Sustainable Production of Metals for Low-Carbon Technologies

Sustainable Production of Metals for Low-Carbon Technologies

Editors

Ilhwan Park
Marthias Silwamba

MDPI

Basel • Beijing • Wuhan • Barcelona • Belgrade • Novi Sad • Cluj • Manchester

Editors
Ilhwan Park
Faculty of Engineering
Hokkaido University
Sapporo
Japan

Marthias Silwamba
Department of Metallurgical
Engineering
University of Zambia
Lusaka
Zambia

Editorial Office
MDPI
St. Alban-Anlage 66
4052 Basel, Switzerland

This is a reprint of articles from the Special Issue published online in the open access journal *Minerals* (ISSN 2075-163X) (available at: www.mdpi.com/journal/minerals/special_issues/SPMLCT).

For citation purposes, cite each article independently as indicated on the article page online and as indicated below:

Lastname, A.A.; Lastname, B.B. Article Title. *Journal Name* **Year**, *Volume Number*, Page Range.

ISBN 978-3-0365-9717-1 (Hbk)
ISBN 978-3-0365-9716-4 (PDF)
doi.org/10.3390/books978-3-0365-9716-4

Contents

About the Editors

Ilhwan Park

Dr. Ilhwan Park is an associate professor at the Division of Sustainable Resources Engineering, Faculty of Engineering, Hokkaido University, Japan. He obtained his Ph.D. in the field of Sustainable Resources Engineering from Hokkaido University, Japan in 2019. After his graduation, he was appointed as an assistant professor at the same University before being promoted to the position of associate professor in 2023. Dr. Park has acted as Editor of Resources Recycling (The Korean Institute of Resources Recycling, KIRR), Associate Editor of Frontiers in Environmental Science, Section Board Member of Metals (MDPI), and Topic Advisory Panel of Minerals (MDPI). His research interests cover the recovery of critical metals/minerals from primary and secondary resources by mineral processing (e.g., flotation, magnetic separation, gravity separation) and hydrometallurgy (e.g., leaching, solvent extraction, cementation). In addition, Dr. Park has been engaged in the development of sustainable techniques for the prevention/control of acid mine/rock drainage (AMD/ARD) and the remediation of contaminated soils and water. From his work, he has published more than 60 papers in leading international journals in the fields of mineral processing, extractive metallurgy, and environmental science.

Marthias Silwamba

Dr. Marthias Silwamba is a highly competent professional in the field of sustainable resources engineering, particularly mineral processing, extractive metallurgy, and solid mine wastes repurposing. With a robust educational background and extensive experience in academia, he has made remarkable strides in research, teaching, and leadership. Throughout his career, he has actively engaged in research projects, publishing numerous papers in reputable journals and presenting his findings at international conferences.

In 2021, he successfully completed his doctoral studies at Hokkaido University in Japan, earning a PhD in Sustainable Resources Engineering. His PhD thesis was focused on developing innovative and sustainable solutions for zinc plant leach residues by reprocessing them for the recovery of valuable metals, thereby addressing present day, pressing critical metals challenges and solving the environmental challenges associated with the leach residues.

Prior to his doctoral studies, Dr. Silwamba pursued a Master of Science degree in Chemical Engineering at the University of the Witwatersrand in South Africa, graduating in 2016. His academic journey commenced with a Bachelor's degree in Metallurgical Engineering from Copperbelt University in Zambia, providing him with a solid foundation in metallurgy and comprehensive knowledge of its principles and practices.

After obtaining MSc (Chemical) in 2016, Dr. Silwamba has served as a lecturer at the University of Zambia, where his dedication to teaching and mentoring students has garnered recognition and respect within the academic community. His commitment to shaping the minds of future engineers has been instrumental in preparing the next generation for success. Beyond academia, he has collaborated with industry partners, providing consultancy services and offering valuable insights into various metallurgical and environmental aspects.

minerals

MDPI

Editorial

Editorial for Special Issue "Sustainable Production of Metals for Low-Carbon Technologies"

Ilhwan Park [1,*] and Marthias Silwamba [2,*]

1 Division of Sustainable Resources Engineering, Faculty of Engineering, Hokkaido University, Sapporo 060-8628, Japan
2 Department of Metallurgical Engineering, School of Mines, University of Zambia, P.O. Box 32379, Lusaka 10101, Zambia
* Correspondence: i-park@eng.hokudai.ac.jp (I.P.); marthias.silwamba@unza.zm (M.S.); Tel.: +81-11-706-6315 (I.P.); +260-979-287-461 (M.S.)

1. Introduction and Scope

In 2015, 193 governments agreed to act on climate change by drastically reducing carbon dioxide (CO_2) emissions as envisaged in the sustainable development goal (SDG) number 13 [1]. The high consumption of metals during the period of industrialization (i.e., since 1900) and the global call for cutting down on CO_2 emissions have led to high demand for metals such as cobalt (Co), copper (Cu), gold (Au), zinc (Zn), lead (Pb), lithium (Li), nickel (Ni), tin (Sn), vanadium (V), rare earth metals, etc. [2]. These metals are critical as the world moves to low-carbon technologies based on renewable energy sources (RESs) and electric vehicles (EVs), and away from the heavy use of fossil fuels. To meet the high forecasted demand, unconventional sources of metals such as low-grade complex ores, seafloor massive sulphides (SMSs), and wastes (e.g., tailings, metallurgical residues, and electronic wastes (e-wastes)) have become very important sources of metals [3,4]. Moreover, these metals should be extracted in a sustainable manner without negatively impacting the environment.

The sustainable extraction of these critical metals begins with studies on the advancements in the beneficiation (i.e., gravity separation, magnetic separation, flotation, etc.) of valuable minerals from unconventional sources. The concentrated minerals are then subjected to advanced extractive metallurgy where unconventional methods (e.g., unconventional hydrometallurgical processes, pyrometallurgical processes, or the combination of the two) are used to obtain metal with lesser or no negative impacts on the environment.

check for updates

Citation: Park, I.; Silwamba, M. Editorial for Special Issue "Sustainable Production of Metals for Low-Carbon Technologies". *Minerals* **2023**, *13*, 88. https://doi.org/10.3390/min13010088

Received: 13 December 2022
Accepted: 16 December 2022
Published: 6 January 2023

2. Contributions

Eleven research articles and one review article were submitted, peer-reviewed, and have been published in this Special Issue of *Minerals* entitled "Sustainable production of metals for low-carbon technologies". The articles cover mineral concentration techniques (flotation and magnetic separation) and hydrometallurgy processes (leaching and recovery of extracted metal). All the articles published have high scientific rigour and will be of great interest to wider readers in the scientific community. The contents of the published articles are briefly summarized as follows.

2.1. Concentration of Valuable Minerals and Metals

2.1.1. Flotation

One unconventional source of critical metals for low-carbon technologies is the mineral tailings that were produced before the advancement of flotation techniques. These tailings are not only important as a secondary resource, but pose serious environmental threats due to the formation of acid mine drainage (AMD) from the oxidation of sulphide minerals and their subsequent dissolution [5,6]. Therefore, the reprocessing of these tailings to recover valuable sulphide minerals solves both the secondary resource concern and the negative impact on the environment. Flotation is one of the methods studied to recover valuable minerals containing critical metals from tailing. The challenges encountered in the use of flotation in the recovery of valuable minerals are that the particle sizes of these minerals sent to tailing dams are too fine for conventional flotation. In a review paper by Bilal et al. [7], the authors was found the challenges of fine particles' flotation from tailings can be addressed by decreasing the bubble size in the flotation vessel (e.g., the use of column flotation, micro- or nano-bubble flotation) and increasing the particle size approaches (e.g., polymer flocculation, shear flocculation, oil agglomeration, and carrier flotation).

The depletion of high-grade and easy-to-process ore has led to researchers exploring complex ores, such as Cu–Zn–Pb sulphide ore, for the recovery of valuable minerals. The challenges encountered when Cu–Zn–Pb sulphide ore is treated for selective flotation (i.e., recovery of each sulphide mineral) include: (1) the unwanted activation of Zn sulphide minerals during Cu sulphide flotation by Pb^{2+} released from anglesite ($PbSO_4$) and (2) the activation of Zn sulphide minerals by dissolved Cu^{2+} as the result of galvanic interaction between Cu sulphide minerals (e.g., chalcopyrite ($CuFeS_2$)) and pyrite (FeS_2) [8,9]. To eliminate the unwanted activation of Zn sulphide by dissolved Pb^{2+} from $PbSO_4$, Aikawa et al. [8] suggested pretreating the Cu–Zn–Pb ore by washing it in ethylene diamine tetra acetic acid (EDTA) to dissolve $PbSO_4$ and the subsequent cementation of dissolved Pb by zero-valent iron (ZVI). The cemented Pb on the surface of ZVI is easily harvested from the pulp by the magnetic separation method [10]. Meanwhile, to control galvanic interaction between $CuFeS_2$ and FeS_2 in flotation cells leading to unwanted activation, Aikawa et al. [9] suggested purging nitrogen, hence reducing the dissolved oxygen and resulting in limiting the anodic dissolution of $CuFeS_2$ when in contact with FeS_2.

Existing flotation plants need simple but effective tools for diagnosis and optimization as the mined sulphide ore becomes more complex and lean of valuable minerals. In this Special Issue, a study by Han et al. [11] showed that linear circuit analysis (LCA) and mass balance (MB) simulation are simple yet powerful tools that can be used to diagnose and optimize the gold ore flotation circuit. From their simulation results, they showed that the current process flotation circuit can be changed to an alternative circuit to obtain the current recovery while improving the gold grade of the concentrate by 128%. The suggested approach of diagnosis and optimization has the potential of being applied to other plants that concentrate various minerals.

The combustion of high-fluorine (F) coal generates large amounts of F-bearing compounds (gases: $HF_{(g)}$, $SiF_{4(g)}$, and $CF_{4(g)}$; solid particulates: $SiF_{6(s)}$ and $CaF_{2(s)}$), which enter into the atmosphere and cause detrimental environmental and ecological consequences. In this Special Issue, Ni et al. [12] devoted their attention to the reduction in the F content in Chinese coal by the combination of gravity separation for coarse coal (>0.5 mm) and flotation for fine coal (<0.5 mm). The initial F content in the studied coal was 347.74 µg/g, which could be reduced to 90.14 µg/g after the proposed combination of gravity-flotation separation process.

2.1.2. Magnetic Separation and Thermal Treatment

In recent years, there has been an astronomical generation of electronic waste (e-waste), and the current outlook points to more generation of this waste. This e-waste is vital in 'urban mining', as it contains substantial amounts of critical metals that should be extracted. However, the liberation and later concentration of metals from e-waste, particularly from printed circuit boards (PCBs), is difficult because of the layered structure of non-metal (i.e., epoxy resin, glass fibre) and valuable metals (e.g., Cu). In this Special Issue, research by Kim et al. [13] showed that pretreatment of PCBs by thermal treatment under a 300 °C air atmosphere reduced the mechanical strength of PCB from 386.36 to 24.26 MPa, and thus copper liberation improved from 9.3 to 100% in the size range of a coarser size fraction (>1400 µm). After thermal treatment, the concentration of Cu was undertaken via electrostatic separation that produced a high-grade Cu concentrate (>90%) with a high recovery (>90%).

Rare earth metals are critical to low-carbon technologies as they are mainly used in the manufacturing of critical components used in generators for wind turbines, traction motors for electric vehicles, and strong permanent magnets. Although these metals are critical for low-carbon technologies, rare earth deposits are found in only a few countries, and they are mostly of very low grades. Therefore, it is vital to beneficiate mined rare earth ores before the extraction of these metals. In this Special Issue, Park et al. [14] studied the beneficiation of low-grade rare earth ore from the Khalzan Buregtei deposit (Mongolia) by magnetic separation. They demonstrated that for fine particle sizes (i.e., −0.5 + 0.1 mm and −106 + 75 μm), dry high-intensity magnetic separation (DHIMS) and wet high-intensity magnetic separation (WHIMS) produced total rare earth oxide (TREO) concentrates with the enrichment ratios of 2.8 and 5.5 and recoveries of 70% and 80%, respectively.

2.2. *Extraction of Critical Metals by Hydrometallurgy*

2.2.1. Alkaline Leaching of Materials and Critical Metal Recovery

Alkaline leaching of metals from concentrate, ores, or waste (mine waste and e-waste) achieves the selective solubilization of amphoteric elements—Al, Pb, Au, and Zn, among others—leaving the iron (Fe), calcium (Ca), and magnesium (Mg) host minerals that may constitute a large percentage of the processed materials. The challenge, however, is that solid–liquid separation by filtration, especially for strong alkaline, is difficult [15]. Silwamba et al. [15] investigated the concurrent recovery of dissolved Pb and Zn by cementation on aluminium metal powder from alkaline pulp (i.e., when zinc leach plant residues were leached using NaOH solution) before filtration. The results showed that around 100% of the dissolved Pb was cemented in the leaching pulp when Al metal powder was added. Meanwhile, less than 2% of dissolved Zn was cemented in leaching pulp. This meant that Pb could be recovered in alkaline leach pulp, and this could eliminate the challenges of the thorough filtration of difficult alkaline pulp.

Another alkaline leaching and metal recovery paper of this Special Issue was by Jeon et al. [16]. They investigated the kinetics of the recovery of Au from ammonium thiosulfate solution by the enhanced cementation technique by galvanic interaction of aluminium (Al; electron donor) and activated carbon (AC; electron mediator) to recover gold (Au). The addition of AC in the Au cementation on Al metal eliminates the challenges of poor Au recovery in the ammonium thiosulfate solution [17]. The reported kinetics results showed that the recovery followed first-order kinetics (i.e., the cementation rate increased with a higher initial concentration of Au, smaller electron donor size, greater both electron donor and mediator quantity, decrease in temperature, and higher shaking speed in the system).

2.2.2. Acid Leaching of Materials and Critical Metals Recovery

Sulphuric acid is one of the most widely used lixiviants used in hydrometallurgical processes for the dissolution of metals from their host materials (minerals) because it is cheap (often produced as a by-product) and can be regenerated. In this Special Issue, Jha et al. [18] used the sulphuric acid leaching of rare earth metals (REMs) from end-of-life nickel metal hydride (NiMH) batteries, which are now generated as waste due to their extensive use in the manufacturing of portable electronic devices as well as electric vehicles. The obtained results showed that more than 90% of REMs (Nd, Ce, and La) dissolved when 2 M H_2SO_4 was used at 75 °C and leached for 60 min in the presence of 10% H_2O_2 (v/v). The dissolved REMs were concentrated by solvent extraction and selective precipitation.

In a similar study, Ahn et al. [19] reported the leaching of Li, Ni, Co, and Mn in sulphuric acid from cathode active materials that have been used in lithium-ion batteries after their carbothermic pretreatment. The obtained results indicated carbothermic treatment enhanced the conventional leaching process using hydrogen peroxide because the leaching efficiencies of Li, Ni, Co, and Mn increased to over 99.9% within 120 min in all samples roasted at 600–900 °C.

The study by Yang et al. [20] was about the recovery of magnesium from ferronickel slag (FNS) to prepare magnesium oxide by leaching in sulfuric acid. The FNS was leached in two stages (i.e., first leached in 2.4 M sulphuric acid and regenerated residues subjected to 2.4 M sulphuric acid leaching again), and the overall leaching efficiency was 95.82%. The dissolved Mg was precipitated as carbonates and later on calcinated to produce MgO of 94.85% purity.

3. Conclusions and Outlook

The papers published under this Special Issue cover a variety of topics of mineral beneficiation and critical metals extraction as well as recovery from a wide variety of materials and ores. The papers reported the advancement made in the beneficiation and extraction of these metals in a sustainable manner. However, there are still many challenges that need to be overcome, especially the cost–benefit study of extracting these critical metals using non-conventional processes and from secondary resources. It is believed that this Special Issue will help to arouse future scientific discussions and debates on the challenges of the production of metals critical to low-carbon technologies.

Funding: The writing of these editorial comments received no external funding.

Acknowledgments: As guest editors, we would like, first of all, to sincerely thank the contributions of high-quality papers with excellent scientific rigour from all the authors. Many thanks go also to all the reviewers who made sure that the papers published met scientific soundness. Furthermore, we wish to thank the Editors of *Minerals* for their continuous support and the *Minerals* Editorial Assistants for their valuable and inexhaustible engagement and support during the preparation of this volume. My special thanks go to journal editors for their support and assistance.

Conflicts of Interest: Authors declare no conflict of interest.

References

1. Hák, T.; Janoušková, S.; Moldan, B. Sustainable Development Goals: A need for relevant indicators. *Ecol. Indic.* **2016**, *60*, 565–573. [CrossRef]
2. Tabelin, C.B.; Park, I.; Phengsaart, T.; Jeon, S.; Villacorte-Tabelin, M.; Alonzo, D.; Yoo, K.; Ito, M.; Hiroyoshi, N. Copper and critical metals production from porphyry ores and E-wastes: A review of resource availability, processing/recycling challenges, socio-environmental aspects, and sustainability issues. *Resour. Conserv. Recycl.* **2021**, *170*, 105610. [CrossRef]
3. Silwamba, M.; Ito, M.; Hiroyoshi, N.; Tabelin, C.B.; Hashizume, R.; Fukushima, T.; Park, I.; Jeon, S.; Igarashi, T.; Sato, T.; et al. Recovery of Lead and Zinc from Zinc Plant Leach Residues by Concurrent Dissolution-Cementation Using Zero-Valent Aluminum in Chloride Medium. *Metals* **2020**, *10*, 531. [CrossRef]
4. Spooren, J.; Binnemans, K.; Björkmalm, J.; Breemersch, K.; Dams, Y.; Folens, K.; González-Moya, M.; Horckmans, L.; Komnitsas, K.; Kurylak, W.; et al. Near-Zero-Waste Processing of Low-Grade, Complex Primary Ores and Secondary Raw Materials in Europe: Technology Development Trends. *Resour. Conserv. Recycl.* **2020**, *160*, 104919. [CrossRef]
5. Tabelin, C.B.; Uyama, A.; Tomiyama, S.; Villacorte-Tabelin, M.; Phengsaart, T.; Silwamba, M.; Jeon, S.; Park, I.; Arima, T.; Igarashi, T. Geochemical audit of a historical tailings storage facility in Japan: Acid mine drainage formation, zinc migration and mitigation strategies. *J. Hazard. Mater.* **2022**, *438*, 129453. [CrossRef]
6. Park, I.; Tabelin, C.B.; Jeon, S.; Li, X.; Seno, K.; Ito, M.; Hiroyoshi, N. A review of recent strategies for acid mine drainage prevention and mine tailings recycling. *Chemosphere* **2019**, *219*, 588–606. [CrossRef] [PubMed]
7. Bilal, M.; Park, I.; Hornn, V.; Ito, M.; Hassan, F.U.; Jeon, S.; Hiroyoshi, N. The Challenges and Prospects of Recovering Fine Copper Sulfides from Tailings Using Different Flotation Techniques: A Review. *Minerals* **2022**, *12*, 586. [CrossRef]
8. Aikawa, K.; Ito, M.; Kusano, A.; Jeon, S.; Park, I.; Hiroyoshi, N. Development of a Sustainable Process for Complex Sulfide Ores Containing Anglesite: Effect of Anglesite on Sphalerite Floatability, Enhanced Depression of Sphalerite by Extracting Anglesite, and Recovery of Extracted Pb^{2+} as Zero-Valent Pb by Cementation Using Zero-Valent Fe. *Minerals* **2022**, *12*, 723. [CrossRef]
9. Aikawa, K.; Ito, M.; Orii, N.; Jeon, S.; Park, I.; Haga, K.; Kamiya, T.; Takahashi, T.; Sunada, K.; Sakakibara, T.; et al. Flotation of Copper Ores with High Cu/Zn Ratio: Effects of Pyrite on Cu/Zn Separation and an Efficient Method to Enhance Sphalerite Depression. *Minerals* **2022**, *12*, 1103. [CrossRef]
10. Silwamba, M.; Ito, M.; Tabelin, C.B.; Park, I.; Jeon, S.; Takada, M.; Kubo, Y.; Hokari, N.; Tsunekawa, M.; Hiroyoshi, N. Simultaneous extraction and recovery of lead using citrate and micro-scale zero-valent iron for decontamination of polluted shooting range soils. *Environ. Adv.* **2021**, *5*, 100115. [CrossRef]

11. Han, S.; Jung, M.; Lee, W.; Kim, S.; Lee, K.; Lim, G.-T.; Jeon, H.-S.; Choi, S.Q.; Han, Y. Diagnosis and Optimization of Gold Ore Flotation Circuit via Linear Circuit Analysis and Mass Balance Simulation. *Minerals* **2021**, *11*, 1065. [CrossRef]
12. Ni, C.; Zhou, S.; Bu, X.; Bilal, M.; Hassan, F.U.; Chen, Y.; Xu, G.; Xie, G. Characterization and Removal Potential of Fluorine in Lignite from a Mine in Shaanxi Province, China: A Case Study. *Minerals* **2022**, *12*, 280. [CrossRef]
13. Kim, B.; Han, S.; Park, S.; Kim, S.; Jung, M.; Park, C.-H.; Jeon, H.-S.; Kim, D.-W.; Han, Y. Optimal Thermal Treatment for Effective Copper Recovery in Waste Printed Circuit Boards by Physical Separation: Influence of Temperature and Gas. *Minerals* **2021**, *11*, 1213. [CrossRef]
14. Park, I.; Kanazawa, Y.; Sato, N.; Galtchandmani, P.; Jha, M.K.; Tabelin, C.B.; Jeon, S.; Ito, M.; Hiroyoshi, N. Beneficiation of Low-Grade Rare Earth Ore from Khalzan Buregtei Deposit (Mongolia) by Magnetic Separation. *Minerals* **2021**, *11*, 1432. [CrossRef]
15. Silwamba, M.; Ito, M.; Hiroyoshi, N.; Tabelin, C.B.; Hashizume, R.; Fukushima, T.; Park, I.; Jeon, S.; Igarashi, T.; Sato, T.; et al. Alkaline Leaching and Concurrent Cementation of Dissolved Pb and Zn from Zinc Plant Leach Residues. *Minerals* **2022**, *12*, 393. [CrossRef]
16. Jeon, S.; Bright, S.; Park, I.; Kuze, A.; Ito, M.; Hiroyoshi, N. A Kinetic Study on Enhanced Cementation of Gold Ions by Galvanic Interactions between Aluminum (Al) as an Electron Donor and Activated Carbon (AC) as an Electron Mediator in Ammonium Thiosulfate System. *Minerals* **2022**, *12*, 91. [CrossRef]
17. Jeon, S.; Tabelin, C.B.; Takahashi, H.; Park, I.; Ito, M.; Hiroyoshi, N. Enhanced cementation of gold via galvanic interactions using activated carbon and zero-valent aluminum: A novel approach to recover gold ions from ammonium thiosulfate medium. *Hydrometallurgy* **2020**, *191*, 105165. [CrossRef]
18. Jha, M.K.; Choubey, P.K.; Dinkar, O.S.; Panda, R.; Jyothi, R.K.; Yoo, K.; Park, I. Recovery of Rare Earth Metals (REMs) from Nickel Metal Hydride Batteries of Electric Vehicles. *Minerals* **2022**, *12*, 34. [CrossRef]
19. Ahn, Y.; Koo, W.; Yoo, K.; Alorro, R.D. Carbothermic Reduction Roasting of Cathode Active Materials Using Activated Carbon and Graphite to Enhance the Sulfuric-Acid-Leaching Efficiency of Nickel and Cobalt. *Minerals* **2022**, *12*, 1021. [CrossRef]
20. Yang, J.; Duan, X.; Liu, L.; Yang, H.; Jiang, X. Recovery of Magnesium from Ferronickel Slag to Prepare Magnesium Oxide by Sulfuric Acid Leaching. *Minerals* **2021**, *11*, 1375. [CrossRef]

minerals MDPI

Review

The Challenges and Prospects of Recovering Fine Copper Sulfides from Tailings Using Different Flotation Techniques: A Review

Muhammad Bilal [1,2,*], Ilhwan Park [3,*], Vothy Hornn [1], Mayumi Ito [3], Fawad Ul Hassan [2], Sanghee Jeon [3] and Naoki Hiroyoshi [3]

1 Division of Sustainable Resources Engineering, Graduate School of Engineering, Hokkaido University, Sapporo 060-8628, Japan; vothytalis1102@yahoo.com
2 Department of Mining Engineering, Balochistan University of Information Technology Engineering and Management Sciences (BUITEMS), Quetta 87300, Pakistan; fawad-hassan@hotmail.com
3 Division of Sustainable Resources Engineering, Faculty of Engineering, Hokkaido University, Sapporo 060-8628, Japan; itomayu@eng.hokudai.ac.jp (M.I.); shjun1121@eng.hokudai.ac.jp (S.J.); hiroyosi@eng.hokudai.ac.jp (N.H.)
* Correspondence: bilalkhan4p@gmail.com (M.B.); i-park@eng.hokudai.ac.jp (I.P.); Tel.: +81-11-706-6315 (I.P.)

Abstract: Flotation is a common mineral processing method used to upgrade copper sulfide ores; in this method, copper sulfide mineral particles are concentrated in froth, and associated gangue minerals are separated as tailings. However, a significant amount of copper is lost into tailings during the processing; therefore, tailings can be considered secondary resources or future deposits of copper. Particle–bubble collision efficiency and particle–bubble aggregate stability determines the recovery of target particles; this attachment efficiency plays a vital role in the selectivity process. The presence of fine particles in the flotation circuit is because of excessive grinding, which is to achieve a higher degree of liberation. Complex sulfide ores of markedly low grade further necessitate excessive grinding to achieve the maximum degree of liberation. In the flotation process, fine particles due to their small mass and momentum are unable to collide with rising bubbles, and their rate of flotation is very slow, further lowering the recovery of target minerals. This collision efficiency mainly depends on the particle–bubble size ratio and the concentration of particles present in the pulp. To overcome this problem and to maintain a favorable particle–bubble size ratio, different techniques have been employed by researchers to enhance particle–bubble collision efficiency either by increasing particle size or by decreasing bubble size. In this article, the mechanism of tailing loss is discussed in detail. In addition, flotation methods for fine particles recovery such as microbubble flotation, column flotation, nanobubble flotation, polymer flocculation, shear flocculation, oil agglomeration, and carrier flotation are reviewed, and their applications and limitations are discussed in detail.

Keywords: flotation; copper sulfides; tailings; carrier flotation; flocculation; oil agglomeration

check for updates

Citation: Bilal, M.; Park, I.; Hornn, V.; Ito, M.; Hassan, F.U.; Jeon, S.; Hiroyoshi, N. The Challenges and Prospects of Recovering Fine Copper Sulfides from Tailings Using Different Flotation Techniques: A Review. *Minerals* **2022**, *12*, 586. https://doi.org/10.3390/min12050586

Academic Editor: Francesco Veglio

Received: 18 March 2022
Accepted: 4 May 2022
Published: 6 May 2022

1. Introduction

Copper (Cu), the second most produced non-ferrous metal after aluminum, has been widely used for various applications due to its unique properties such as excellent electrical and thermal conductivities, high corrosion resistance, and outstanding malleability and ductility [1,2]. The demand for Cu is projected to be continuously increasing because it is considered one of 17 materials/metals identified as critical materials/metals for realizing a low-carbon future [3,4]. Low-carbon technologies require more Cu than conventional fossil fuel-based technologies; for example, a typical 3 MW wind turbine requires 4.7 t of Cu, photovoltaic (PV) cells need ~1% Cu, and electric vehicles (EVs) use 1.7–11 times more Cu than conventional cars [3–7]. Thus, the establishment of a robust and sustainable supply chain of Cu is an important issue for humanity.

Copper sulfides are the primary source of metallic copper, which makes up 80% of the copper resources [8]. The major copper sulfide ores include chalcopyrite ($CuFeS_2$), bornite (Cu_5FeS_4), covellite (CuS), and chalcocite (Cu_2S). The copper concentrates from sulfide ores are usually produced by flotation and then processed by pyrometallurgical or hydrometallurgical methods to extract the copper [3,9,10].

Froth flotation is widely used to separate sulfide minerals from gangue minerals such as quartz (SiO_2), based on the surface properties of minerals. The principle of the flotation process is illustrated in Figure 1. In flotation, ground ore particles are suspended in water in the flotation cell and conditioned with various reagents (e.g., collector, frother, activator, depressant, and pH adjustor) to control the hydrophobicities of target and gangue minerals [11–14]. Afterward, air bubbles are induced into the pulp (mineral/water suspension), and hydrophobic copper mineral particles are attached to the bubble surface and recovered into the froth layer as a copper concentrate [15,16]. Hydrophilic gangue minerals such as quartz are not attached to air bubbles and remain in the pulp as a by-product called tailings.

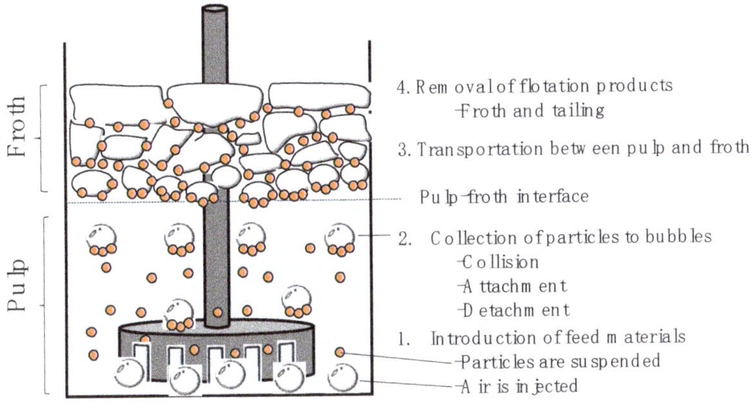

Figure 1. Flotation mechanism.

Increasing demand for energy and metals around the world has led to an acute depletion of high-grade ore deposits. As ore grades decline, more challenges arise in mineral processing, including the increasing complexity of mineralogy and the decreasing grain size of mineral particles [17,18]. The first strategy to overcome the implications of complex mineralogy is to decrease the particle size in order to achieve a high degree of liberation, which produces fine to ultrafine feeds for flotation circuits. As ultrafine grinding requires a lot of energy, the second strategy includes the use of a rougher circuit for processing of coarser particles (>250 μm) followed by regrinding rougher flotation concentrates to save energy costs [19]. In either case, flotation circuits are required to process much finer or coarser feeds nowadays.

In the actual flotation process, a significant quantity of copper sulfide particles (finest and coarsest fractions) is lost in tailings, which are disposed of into the tailings storage facilities (TSFs). Since the conventional copper ores have been becoming depleted, copper tailings are gaining increasing attention as potential resources. For example, the amount of copper in TSFs is estimated to be of the magnitude of 0.13 Gt, equivalent to ~15% of the current geogenic Cu reserves [20–22]. The recovery of copper sulfides from TSFs will not only add economic value, but also protect the environment because the oxidation of copper sulfides results in the release of protons and heavy metal ions that contaminate surrounding ecosystems [23].

Efficiency of the flotation process is mainly dependent on particle size. Fine particles are unable to collide with air bubbles [24] and coarse particles are unable to attach to air bubbles; hence, they are lost into tailings. The presence of fine particles in flotation circuits is largely the result of excessive grinding, performed to achieve a higher degree of mineral liberation.

As discussed, the tailings mostly contain the finest and coarsest size fractions. A few review papers have addressed fine and coarse particle flotation [25–29]. Some previous studies have mainly focused on the use of different types of flotation cells [18,27,28]. A detailed discussion on the development of flotation cells to address the issue of recovering fine/ultrafine and coarse particles is out of the scope of the present work. An excellent review by Hassanzadeh et al. [27,28] on this topic is recommended to the readers, while other review articles have mainly focused on flotation kinetics [25,26,29]. However, to the best of our knowledge, there is a lack of thorough review on the comparison of different flotation techniques with a focus on the recovery of fine copper sulfides from tailings. Therefore, the present article reviews the application of different flotation techniques to recover fine copper mineral particles from tailings. In the first part of this review paper, the mechanism of tailing loss is discussed in detail. After this, different flotation techniques such as column flotation, microbubble flotation, nanobubble flotation, polymer flocculation, shear flocculation, oil agglomeration, and carrier flotation are reviewed, and their applications on tailings and limitations are discussed in detail. This review aims to provide researchers who are working in the area of mineral processing with rough information about fine particle flotation, and shed some light on developed flotation techniques for fine particle separation from tailings.

2. Tailing Loss

2.1. Tailing Amount

Tailings from the flotation process account for ~98% of total ore fed to the concentrator and are usually stored in large tailing dams near the mine [30]. Flotation tailings are termed as the rejected product of the flotation process, and according to the reported literature data, 90–95% of plant tailings are produced from the flotation process [31–33]. Flotation tailings mainly contain gangue minerals, while a significant amount of valuable copper sulfide particles is also contained. As shown in Table 1, it has been reported that around 10–20% of copper in the ore is lost in the tailings. The copper grade in the tailings is 0.04–0.2%, which is very low in comparison with standard copper ores containing over 0.5% of copper. Because of this, low-grade tailings are not considered an economically valuable copper source at present; however, when copper demand increases and methods to recover copper from tailings are developed, tailings can be considered as a secondary deposit or future resource of copper.

2.2. Mechanism

The main operation of copper mineral processing is generally composed of two steps: (i) comminution to liberate valuable copper minerals from non-valuable gangue minerals, and (ii) flotation to separate copper minerals from gangue minerals. After flotation, copper concentrates are sent to the smelter for the production of highly purified copper cathodes (>99.99% Cu), while tailings are disposed of in tailing dams near mine sites.

In an ideal mineral processing of copper ores, all copper mineral grains must be liberated from gangue minerals in the comminution step (crushing and grinding) before separation. In an actual comminution step, perfect liberation is difficult because the process consumes a lot of energy and generates excessively fine particles, which cannot be recovered in the separation step [34]. Because of these reasons, as shown in Figure 2, actual ground ores are a mixture of liberated (fine) copper mineral and gangue grains, and unliberated (coarse) copper minerals locked with gangue minerals.

Table 1. Industrial flotation data from copper concentrators [30].

Concentrator	Ore Treated, Tonnes/Year	Concentrate, Tonnes/Year	Cu Recovery to Concentrate, %	Cu Loss to Tailings, %	Tailings Grade, % Cu
El Soldado, Chile	7,700,000 (2010)	67,000	-	-	0.18
Los Bronces, Chile	20,500,000 (2010)	676,000	-	-	0.133
Africa (open Pit)	4,000,000	35,000	80	20	0.018
Africa (under ground)	900,000	21,000	85	15	0.04
Mantos Blancos Chile (open pit)	4,500,000	125,000	89	11	0.12
Cerro Verde, Peru	39,000,000	750,000	88	12	0.075
Sierrita, USA	37,000,000	325,000	84.3	15.7	0.041
Ray, USA	10,000,000	150,000	90	10	0.035
Dexing, Jiangxi Province, China	13,500,000	150,000	85.5	14.5	0.084
Baiyin, Gansu Province, China	3,600,000	46,000	83	17	0.28
Hujiayu, Shanxi Province, China	750,000	6800	96.05	3.95	-
Fenghaungshan, Anhui Province, China	600,000	7200	93.62	6.38	0.084

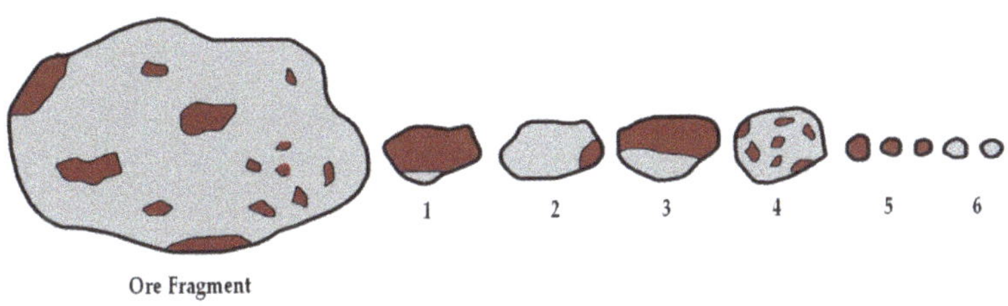

Figure 2. Illustration of an ore fragment (mineralization in dark red) and the products of comminution: (1) highly liberated particle, (2) poorly liberated or gangue-rich particle, (3,4) moderately liberated particles (middlings), (5) liberated fine mineral particles, and (6) fine gangue particles.

In the separation process, the behavior of unliberated grains depends on the copper mineral portion in the grain; that is, when the copper portion is large, an unliberated particle may act as a (liberated) copper mineral; when the copper portion is small, it may behave as gangue mineral. Because of this, a part of copper minerals in unliberated grains is lost into tailings together with gangue minerals. This is one of the reasons (mechanisms) for copper loss in the tailings.

Another case of copper loss occurs even for fine liberated particles during separation using flotation. In flotation, mineral particles suspended in water collide and attach to air bubbles induced into the flotation cell, and then they float to the surface of the water as a froth product. The flotation efficiency of mineral particles depends on particle size. Research on the relationship between particle size and floatability began in the early 1930s in work presented by Gaudin et al. (1931) [35], which showed that coarse and fine particles are more difficult to recover than intermediate-sized particles. In the case of chalcopyrite flotation, it has been reported that copper recovery to the froth was around 95% at particle sizes of 10–150 μm, while the recoveries decreased to 50% when particle size decreased from 20 μm to 3 μm [26,36]. The flotation process is more efficient in the particle size range of approximately 10–100 μm [24,37,38], and it is usually poor with an increased percentage of ultrafine particles (<10 μm) in the system [39–41]. The majority of copper lost into flotation tailings at operating concentration plants occurs for particle sizes smaller than 20 μm and larger than 105 μm. In general, liberating particles can be increased by decreasing particle size (which increases mineral surface area) [42]. In flotation, a particle's lower size limit varies for different minerals [39], and some studies have reported good recovery of ultrafine particles [24,43]. Michael et al. [35] reported the effects of particle size on the distribution of copper to froth and tailings during conventional flotation. Recoveries of copper to froth were maximum at particle sizes between 30 and 150 μm, with a distinct drop in recovery outside of this range, resulting in a loss of ultrafine and coarse particles into tailings (Figure 3). It is evident from Figure 3 that a substantial amount of valuable minerals is lost into the tailings in both the finest and coarsest fractions. The reason for the dependence of flotation efficiency on particle size is described below.

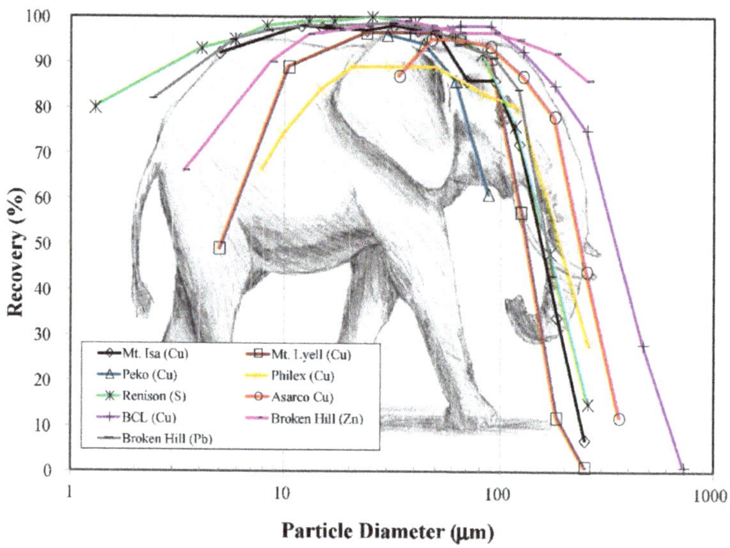

Figure 3. Conventional flotation data for industrial sulfide flotation circuits [44].

In flotation, air bubbles are injected into a mineral particle/water suspension, which rise by buoyancy and collide with suspended mineral particles. If the particle is hydrophobic, it is stably attached to air bubble and is not detached from the bubble. This captured particle is then floated with air bubbles and recovered as a froth product.

Figure 4 illustrates the process of particles being captured by a rising air bubble. The probability of particles (P) being captured by bubbles during the flotation process in the cell can be expressed as follows [40,45]:

$$P = P_c \times P_a \times (1 - P_d) \quad (1)$$

where P_c, P_a, and P_d represent the probabilities of collision, attachment, and detachment, respectively (Figure 4). The collision between fine particles and the rising bubbles in mechanical flotation cells becomes poor because of the smaller mass and low momentum of fine particles [24,46–48]; fine particles follow fluid streamlines around the bubble.

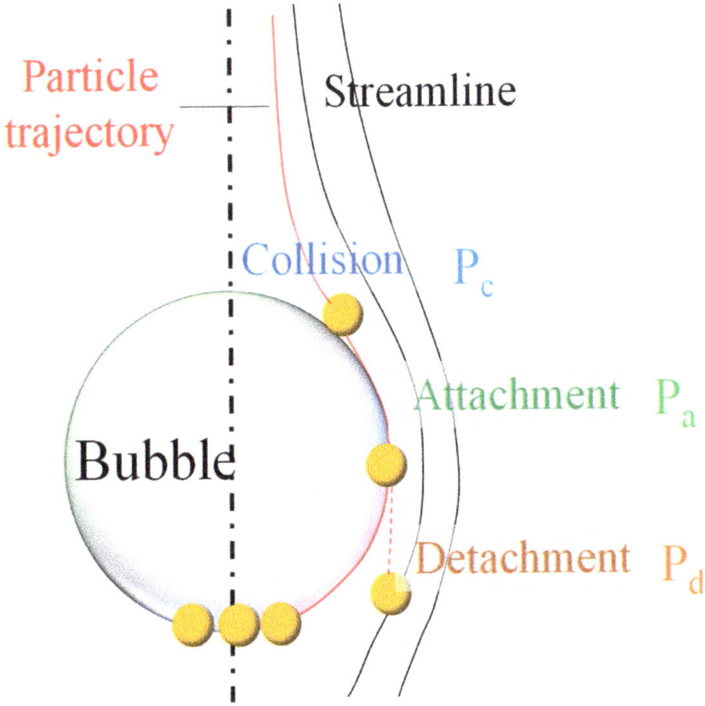

Figure 4. Schematic representation of particle–bubble collision, attachment, and detachment. The thick lines represent particle trajectories whilst the thin lines represent the fluid streamlines (reprinted with permission from [25], copyright 2022 Elsevier).

Collision efficiency is linked with the size ratio of the mineral particle and air bubble.

$$P_c = A\left(\frac{D_p}{D_b}\right)^n \quad (2)$$

where D_p and D_b represent the diameters of particle and bubble, respectively, and A and n are empirical flow regime constants.

Attachment probability (P_a) and detachment probability (P_d) are mainly functions of the particle surface properties. For hydrophobic minerals such as chalcopyrite, the values of P_a and P_d could be assumed to be 1 and 0, respectively. In this case, P is mainly dependent on P_c. Based on Equation (2), the collision probability (P_c) is directly proportional to the particle to bubble diameter ratio. With a significant decrease in the diameter of particles ($D_p << D_b$), the collision probability is substantially decreased, which explains the very low recovery rates of fine particles. For excessively coarse particles, collision probability (P_c) is high; however, attached particles are easily detached from air bubbles by turbulent flow, i.e., P_d is high [49–51]. This is because of the relatively small specific surface area of coarse particles, which reduces particle–bubble attachment strength. This is one of the reasons for the low flotation recovery of coarse particles into the froth. Moreover, because of poor suspension and low dispersion of coarse particles, particle–bubble collision efficiency is limited, which reduces the flotation kinetics rate and final recovery.

As discussed above, there are two major reasons for tailing loss of copper in the flotation process: (i) tailing loss of ultrafine particles due to the excessively low collision efficiency between particles and air bubbles in flotation cells; and (ii) coarse unliberated copper minerals associated with gangue minerals, which are easily detached from an air bubble in flotation.

For recovering copper from tailings, regrinding is needed to liberate copper minerals from locked grains. As a result, liberated but fine copper mineral particles are formed. This necessitates the development of flotation techniques to recover fine copper mineral particles from tailings. Numerous flotation methods have been developed to enhance particle–bubble collision efficiency and improve fine particle recovery either by increasing the particle size or decreasing the bubble size [43,52]. In the following sections, these techniques are reviewed concerning their applications towards tailings, which contain mostly fine mineral particles.

3. Methods for Copper Recovery from Tailings

3.1. Decreasing Bubble Size Approach

3.1.1. Column Flotation

The column flotation cell (CFC) was invented in the 1960s but did not gain wide acceptance in the processing of base metals until the early 1980s [28]. In column flotation, ore is introduced near the top of the column and sparging air from bottom, creating a deep froth zone. Wash water is added from the top to remove fine entrained particles [53]. Flotation circuits use multiple flotation cells in series to recover the maximum amount of desired minerals. Circuits consist of a set of cells called rougher, cleaner, and scavenger, as illustrated in Figure 5. Rougher flotation cells are usually mechanical type and appear first in the circuit, to increase the recovery of the valuable minerals. To minimize valuable mineral loss from final tailings, tailings from rougher cells are fed to scavenger cells, and the froth from the rougher cells is reground to liberate the middlings. This product is then fed to the cleaner cells. For cleaner flotation, column-type flotation cells that produce smaller bubbles than mechanical flotation units are frequently used [54,55]. By using smaller bubbles, the collision efficiency of particles and bubbles increases, resulting in higher copper recovery for fines in column flotation.

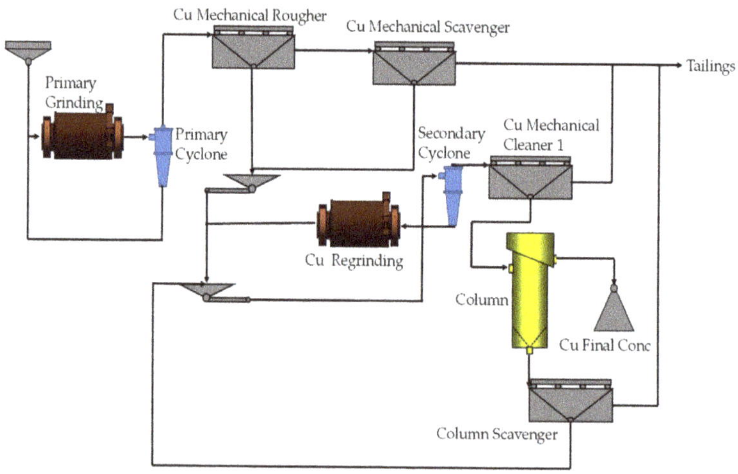

Figure 5. A simplified flotation circuit.

The size of air bubble in the column flotation cell depends strongly on the bubble generator/sparger. Different types of bubble generators/spargers are used in column flotation. Depending on the technique used to create the bubbles, there are three types of spargers. In one type, the air is pumped into a porous material, such as rubber or cloth, to produce bubbles. However, such spargers have durability issues (wear and tear), pore blockage by particles, and difficulties in consistently maintaining optimal bubble size. In the jetting type, air and water are forced through a small circular opening at high pressure [54,56]. Lastly, the sparger type utilizes the shearing motion of a slurry-air mixture to generate fine bubbles after passing through a specially designed tube [56].

An in-line mixer bubble generator produced relatively large bubbles up to a diameter of 0.35 mm, according to laboratory studies [57]. In a comparison of Microcel^{TM} and jet-type spargers, it was found that the former produced smaller bubbles, thereby increasing the recovery [58]. However, the mean diameter of bubbles produced by these two types of spargers (1−3 mm) is still relatively large. Considering this, according to Equation (2), for particles with a size <10 μm, the ideal bubble size is <150 μm.

The advantages of these devices include minimal installation costs, low operating costs, and the ability to produce significantly higher-grade concentrates than conventional cells. Even though column cells have been successful in industrial installations, they still have two inherent flaws. The first problem is that coarse particles have high settling velocities, leading to a shorter residence time in the cell, making them hard to recover. Furthermore, fine particles generally do not possess enough inertia to overcome the streamlines around bubbles and are less likely to collide with them [59].

3.1.2. Microbubble Flotation

As already discussed in the above section, one of the possible ways to overcome the difficulties associated with fine-particle flotation is to reduce bubble size to a few micrometers. Microbubbles are defined as air bubbles with a diameter range of 1−100 μm [60,61]. To generate microbubbles, several techniques are used (Table 2), e.g., electro-flotation, dissolved-air flotation, hydrodynamic cavitation, and dispersed-air flotation. Electro-flotation (or electrolytic flotation) has good controllability of bubble size, but it is more suitable for small-scale applications due to its low throughput and high energy consumption [62–64]. Due to microbubble stability problems in agitated flotation cells, microbubble flotation is primarily carried out in column-type flotation cells [65]. Laboratory experiments were performed using this technique to recover copper, zinc, and lead from a complex sulfide ore of <5 μm size [60,66].

Even though microbubble flotation is promising, it has several drawbacks [25]. When the bubbles are too small (<200 μm), their collision efficiency with fine particles is high, however, they will float too slowly due to their low buoyancy. On the other hand, if bubbles are large (>500 μm), they will float fast; however, they have low collision efficiency with fine particles [67]. Due to the slow speed of the microbubbles rising with attached mineral particles, longer residence times are required in the flotation circuit. Furthermore, the microbubbles cause a higher water recovery, thereby increasing gangue mineral entrainment, which reduces the selectivity [39,68].

Table 2. Studies on the use of microbubbles in flotation.

Name	Minerals Studied	Process/Effects	References
Electroflotation	Chalcopyrite (<20 μm)	Bubble generation by electrolysis.	[69]
		Improved recovery by electrolytic oxygen.	
Charged microbubble/colloid gas aphrons (CGAs) microbubble	CuO and SiO_2 (<10 μm)	CGAs generated using high speed impeller.	[70]
		The grade of CuO improved from 59% to 82%. Recovery of CuO improved from 58% to 77%.	

Table 2. *Cont.*

Name	Minerals Studied	Process/Effects	References
Microbubble flotation	Pyrite (FeS_2), chalcopyrite ($CuFeS_2$), galena (PbS), and sphalerite (ZnS)	Bubbles generated using Silica Porous Glass (SPG) filter.	[66]
		Microbubble flotation of ultrafine sulfide minerals.	
Microbubble flotation	Galena (PbS) and sphalerite (ZnS)	Bubbles generated using Silica Porous Glass (SPG) filter.	[71]
		Collector-less microbubble flotation using sodium hydrosulfide.	
Nano-microbubble	Chalcopyrite fines (14−38 μm) Ultrafine particles (5−14 μm)	Bubble size (D_{90} = 100 μm) generated by hydrodynamic cavitation.	[72]
		Improved recovery of fine and ultrafine particles (16−21%).	

3.1.3. Nanobubble Flotation

Flotation of fine particles can be enhanced by nanobubbles. Parker et al. were the first to propose nanobubbles to improve the flotation of fines [73]. Nanobubbles (NBs), also called ultrafine bubbles, are gas cavities with a diameter of <1 μm [74–76]. As shown in Figure 6, nanobubbles are adsorbed on fine mineral particles when they are pre-treated with them, forming nanobubble−particle aggregates and, thus, increasing the apparent particle size and improving collision efficiency [77–79]. In addition to enhancing the fine particle hydrophobicity and forming aggregates, NBs also act as secondary collectors [80–82].

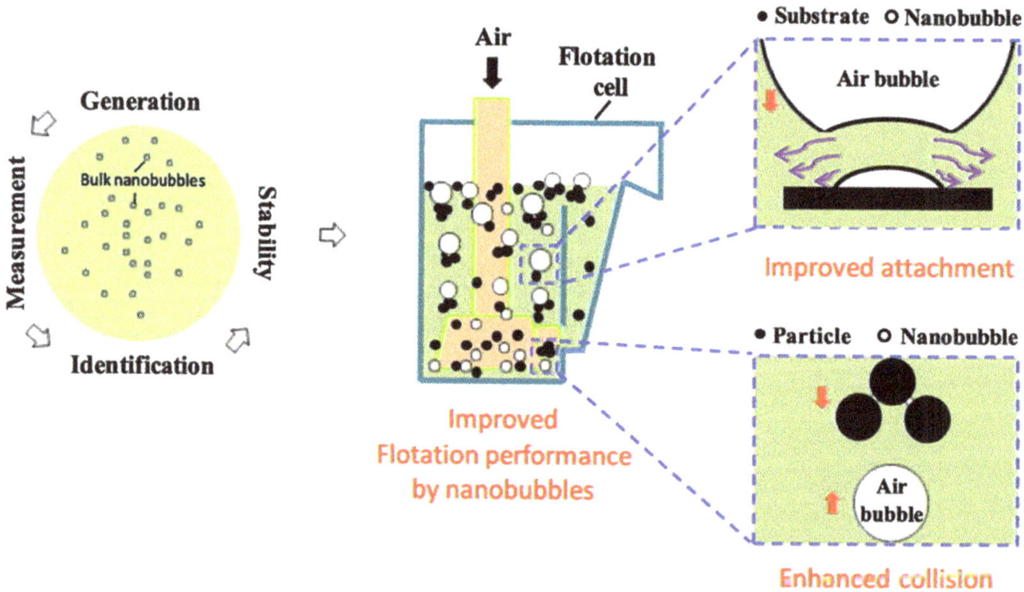

Figure 6. Schematic mechanism of nanobubble flotation (reprinted with permission from Li et al. [83], copyright (2022) Elsevier).

NBs are produced by different methods such as hydrodynamic cavitation, gas over-saturation, power ultrasound, electrolysis, chemical reaction, solvent exchange, temperature exchange, and membrane modules [83,84]. The generation of NBs is a complex physicochemical process influenced by several factors, including temperature, electrolyte concentration, dissolved gas concentration in solution, and the type and concentration of reagents [84–90]. Studies have shown that nanobubbles can significantly boost process efficiency by enhancing coarse bubble attachment onto adhered NBs and by aggregating the problematic fine particles [72,77,80,89,91]. The NBs are considered to have significant potential in mineral processing due to the better recovery of ultrafine particles [92].

Flotation kinetics and recovery are reportedly higher for a fine fraction (<38 μm), while the grade and selectivity index tend to be higher for coarse size fraction (<150 μm) [93]. Ahmadi et al. [72] confirmed that the flotation recovery of fine (14–38 μm) and ultrafine (5–14 μm) chalcopyrite particles with NBs increased by approximately 16–21%. With surfactants (collector and frother), nanobubbles may cause a decrease in grade, primarily due to increased entrainment [93,94]. Nanobubbles should be used in the rougher and scavenger stages, where recovery is of primary importance, while the cleaner stage mainly controls the grade. Optimizing the hydrodynamic conditions and surfactant types is required to reduce the amount of water recovery associated with NBs, which will reduce the entrainment rate [94].

3.2. *Increasing Particle Size Approaches*

Theoretically and experimentally, it has been observed that the flotation rate elevates when the particle size is large [39,95–99]. As mentioned above, the agglomeration of fine mineral particles can be an effective method for improving collision efficiency between mineral particles and air bubbles during flotation. The recovery of fines by using this approach has been studied extensively, and several techniques have been developed. These techniques mainly include polymer flocculation, shear flocculation, oil agglomeration, and carrier flotation. Table 3 provides a brief overview of these techniques, and Figure 7 presents a schematic diagram of the governing mechanisms.

Table 3. A brief description of different agglomeration techniques.

Method	Reagents and Material	Types of Interaction
Polymer flocculation	Polymer, surfactant	Hydrophobic, chemical, hydroxyl functional group, electrostatic
Shear flocculation	Surfactant	Hydrophobic interaction
Oil agglomeration	Oil, surfactant	Capillary forces
Carrier flocculation	Surfactant, carrier material	Hetero coagulation, hydrophobic interaction

3.2.1. Polymer Flocculation

In flocculation, organic polymers with a high molecular weight (>1 million) are used to form a molecular bridge between particles [100–102]. A polymer induces flocculation of mineral particles by attaching itself to two or more particles, bridging the particles. If polymers are adsorbed on a specific mineral surface, selective flocculation occurs [103,104]. Selective flocculation is usually divided into feed preparation, selective adsorption of flocculant (polymers) on specific minerals, floc formation, and conditioning. For separating flocs from dispersed particles, flotation or sedimentation can be used [104].

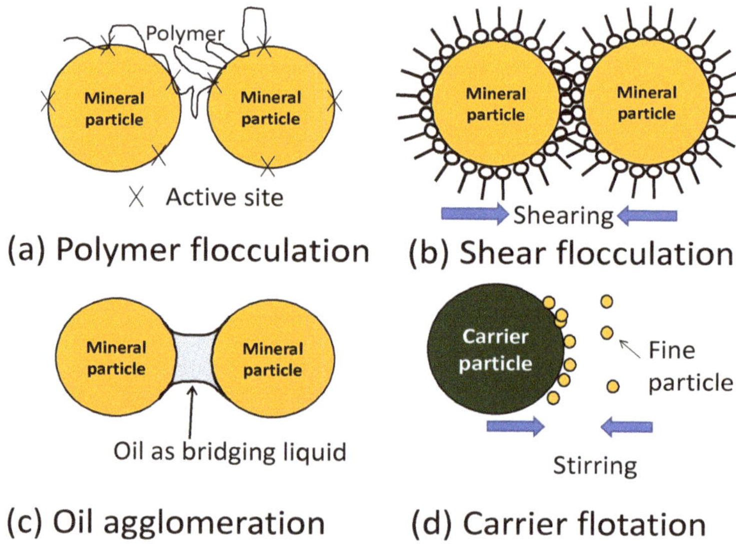

Figure 7. Schematic representation of polymer flocculation, shear flocculation, oil agglomeration, and carrier flotation.

Usually, flocculation is possible with a small amount of flocculant. This is an advantage of the flocculation technique, in comparison with other techniques discussed below. However, selective flocculation has proven to be a challenging technique [104]. Cellulose-type compounds are typical flocculants, and their effects on the flocculation of ground chalcopyrite, pyrite, and quartz were investigated using silylated cellulose nanofibers (SiCNF) [105]. The results showed that all minerals are flocculated in the presence of SiCNF, suggesting that cellulose-type compounds cannot be used for the selective flocculation of typical copper ores containing chalcopyrite, pyrite, and quartz. In another study [106], synthesized polymer xanthate was used as a flocculant before the flotation of complex sulfide ores, demonstrating good recovery. However, the grade of the concentrate was very low, indicating that the flocculant was non-selective.

3.2.2. Shear Flocculation

Shear flocculation is a technique to improve flotation kinetics. Mineral particles suspended in water have an electrical surface charge, and this causes a repulsive energy barrier between the homogeneously charged particles, suppressing their flocculation. In shear flocculation, kinetic energy is mechanically imparted to hydrophobic particles to make them collide together, in a mechanical mixing tank. High intensity stirring of hydrophobic mineral particles overcomes the energy barrier causing repulsion due to homogeneously charged particles [57,107,108]. Warren initially proposed this technique [108], concluding that, subject to provision of sufficient kinetic energy to the system, using agitation overcomes the energy barrier, forming flocs/aggregates of hydrophobic particles. Warren proposed that shear flocculation occurs in three steps: (1) in the turbulent movement, the average collision energy is much higher than thermal energy, which allows colloidal particles to approach each other; (2) the formation of flocs/aggregates is preferred with the energy of a "hydrophobic bond" in the event of a collision in direct contact between hydrophobic particles; (3) the resistance to the thinning and removal of the water layers separating the approaching particles is likely less in the case of hydrophobic particles than with colliding hydrophilic particles [109].

In a study conducted by Tabosa et al. [52], it was observed that high-intensity conditioning had less effect when the particle size of copper particles was less than 15 μm.

However, for particle sizes >15 μm, high-intensity conditioning increased copper recovery. The flotation of galena or sphalerite with a particle size less than 20 μm after shear flocculation has been reported [110]. Over 80% recovery was obtained for both galena and sphalerite when shear flocculation was applied. Shear flocculation also requires a precise control of the hydrodynamics of the system. In the flocculation process, there is a fine balance between the amount of shear force that initiates the process and the amount of shear that disintegrates aggregates beyond a certain point. Currently, shear flocculation is not frequently used in mineral processing, despite its potential. This may be due to the difficulty in controlling shear flocculation, selectivity issues, and high energy demands.

3.2.3. Oil Agglomeration

The oil agglomeration technique was developed in the early 1920s for coal cleaning; it has been modified and improved many times since then [111]. This technique uses oil as a bridging liquid to aggregate minerals [46–48,112,113]. Kerosene, vegetable oil, and diesel oil are used as bridging liquids for agglomeration to increase apparent particle size, leading to improved flotation of ultrafine particles [114]. The capillary force of the bridging liquid between particles causes an attraction force between particles [111]. The application of this technique has been researched for the purification of coal [111,115–121], recovery of gold [122,123], molybdenite [114], and chalcopyrite [46–48].

Dosage and type of oil, hydrophobicity of minerals, and agitation strength are the key parameters for oil agglomeration [117,124–127]. Oil is expensive and required in large amounts, which need to be reduced to improve economic viability. This is a major disadvantage associated with oil agglomeration for industrial-scale operations [112].

To reduce the amount of oil used in the agglomeration, the use of oil–water emulsions has been proposed [112]. Using emulsifiers decreases oil droplet size and increases the surface area [128–130]. In a recent study [46], emulsified oil agglomeration flotation was conducted on a fine chalcopyrite/quartz mixture by varying different parameters: agitation strength, collector dosage, and amount of oil. Emulsified oil agglomeration flotation significantly improved copper (Cu) recovery from around 45% without agglomeration to over 80% with agglomeration, as well as separation efficiency from around 30% without agglomeration to 70% with agglomeration.

It was observed that strong agitation strength is required to produce a large number of oil droplets, which is important for producing a larger agglomerate of mineral particles. However, this process consumes a lot of energy, making it expensive and difficult to incorporate into flotation circuits. Later, a new method was developed to overcome this problem by using emulsified oil stabilized by a surfactant to produce a stable oil–water emulsion that does not require strong agitation strength [47]; this method can be easily integrated into the existing flotation circuits.

A major problem of oil agglomeration is the amount of oil and high agitation speed required. This problem can be overcome when emulsified oil is used: small oil droplets become stable [131], and strong agitation is not needed. However, when the number of copper minerals becomes smaller, agglomerate size becomes small, and the flotation rate is low. Therefore, it is difficult to use oil agglomeration for Cu recovery from tailings with a very minute amount of copper mineral particles (Figure 8).

3.2.4. Carrier Flotation

Carrier flotation is a method for improving fine particle recovery in flotation. In this method, coarse carrier particles are added to the flotation cell and agitated together with fine mineral particles. The fine particles are then attached to the surface of coarse carrier particles and floated together.

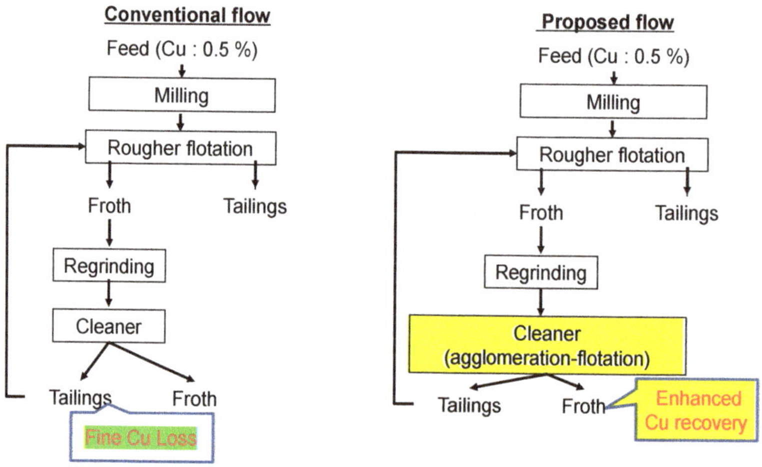

Figure 8. Application of agglomeration to flotation circuit.

The carrier flotation technique was initially developed to remove impurities from kaolin clay, using coarse limestone as a seed or carrier [132]. Normally, coarse particles are present in the ore, so adding extra coarse particles is not desirable in that case. If coarse particles are already present, they may act as a seed, in which case introducing extra carrier particles may not work [132,133]. When fine particles are attached to coarse particles (carrier or seed) of the same mineral, the process is termed autogenous carrier flotation [134]. Heterogenous carrier flotation uses coarse carrier particles that are different from the target fine minerals. Autogenous carrier flotation was first carried out on wolframite [135], and later used for coal [136]. With heterogeneous carrier flotation, numerous research works have been reported. For example, Jorge Rubio and Heinz used polypropylene treated with oleic acid as a carrier for different minerals [137].

Requirements for carrier flotation are the presence of coarse particles (carrier or seed), and hydrophobic coarse and fine particles [107]. The collision rate between fine and coarse particles is comparatively greater than fine and fine particles; also, fine particles attach well to the surface of coarse particles [134,138–140]. The schematic mechanism of carrier flotation is presented in Figure 9. It can be seen that when coarse carrier particles are agitated together with fine mineral particles, the fine particles are selectively attached to the surface of carrier particles.

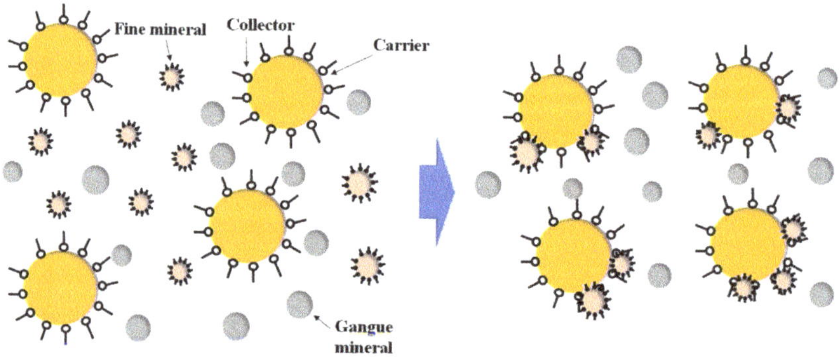

Figure 9. Schematic mechanism of carrier flotation.

In carrier flotation, agitation time and intensity; pulp density; carrier particle size; coarse–fine particle ratio; and chemical factors such as pH, collector dosage, the extent of hydrophobicity, the surface charge of the particle, and geometry of the cell are important variables, affecting the recovery of fines. Among these variables, the coarse–fine ratio may be the most important.

When the carrier (coarse particle) amount is increased, fine particle recovery is also increased [139]. However, overly large amounts of coarse particles are harmful. If the amount of coarse particles is too large, collision rates between coarse particles become high, which may cause detachment of already adhered fine particles [107].

In a study by Li et al. [141], interactions between coarse and fine hematite particles were studied. When the amounts of coarse and fine hematite particles were equal, attachment of fines to coarse carrier particles was maximized, causing maximum flotation recovery of fines.

It has been reported that carrier flotation is an effective method for improving the recovery of fine copper grains, where researchers investigated the autogenous carrier flotation technique using coarse chalcopyrite particles [142]. The fine chalcopyrite recovery was improved from 25% without a carrier to 80% with a carrier. The particle size distribution results as well as extended Derjaguin–Landau–Verwey–Overbeek (EDLVO) calculations confirmed the attachment of fine particles to coarse carrier particles.

As already discussed, considering tailings as future deposits, polymer flocculation, oil agglomeration, and shear flocculation are difficult to apply to copper recovery from tailings. This is mainly because of the low copper grade in tailings: floc/aggregate size depends on the number of copper minerals, and the size is limited when the amount of copper mineral is small.

Tailings consist primarily of gangue minerals, with a minute amount of valuable/target minerals, so the above-mentioned methods are unlikely to efficiently recover fine minerals from tailings. On the other hand, in carrier flotation, aggregate size is mainly dependent on the carrier size. This gives carrier flotation (autogenous carrier flotation) an advantage over other mentioned techniques (polymer flocculation, shear flocculation, and oil agglomeration) for recovering fine particles from tailings. However, an alternate carrier is required because of the non-availability of coarse chalcopyrite particles at the tailings site. Pyrite may be a good choice because of its abundance and availability at mine sites. The effectiveness of pyrite as a carrier for finely ground chalcopyrite particles was recently reported by Bilal et al. [143]. In this study [143], Cu^{2+}-activated pyrite particles were used as the carrier for recovering finely ground chalcopyrite particles. The wet sieving method (using ethanol spraying) was used to separate fine fractions from coarse fractions. However, when integrating this method into existing flotation circuits on an industrial scale, a post-flotation method was still not proposed. Therefore, the development of a post-flotation process to separate attached fine copper minerals from carriers (Cu^{2+}-activated pyrite) is needed.

4. Summary

In mineral processing at copper mines, around 10−20% of copper is lost into tailings. This is mainly because of the presence of overly fine copper mineral particles and unliberated copper mineral particles, which cannot be recovered in flotation. To recover copper from tailings, regrinding is required to liberate locked copper mineral particles, producing fine grains. This necessitates the development of flotation techniques to recover fine copper minerals from tailings. To this end, the present review addresses the challenges and prospects of recovering fine copper sulfides from tailings using different flotation techniques. Several techniques have been developed to recover fine mineral particles either by increasing particle size or by decreasing bubble size. The following findings are highlighted:

- Evaluating literature data shows that to recover fine particles and to maintain the required particle–bubble diameter ratio, bubble size needs to be reduced or particle size needs to be increased.

- The mean diameter of bubbles produced in column cells (1−3 mm) is still relatively large, and fine particles are less likely to collide with rising bubbles.
- In addition to enhancing fine particle recovery by forming aggregates, NBs also act as secondary collectors, thereby reducing collector and frother consumption.
- Microbubbles rising with attached mineral particles lead to longer residence times in the flotation circuit due to their slow speed. Moreover, the microbubbles increase water recovery, thereby increasing gangue mineral entrainment and reducing selectivity.
- Polymer flocculation is a promising method for increasing the apparent particle size, but this method is less selective.
- Shear flocculation requires a comparatively high amount of energy compared to other aggregation methods.
- Oil agglomeration requires a large amount of oil and agitation strength; however, this problem can be overcome when emulsified oil stabilized by a surfactant is used.
- Carrier flotation utilizes hydrophobic interactions, and fine particles are selectively attached onto the surface of coarse carrier particles.
- Polymer flocculation, shear flocculation, and oil agglomeration can generate larger flocs/aggregates from fine particles, thereby improving fine particle recovery. However, none of these techniques can be applied to recovering copper from tailings because the amount of copper minerals in tailings is small, causing a limitation of floc/aggregate size. Their applicability in mineral flotation has also been limited due to the lack of selectivity in aggregation.
- Carrier flotation is more suitable for copper recovery from tailings because agglomerate size is mainly dependent on carrier size itself, and can be applied even if the amount of fine particles is limited.

In light of the above conclusions and given information in the literature, the following future studies are recommended:

- Microbubble flotation and nanobubble flotation can improve the flotation rate and recovery of fine particles; however, they cause a higher water recovery, which causes increased entrainment of gangue minerals. Researchers have rarely addressed NB's impact on grade, separation efficiency, and selectivity of separation. More research is needed in this regard.
- The size of the agglomerate depends on the number of valuable mineral particles available in the feed. This gives the carrier flotation an advantage over the other agglomeration techniques because, in carrier flotation, agglomerate size is decided by the carrier itself; however, all these techniques have been applied on high-grade ores. The efficiency of these methods needs to be checked on real samples, i.e., tailings and low-grade ores.
- Autogenous carrier flotation can be integrated into the existing flotation circuits to minimize the loss of fine copper mineral particles, but the amount of carrier particles needs to be optimized for industrial scale operations.
- Research has shown that Cu^{2+}-activated pyrite can be used as a carrier; however, there is no study on the post-flotation separation of fine particles attached to Cu^{2+}-activated pyrite particles. Therefore, further studies are recommended in this regard.

Author Contributions: Conceptualization, M.B., I.P., M.I., F.U.H., V.H. and N.H.; writing—original draft preparation, M.B.; writing—review and editing, M.B., I.P., M.I., F.U.H., S.J. and N.H.; visualization, M.B. All authors have read and agreed to the published version of the manuscript.

Funding: This research received no external funding.

Conflicts of Interest: The authors declare no conflict of interest.

References

1. Schlesinger, M.E.; King, M.J.; Sole, K.C.; Davenport, W.G. (Eds.) Chapter 2—Production and Use. In *Extractive Metallurgy of Copper*, 5th ed.; Elsevier: Oxford, UK, 2011; pp. 13–30.
2. Tran, T.Q.; Chinnappan, A.; Lee, J.K.; Loc, N.H.; Tran, L.T.; Wang, G.; Kumar, V.V.; Jayathilaka, W.A.D.M.; Ji, D.; Doddamani, M.; et al. 3D Printing of Highly Pure Copper. *Metals* **2019**, *9*, 756. [CrossRef]
3. Tabelin, C.B.; Park, I.; Phengsaart, T.; Jeon, S.; Villacorte-Tabelin, M.; Alonzo, D.; Yoo, K.; Ito, M.; Hiroyoshi, N. Copper and critical metals production from porphyry ores and E-wastes: A review of resource availability, processing/recycling challenges, socio-environmental aspects, and sustainability issues. *Resour. Conserv. Recycl.* **2021**, *170*, 105610. [CrossRef]
4. Hund, K.; La Porta, D.; Fabregas, T.P.; Laing, T.; Drexhage, J. *Minerals for Climate Action: The Mineral Intensity of the Clean Energy Transition*; World Bank Group: Washington, DC, USA, 2020.
5. Park, I. Advances in Selective Flotation and Leaching Process in Metallurgy. *Metals* **2022**, *12*, 144. [CrossRef]
6. Park, I.; Kanazawa, Y.; Sato, N.; Galtchandmani, P.; Jha, M.K.; Tabelin, C.B.; Jeon, S.; Ito, M.; Hiroyoshi, N. Beneficiation of Low-Grade Rare Earth Ore from Khalzan Buregtei Deposit (Mongolia) by Magnetic Separation. *Minerals* **2021**, *11*, 1432. [CrossRef]
7. Copper Drives Electric Vehicles. Available online: https://www.copper.org/publications/pub_list/pdf/A6191-ElectricVehicles-Factsheet.pdf (accessed on 15 March 2022).
8. Schlesinger, M.E.; King, M.J.; Sole, K.C.; Davenport, W.G. (Eds.) Chapter 1—Overview. In *Extractive Metallurgy of Copper*, 5th ed.; Elsevier: Oxford, UK, 2011; pp. 1–12.
9. de Lima, G.F.; de Oliveira, C.; de Abreu, H.A.; Duarte, H.A. Water Adsorption on the Reconstructed (001) Chalcopyrite Surfaces. *J. Phys. Chem. C* **2011**, *115*, 10709–10717. [CrossRef]
10. Park, I.; Hong, S.; Jeon, S.; Ito, M.; Hiroyoshi, N. A Review of Recent Advances in Depression Techniques for Flotation Separation of Cu–Mo Sulfides in Porphyry Copper Deposits. *Metals* **2020**, *10*, 1269. [CrossRef]
11. Park, I.; Hong, S.; Jeon, S.; Ito, M.; Hiroyoshi, N. Flotation Separation of Chalcopyrite and Molybdenite Assisted by Microencapsulation Using Ferrous and Phosphate Ions: Part I—Selective Coating Formation. *Metals* **2020**, *10*, 1667. [CrossRef]
12. Park, I.; Hong, S.; Jeon, S.; Ito, M.; Hiroyoshi, N. Flotation Separation of Chalcopyrite and Molybdenite Assisted by Microencapsulation Using Ferrous and Phosphate Ions: Part II—Flotation. *Metals* **2021**, *11*, 439. [CrossRef]
13. Aikawa, K.; Ito, M.; Kusano, A.; Park, I.; Oki, T.; Takahashi, T.; Furuya, H.; Hiroyoshi, N. Flotation of Seafloor Massive Sulfide Ores: Combination of Surface Cleaning and Deactivation of Lead-Activated Sphalerite to Improve the Separation Efficiency of Chalcopyrite and Sphalerite. *Metals* **2021**, *11*, 253. [CrossRef]
14. Bulatovic, S.M. (Ed.) 12—Flotation of Copper Sulfide Ores. In *Handbook of Flotation Reagents*; Elsevier: Amsterdam, The Netherlands, 2007; pp. 235–293.
15. Yuan, X.M.; Palsson, B.I.; Forssberg, K.S.E. Statistical interpretation of flotation kinetics for a complex sulphide ore. *Miner. Eng.* **1996**, *9*, 429–442. [CrossRef]
16. Sripriya, R.; Rao, P.V.T.; Choudhury, B.R. Optimisation of operating variables of fine coal flotation using a combination of modified flotation parameters and statistical techniques. *Int. J. Miner. Processing* **2003**, *68*, 109–127. [CrossRef]
17. Peng, Y.; Bhambhani, T. Preface to the MME Special Focus Issue on Managing Gangue Minerals. *Min. Metall. Explor.* **2021**, *38*, 669–671. [CrossRef]
18. Chen, J.; Chimonyo, W.; Peng, Y. Flotation behaviour in reflux flotation cell—A critical review. *Miner. Eng.* **2022**, *181*, 107519. [CrossRef]
19. Fosu, S.; Awatey, B.; Skinner, W.; Zanin, M. Flotation of coarse composite particles in mechanical cell vs. the fluidised-bed separator (The HydroFloat™). *Miner. Eng.* **2015**, *77*, 137–149. [CrossRef]
20. Suppes, R.; Heuss-Aßbichler, S. Resource potential of mine wastes: A conventional and sustainable perspective on a case study tailings mining project. *J. Clean. Prod.* **2021**, *297*, 126446. [CrossRef]
21. Kapur, A.; Graedel, T.E. Copper Mines Above and Below the Ground. *Environ. Sci. Technol.* **2006**, *40*, 3135–3141. [CrossRef] [PubMed]
22. Ober, J.A. *Mineral Commodity Summaries 2018*; USGS: Reston, VA, USA, 2018; p. 204.
23. Park, I.; Tabelin, C.B.; Jeon, S.; Li, X.; Seno, K.; Ito, M.; Hiroyoshi, N. A review of recent strategies for acid mine drainage prevention and mine tailings recycling. *Chemosphere* **2019**, *219*, 588–606. [CrossRef]
24. Trahar, W.J. A rational interpretation of the role of particle size in flotation. *Int. J. Miner. Processing* **1981**, *8*, 289–327. [CrossRef]
25. Miettinen, T.; Ralston, J.; Fornasiero, D. The limits of fine particle flotation. *Miner. Eng.* **2010**, *23*, 420–437. [CrossRef]
26. De, F.; Gontijo, C.; Fornasiero, D.; Ralston, J. The limits of fine and coarse particle flotation. *Can. J. Chem. Eng.* **2007**, *85*, 739–747. [CrossRef]
27. Hassanzadeh, A.; Safari, M.; Hoang, D.H. Fine, Coarse and Fine-Coarse Particle Flotation in Mineral Processing with A Particular Focus on The Technological Assessments. In Proceedings of the 2nd International Conference on Mineral Science, Online, 1–15 March 2021; pp. 1–15.
28. Hassanzadeh, A.; Safari, M.; Hoang, D.H.; Khoshdast, H.; Albijanic, B. and Kowalczuk, P.B. Technological assessments on recent developments in fine and coarse particle flotation systems. *Miner. Eng.* **2022**, *180*, 107509. [CrossRef]
29. Farrokhpay, S.; Filippov, L.; Fornasiero, D. Flotation of Fine Particles: A Review. *Miner. Processing Extr. Metall. Rev.* **2021**, *42*, 473–483. [CrossRef]

30. Schlesinger, M.E.; King, M.J.; Sole, K.C.; Davenport, W.G. (Eds.) Chapter 4—Production of Cu Concentrate from Finely Ground Cu Ore. In *Extractive Metallurgy of Copper*, 5th ed.; Elsevier: Oxford, UK, 2011; pp. 51–71.
31. Han, B.; Altansukh, B.; Haga, K.; Stevanović, Z.; Jonović, R.; Avramović, L.; Urosević, D.; Takasaki, Y.; Masuda, N.; Ishiyama, D.; et al. Development of copper recovery process from flotation tailings by a combined method of high–pressure leaching–solvent extraction. *J. Hazard. Mater.* **2018**, *352*, 192–203. [CrossRef] [PubMed]
32. Antonijević, M.M.; Dimitrijević, M.D.; Stevanović, Z.O.; Serbula, S.M.; Bogdanovic, G.D. Investigation of the possibility of copper recovery from the flotation tailings by acid leaching. *J. Hazard. Mater.* **2008**, *158*, 23–34. [CrossRef] [PubMed]
33. Hansen, H.K.; Yianatos, J.B.; Ottosen, L.M. Speciation and leachability of copper in mine tailings from porphyry copper mining: Influence of particle size. *Chemosphere* **2005**, *60*, 1497–1503. [CrossRef]
34. Bagster, D.F.; McIlvenny, J.D. Studies in the selective flocculation of hematite from gangue using high molecular weight polymers. Part 1: Chemical factors. *Int. J. Miner. Processing* **1985**, *14*, 1–20. [CrossRef]
35. Mankosa, M.J.; Kohmuench, J.N.; Christodoulou, L.; Luttrell, G.H. Recovery of values from a porphory copper tailings stream. In Proceedings of the XXVIII International Mineral Processing Congress, Quebec City, QC, Canada, 11–15 September 2016; pp. 11–15.
36. Shergold, H. Flotation in mineral processing. In *The Scientific Basis of Flotation*; Springer: Berlin/Heidelberg, Germany, 1984; Volume 75, pp. 229–287.
37. Jameson, G.J. The effect of surface liberation and particle size on flotation rate constants. *Miner. Eng.* **2012**, *36*, 132–137. [CrossRef]
38. Schubert, H. On the optimization of hydrodynamics in fine particle flotation. *Miner. Eng.* **2008**, *21*, 930–936. [CrossRef]
39. Trahar, W.J.; Warren, L.J. The flotability of very fine particles—A review. *Int. J. Miner. Processing* **1976**, *3*, 103–131. [CrossRef]
40. Sivamohan, R. The problem of recovering very fine particles in mineral processing—A review. *Int. J. Miner. Processing* **1990**, *28*, 247–288. [CrossRef]
41. Leistner, T.; Embrechts, M.; Leißner, T.; Chehreh Chelgani, S.; Osbahr, I.; Möckel, R.; Peuker, U.A.; Rudolph, M. A study of the reprocessing of fine and ultrafine cassiterite from gravity tailing residues by using various flotation techniques. *Miner. Eng.* **2016**, *96*, 94–98. [CrossRef]
42. Asghari, M.; Nakhaei, F.; VandGhorbany, O. Copper recovery improvement in an industrial flotation circuit: A case study of Sarcheshmeh copper mine. *Energy Sources Part A Recovery Util. Environ. Eff.* **2019**, *41*, 761–778. [CrossRef]
43. Leistner, T.; Peuker, U.A.; Rudolph, M. How gangue particle size can affect the recovery of ultrafine and fine particles during froth flotation. *Miner. Eng.* **2017**, *109*, 1–9. [CrossRef]
44. Mankosa, M.J.; Kohmuench, J.N.; Christodoulou, L.; Yan, E.S. Improving fine particle flotation using the StackCell™ (raising the tail of the elephant curve). *Miner. Eng.* **2018**, *121*, 83–89. [CrossRef]
45. Yoon, R.H. The role of hydrodynamic and surface forces in bubble–particle interaction. *Int. J. Miner. Processing* **2000**, *58*, 129–143. [CrossRef]
46. Hornn, V.; Ito, M.; Shimada, H.; Tabelin, C.B.; Jeon, S.; Park, I.; Hiroyoshi, N. Agglomeration-Flotation of Finely Ground Chalcopyrite and Quartz: Effects of Agitation Strength during Agglomeration Using Emulsified Oil on Chalcopyrite. *Minerals* **2020**, *10*, 380. [CrossRef]
47. Hornn, V.; Ito, M.; Shimada, H.; Tabelin, C.B.; Jeon, S.; Park, I.; Hiroyoshi, N. Agglomeration–Flotation of Finely Ground Chalcopyrite Using Emulsified Oil Stabilized by Emulsifiers: Implications for Porphyry Copper Ore Flotation. *Metals* **2020**, *10*, 912. [CrossRef]
48. Hornn, V.; Ito, M.; Yamazawa, R.; Shimada, H.; Tabelin, C.B.; Jeon, S.; Park, I.; Hiroyoshi, N. Kinetic Analysis for Agglomeration-Flotation of Finely Ground Chalcopyrite: Comparison of First Order Kinetic Model and Experimental Results. *Mater. Trans.* **2020**, *61*, 1940–1948. [CrossRef]
49. Nguyen-Van, A. The Collision between Fine Particles and Single Air Bubbles in Flotation. *J. Colloid Interface Sci.* **1994**, *162*, 123–128. [CrossRef]
50. Tortorelli, J.P.; Craven, J.W.; Toguri, J.M.; Dobby, G.S.; Agar, G.E. The effect of external gas/slurry contact on the flotation of fine particles. *Miner. Eng.* **1997**, *10*, 1127–1138. [CrossRef]
51. Nguyen, A.; Schulze, H.J. *Colloidal Science of Flotation*; CRC Press: Boca Raton, FL, USA, 2003; Volume 118.
52. Tabosa, E.; Rubio, J. Flotation of copper sulphides assisted by high intensity conditioning (HIC) and concentrate recirculation. *Miner. Eng.* **2010**, *23*, 1198–1206. [CrossRef]
53. Yianatos, J.; Bucarey, R.; Larenas, J.; Henríquez, F.; Torres, L. Collection zone kinetic model for industrial flotation columns. *Miner. Eng.* **2005**, *18*, 1373–1377. [CrossRef]
54. Finch, J.A. Column flotation: A selected review—Part IV: Novel flotation devices. *Miner. Eng.* **1995**, *8*, 587–602. [CrossRef]
55. Rodrigues, R.T.; Rubio, J. DAF—Dissolved air flotation: Potential applications in the mining and mineral processing industry. *Int. J. Miner. Processing* **2007**, *82*, 1–13. [CrossRef]
56. Fuerstenau, M.C.; Jameson, G.J.; Yoon, R.-H. *Froth Flotation: A Century of Innovation*; SME: Littleton, CO, USA, 2007.
57. Patil, D.P.; Andrews, J.R.G.; Uhlherr, P.H.T. Shear flocculation—Kinetics of floc coalescence and breakage. *Int. J. Miner. Processing* **2001**, *61*, 171–188. [CrossRef]
58. Pyecha, J.; Lacouture, B.; Sims, S.; Hope, G.; Stradling, A. Evaluation of a Microcel™ sparger in the Red Dog column flotation cells. *Miner. Eng.* **2006**, *19*, 748–757. [CrossRef]

59. Zahab Nazouri, A.; Shojaei, V.; Khoshdast, H.; Hassanzadeh, A. Hybrid CFD-experimental investigation into the effect of sparger orifice size on the metallurgical response of coal in a pilot-scale flotation column. *Int. J. Coal Prep. Util.* **2022**, *42*, 349–368. [CrossRef]
60. Vothy, H. Development of Agglomeration-Flotation for Finely Ground Copper Sulfides. Ph.D. Thesis, Hokkaido University, Sapporo, Japan, 2020.
61. Terasaka, K. Leading edge in fine bubble technology. *J. Jpn. Inst. Energy* **2014**, *93*, 1022–1024.
62. Ketkar, D.R.; Mallikarjunan, R.; Venkatachalam, S. Electroflotation of quartz fines. *Int. J. Miner. Processing* **1991**, *31*, 127–138. [CrossRef]
63. Rubio, J.; Souza, M.L.; Smith, R.W. Overview of flotation as a wastewater treatment technique. *Miner. Eng.* **2002**, *15*, 139–155. [CrossRef]
64. Kyzas, G.Z.; Matis, K.A. Electroflotation process: A review. *J. Mol. Liq.* **2016**, *220*, 657–664. [CrossRef]
65. Yoon, R.H. Microbubble flotation. *Miner. Eng.* **1993**, *6*, 619–630. [CrossRef]
66. Kusaka, E. Microbubble flotation of ultrafine divided sulfide mineral. *Min. Mater. Processing Inst. Jpn.* **2014**, *1*. (In Japanese)
67. Farrokhpay, S.; Filippova, I.; Filippov, L.; Picarra, A.; Rulyov, N.; Fornasiero, D. Flotation of fine particles in the presence of combined microbubbles and conventional bubbles. *Miner. Eng.* **2020**, *155*, 106439. [CrossRef]
68. Liu, Q.; Wannas, D. The role of polymeric-depressant-induced flocculation in fine particle flotation. In Proceedings of the UBC Mcgill Biennial International Symposium on Fundamentals of Mineral Processing, Toronto, ON, Canada, 22–25 August 2004.
69. Bhaskar Raju, G.; Khangaonkar, P.R. Electro-flotation of chalcopyrite fines. *Int. J. Miner. Processing* **1982**, *9*, 133–143. [CrossRef]
70. Waters, K.E.; Hadler, K.; Cilliers, J.J. The flotation of fine particles using charged microbubbles. *Miner. Eng.* **2008**, *21*, 918–923. [CrossRef]
71. Murao, K. Fundamental study on the collectorless microbubble flotation using sodium hydrosulfide. *Min. Mater. Processing Inst. Jpn.* **2014**, *1*.
72. Ahmadi, R.; Khodadadi, D.A.; Abdollahy, M.; Fan, M. Nano-microbubble flotation of fine and ultrafine chalcopyrite particles. *Int. J. Min. Sci. Technol.* **2014**, *24*, 559–566. [CrossRef]
73. Parker, J.L.; Claesson, P.M.; Attard, P. Bubbles, cavities, and the long-ranged attraction between hydrophobic surfaces. *J. Phys. Chem.* **1994**, *98*, 8468–8480. [CrossRef]
74. Uchida, T.; Oshita, S.; Ohmori, M.; Tsuno, T.; Soejima, K.; Shinozaki, S.; Take, Y.; Mitsuda, K. Transmission electron microscopic observations of nanobubbles and their capture of impurities in wastewater. *Nanoscale Res. Lett.* **2011**, *6*, 295. [CrossRef]
75. Zimmerman, W.B.; Tesař, V.; Bandulasena, H.C.H. Towards energy efficient nanobubble generation with fluidic oscillation. *Curr. Opin. Colloid Interface Sci.* **2011**, *16*, 350–356. [CrossRef]
76. Wang, Q.; Zhao, H.; Qi, N.; Qin, Y.; Zhang, X.; Li, Y. Generation and Stability of Size-Adjustable Bulk Nanobubbles Based on Periodic Pressure Change. *Sci. Rep.* **2019**, *9*, 1118. [CrossRef] [PubMed]
77. Calgaroto, S.; Azevedo, A.; Rubio, J. Flotation of quartz particles assisted by nanobubbles. *Int. J. Miner. Processing* **2015**, *137*, 64–70. [CrossRef]
78. Azevedo, A.; Etchepare, R.; Calgaroto, S.; Rubio, J. Aqueous dispersions of nanobubbles: Generation, properties and features. *Miner. Eng.* **2016**, *94*, 29–37. [CrossRef]
79. Zhou, S.; Wang, X.; Bu, X.; Wang, M.; An, B.; Shao, H.; Ni, C.; Peng, Y.; Xie, G. A novel flotation technique combining carrier flotation and cavitation bubbles to enhance separation efficiency of ultra-fine particles. *Ultrason. Sonochemistry* **2020**, *64*, 105005. [CrossRef] [PubMed]
80. Fan, M.; Tao, D.; Honaker, R.; Luo, Z. Nanobubble generation and its applications in froth flotation (part II): Fundamental study and theoretical analysis. *Min. Sci. Technol. China* **2010**, *20*, 159–177. [CrossRef]
81. Han, H.; Liu, A.; Wang, H. Effect of Hydrodynamic Cavitation Assistance on Different Stages of Coal Flotation. *Minerals* **2020**, *10*, 221. [CrossRef]
82. Atluri, V.; Gao, Y.; Wang, X.; Pan, L.; Miller, J.D. The Influence of Polysaccharides on Film Stability and Bubble Attachment at the Talc Surface. *Min. Metall. Explor.* **2019**, *36*, 71–80. [CrossRef]
83. Li, C.; Zhang, H. A review of bulk nanobubbles and their roles in flotation of fine particles. *Powder Technol.* **2022**, *395*, 618–633. [CrossRef]
84. Nazari, S.; Hassanzadeh, A.; He, Y.; Khoshdast, H.; Kowalczuk, P.B. Recent Developments in Generation, Detection and Application of Nanobubbles in Flotation. *Minerals* **2022**, *12*, 462. [CrossRef]
85. Cho, S.-H.; Kim, J.-Y.; Chun, J.-H.; Kim, J.-D. Ultrasonic formation of nanobubbles and their zeta-potentials in aqueous electrolyte and surfactant solutions. *Colloids Surf. A Physicochem. Eng. Asp.* **2005**, *269*, 28–34. [CrossRef]
86. Najafi, A.S.; Drelich, J.; Yeung, A.; Xu, Z.; Masliyah, J. A novel method of measuring electrophoretic mobility of gas bubbles. *J. Colloid Interface Sci.* **2007**, *308*, 344–350. [CrossRef] [PubMed]
87. Yang, S.; Dammer, S.M.; Bremond, N.; Zandvliet, H.J.W.; Kooij, E.S.; Lohse, D. Characterization of Nanobubbles on Hydrophobic Surfaces in Water. *Langmuir* **2007**, *23*, 7072–7077. [CrossRef] [PubMed]
88. Ushikubo, F.Y.; Furukawa, T.; Nakagawa, R.; Enari, M.; Makino, Y.; Kawagoe, Y.; Shiina, T.; Oshita, S. Evidence of the existence and the stability of nano-bubbles in water. *Colloids Surf. A Physicochem. Eng. Asp.* **2010**, *361*, 31–37. [CrossRef]
89. Fan, M.; Tao, D.; Honaker, R.; Luo, Z. Nanobubble generation and its applications in froth flotation (part III): Specially designed laboratory scale column flotation of phosphate. *Min. Sci. Technol. China* **2010**, *20*, 317–338. [CrossRef]

90. Tuziuti, T.; Yasui, K.; Kanematsu, W. Influence of addition of degassed water on bulk nanobubbles. *Ultrason. Sonochemistry* **2018**, *43*, 272–274. [CrossRef]
91. Sobhy, A.; Tao, D. Nanobubble column flotation of fine coal particles and associated fundamentals. *Int. J. Miner. Processing* **2013**, *124*, 109–116. [CrossRef]
92. Azevedo, A.; Oliveira, H.; Rubio, J. Bulk nanobubbles in the mineral and environmental areas: Updating research and applications. *Adv. Colloid Interface Sci.* **2019**, *271*, 101992. [CrossRef]
93. Chipakwe, V.; Sand, A.; Chelgani, S.C. Nanobubble assisted flotation separation of complex Pb–Cu–Zn sulfide ore—Assessment of process readiness. *Sep. Sci. Technol.* **2022**, *57*, 1351–1358. [CrossRef]
94. Chipakwe, V.; Jolsterå, R.; Chelgani, S.C. Nanobubble-Assisted Flotation of Apatite Tailings: Insights on Beneficiation Options. *ACS Omega* **2021**, *6*, 13888–13894. [CrossRef]
95. Sutherland, K.L. Physical Chemistry of Flotation. XI. Kinetics of the Flotation Process. *J. Phys. Colloid Chem.* **1948**, *52*, 394–425. [CrossRef] [PubMed]
96. Crawford, R.; Ralston, J. The influence of particle size and contact angle in mineral flotation. *Int. J. Miner. Processing* **1988**, *23*, 1–24. [CrossRef]
97. Ralston, J.; Dukhin, S.S. The interaction between particles and bubbles. *Colloids Surf. A Physicochem. Eng. Asp.* **1999**, *151*, 3–14. [CrossRef]
98. Duan, J.; Fornasiero, D.; Ralston, J. Calculation of the flotation rate constant of chalcopyrite particles in an ore. *Int. J. Miner. Processing* **2003**, *72*, 227–237. [CrossRef]
99. Pyke, B.; Fornasiero, D.; Ralston, J. Bubble particle heterocoagulation under turbulent conditions. *J. Colloid Interface Sci.* **2003**, *265*, 141–151. [CrossRef]
100. Hogg, R. Flocculation and dewatering. *Int. J. Miner. Processing* **2000**, *58*, 223–236. [CrossRef]
101. Pearse, M.J. Historical use and future development of chemicals for solid–liquid separation in the mineral processing industry. *Miner. Eng.* **2003**, *16*, 103–108. [CrossRef]
102. Usher, S.P.; Spehar, R.; Scales, P.J. Theoretical analysis of aggregate densification: Impact on thickener performance. *Chem. Eng. J.* **2009**, *151*, 202–208. [CrossRef]
103. Phiri, T.; Tepa, C.; Nyati, R. Effect of Desliming on Flotation Response of Kansanshi Mixed Copper Ore. *J. Miner. Mater. Charact. Eng.* **2019**, *7*, 20. [CrossRef]
104. Wills, B.A.; Finch, J.A. (Eds.) Chapter 12—Froth Flotation. In *Wills' Mineral Processing Technology*, 5th ed.; Butterworth-Heinemann: Boston, UK, 2016; pp. 265–380.
105. Zhou, H.; Geng, L.; Zhang, Y.; Yang, Z.; He, K.; Xie, F. Selective flotation separation of chalcopyrite and sphalerite by thermal pretreatment under air atmosphere. *Physicochem. Probl. Miner. Process.* **2021**, *57*, 305–314. [CrossRef]
106. Mandre, N.R.; Panigrahi, D. Studies on selective flocculation of complex sulphides using cellulose xanthate. *Int. J. Miner. Processing* **1997**, *50*, 177–186. [CrossRef]
107. Subrahmanyam, T.V.; Forssberg, K.S.E. Fine particles processing: Shear-flocculation and carrier flotation—A review. *Int. J. Miner. Processing* **1990**, *30*, 265–286. [CrossRef]
108. Bilgen, S.; Wills, B.A. Shear flocculation—A review. *Miner. Eng.* **1991**, *4*, 483–487. [CrossRef]
109. Warren, L.J. Shear-flocculation of ultrafine scheelite in sodium oleate solutions. *J. Colloid Interface Sci.* **1975**, *50*, 307–318. [CrossRef]
110. Song, S.; Lopez-Valdivieso, A.; Reyes-Bahena, J.L.; Lara-Valenzuela, C. Floc flotation of galena and sphalerite fines. *Miner. Eng.* **2001**, *14*, 87–98. [CrossRef]
111. Mehrotra, V.P.; Sastry, K.V.S.; Morey, B.W. Review of oil agglomeration techniques for processing of fine coals. *Int. J. Miner. Processing* **1983**, *11*, 175–201. [CrossRef]
112. Bensley, C.N.; Swanson, A.R.; Nicol, S.K. The effect of emulsification on the selective agglomeration of fine coal. *Int. J. Miner. Processing* **1977**, *4*, 173–184. [CrossRef]
113. Wu, X.Q.; Monhemius, A.J.; Gochin, R.J. Quantitative assessment of hydrophobic agglomeration performance. *Miner. Eng.* **2005**, *18*, 567–573. [CrossRef]
114. Jiangang, F.; Kaida, C.; Hui, W.; Chao, G.; Wei, L. Recovering molybdenite from ultrafine waste tailings by oil agglomerate flotation. *Miner. Eng.* **2012**, *39*, 133–139. [CrossRef]
115. Laskowski, J.S.; Yu, Z. Oil agglomeration and its effect on beneficiation and filtration of low-rank/oxidized coals. *Int. J. Miner. Processing* **2000**, *58*, 237–252. [CrossRef]
116. Gray, M.L.; Champagne, K.J.; Soong, Y.; Finseth, D.H. Parametric study of the column oil agglomeration of fly ash. *Fuel* **2001**, *80*, 867–871. [CrossRef]
117. Aktaş, Z. Some factors affecting spherical oil agglomeration performance of coal fines. *Int. J. Miner. Processing* **2002**, *65*, 177–190. [CrossRef]
118. Alonso, M.I.; Valdés, A.F.; Martínez-Tarazona, R.M.; Garcia, A.B. Coal recovery from fines cleaning wastes by agglomeration with colza oil: A contribution to the environment and energy preservation. *Fuel Processing Technol.* **2002**, *75*, 85–95. [CrossRef]
119. Cebeci, Y.; Sönmez, İ. The investigation of coal–pyrite/lignite concentration and their separation in the artificial mixture by oil agglomeration. *Fuel* **2002**, *81*, 1139–1146. [CrossRef]
120. Cebeci, Y.; Sönmez, İ. Application of the Box-Wilson experimental design method for the spherical oil agglomeration of coal. *Fuel* **2006**, *85*, 289–297. [CrossRef]

121. Sahinoglu, E.; Uslu, T. Amenability of Muzret bituminous coal to oil agglomeration. *Energy Convers. Manag.* **2008**, *49*, 3684–3690. [CrossRef]
122. Moses, L.B.; Petersen, F.W. Flotation as a separation technique in the coal gold agglomeration process. *Miner. Eng.* **2000**, *13*, 255–264. [CrossRef]
123. Sen, S.; Seyrankaya, A.; Cilingir, Y. Coal–oil assisted flotation for the gold recovery. *Miner. Eng.* **2005**, *18*, 1086–1092. [CrossRef]
124. Alonso, M.I.; Valdés, A.F.; Martínez-Tarazona, R.M.; Garcia, A.B. Coal recovery from coal fines cleaning wastes by agglomeration with vegetable oils: Effects of oil type and concentration. *Fuel* **1999**, *78*, 753–759. [CrossRef]
125. Bos, J.L.; Quast, K.B. Effects of oils and lubricants on the flotation of copper sulphide minerals. *Miner. Eng.* **2000**, *13*, 1623–1627. [CrossRef]
126. Slaghuis, J.H.; Ferreira, L.C. Selective spherical agglomeration of coal: An amended mechanism of agglomerate formation and growth and its effect on product quality. *Fuel* **1987**, *66*, 1427–1430. [CrossRef]
127. Wheelock, T.D.; Milana, G.; Vettor, A. The role of air in oil agglomeration of coal at a moderate shear rate. *Fuel* **1994**, *73*, 1103–1107. [CrossRef]
128. Sahinoglu, E.; Uslu, T. Use of ultrasonic emulsification in oil agglomeration for coal cleaning. *Fuel* **2013**, *113*, 719–725. [CrossRef]
129. van Netten, K.; Moreno-Atanasio, R.; Galvin, K.P. Fine Particle Beneficiation through Selective Agglomeration with an Emulsion Binder. *Ind. Eng. Chem. Res.* **2014**, *53*, 15747–15754. [CrossRef]
130. Zhang, Q.; Niu, C.; Bu, X.; Bilal, M.; Ni, C.; Peng, Y. Enhancement of Flotation Performance of Oxidized Coal by the Mixture of Laurylamine Dipropylene Diamine and Kerosene. *Minerals* **2021**, *11*, 1271. [CrossRef]
131. Gao, J.; Bu, X.; Zhou, S.; Wang, X.; Bilal, M.; Hassan, F.U.; Hassanzadeh, A.; Xie, G.; Chelgani, S.C. Pickering emulsion prepared by nano-silica particles—A comparative study for exploring the effect of various mechanical methods. *Ultrason. Sonochemistry* **2022**, *83*, 105928. [CrossRef]
132. Ateşok, G.; Boylu, F.; Çelik, M.S. Carrier flotation for desulfurization and deashing of difficult-to-float coals. *Miner. Eng.* **2001**, *14*, 661–670. [CrossRef]
133. Warren, L.J.T.I.M.M.; Sect, C. Slime coating and shear flocculation in the scheelite-sodium oleate system. *Min. Proc. Ext. Metall.* **1975**, *84*, 99–104.
134. Valderrama, L.; Rubio, J. High intensity conditioning and the carrier flotation of gold fine particles. *Int. J. Miner. Processing* **1998**, *52*, 273–285. [CrossRef]
135. Hu, W.; Wang, D.; Jin, H. Flotation of Wolframite slime-practice and technological innovation. In *CIM BULLETIN*; CIM: Calgary, AB, Canada, 1982; Volume 75, p. 78.
136. Hu, W.; Wang, D.; Qu, G. Autogenous carrier flotation. In Proceedings of the International Mineral Processing Congress, Stockholm, Sweden, 5–10 June 1988; pp. 445–452.
137. Rubio, J.; Hoberg, H. The process of separation of fine mineral particles by flotation with hydrophobic polymeric carrier. *Int. J. Miner. Processing* **1993**, *37*, 109–122. [CrossRef]
138. Dianzuo, W.; Guanzhou, Q.; Weibai, H. The effect of carrier—Promoting aggregation of coarse particles in fine particle flotation. In *Production and Processing of Fine Particles*; Plumpton, A.J., Ed.; Pergamon: Amsterdam, The Netherlands, 1988; pp. 309–316.
139. Lange, A.G.; Skinner, W.M.; Smart, R.S.C. Fine: Coarse particle interactions and aggregation in sphalerite flotation. *Miner. Eng.* **1997**, *10*, 681–693. [CrossRef]
140. Hu, Y.; Qiu, G.; Miller, J.D. Hydrodynamic interactions between particles in aggregation and flotation. *Int. J. Miner. Processing* **2003**, *70*, 157–170. [CrossRef]
141. Li, D.; Yin, W.; Liu, Q.; Cao, S.; Sun, Q.; Zhao, C.; Yao, J. Interactions between fine and coarse hematite particles in aqueous suspension and their implications for flotation. *Miner. Eng.* **2017**, *114*, 74–81. [CrossRef]
142. Bilal, M.; Ito, M.; Koike, K.; Hornn, V.; Ul Hassan, F.; Jeon, S.; Park, I.; Hiroyoshi, N. Effects of coarse chalcopyrite on flotation behavior of fine chalcopyrite. *Miner. Eng.* **2021**, *163*, 106776. [CrossRef]
143. Bilal, M.; Ito, M.; Akishino, R.; Bu, X.; Ul Hassan, F.; Park, I.; Jeon, S.; Aikawa, K.; Hiroyoshi, N. Heterogenous carrier flotation technique for recovering finely ground chalcopyrite particles using coarse pyrite particles as a carrier. *Miner. Eng.* **2022**, *180*, 107518. [CrossRef]

minerals MDPI

Article

Flotation of Copper Ores with High Cu/Zn Ratio: Effects of Pyrite on Cu/Zn Separation and an Efficient Method to Enhance Sphalerite Depression

Kosei Aikawa [1,*], Mayumi Ito [2], Nodoka Orii [1], Sanghee Jeon [2], Ilhwan Park [2], Kazutoshi Haga [3], Taro Kamiya [4], Tatsuru Takahashi [5], Kazuya Sunada [6], Taisuke Sakakibara [7], Tatsuhiro Ono [4], Refilwe S. Magwaneng [4] and Naoki Hiroyoshi [2]

1 Division of Sustainable Resources Engineering, Graduate School of Engineering, Hokkaido University, Sapporo 060-8628, Japan
2 Division of Sustainable Resources Engineering, Faculty of Engineering, Hokkaido University, Sapporo 060-8628, Japan
3 Department of Earth Resource Engineering and Environmental Science, Akita University, Akita 010-0865, Japan
4 Metals Technology Center, Japan Oil, Gas and Metals National Corporation (JOGMEC), Kosaka 017-0202, Japan
5 Seafloor Mineral Resources Research & Development Division, Seafloor Mineral Resources Department, Japan Oil, Gas and Metals National Corporation (JOGMEC), Tokyo 105-0001, Japan
6 Finance Division, Metals Development Department, Japan Oil, Gas and Metals National Corporation (JOGMEC), Tokyo 105-0001, Japan
7 Metals Technology Division, Metals Development Department, Japan Oil, Gas and Metals National Corporation (JOGMEC), Tokyo 105-0001, Japan
* Correspondence: k-aikawa@frontier.hokudai.ac.jp; Tel.: +81-11-706-6315

Citation: Aikawa, K.; Ito, M.; Orii, N.; Jeon, S.; Park, I.; Haga, K.; Kamiya, T.; Takahashi, T.; Sunada, K.; Sakakibara, T.; et al. Flotation of Copper Ores with High Cu/Zn Ratio: Effects of Pyrite on Cu/Zn Separation and an Efficient Method to Enhance Sphalerite Depression. *Minerals* **2022**, *12*, 1103. https://doi.org/10.3390/min12091103

Academic Editor: Andrea Gerson

Received: 28 July 2022
Accepted: 27 August 2022
Published: 30 August 2022

Abstract: Porphyry copper deposits are important sources of copper and typically processed by flotation to produce copper concentrates. As mining areas become deeper, the amounts of impurities, such as sphalerite, can be increased in copper ores, so the appropriate depression of sphalerite floatability should be achieved to obtain saleable copper concentrates. In this study, the flotation behaviors of chalcopyrite and sphalerite in model samples mimicking copper ores with high Cu/Zn ratios (i.e., the ratio of chalcopyrite/sphalerite = 13:1) were investigated with zinc sulfate as a depressant for sphalerite. In addition, the effect of pyrite—a major gangue mineral in copper ores—on the depression of sphalerite floatability with zinc sulfate was examined. When sphalerite and chalcopyrite coexisted, the floatability of the former was effectively depressed by zinc sulfate (Zn recovery: <12%), whereas the presence of pyrite promoted the release of Cu^{2+} due to the galvanic interaction with chalcopyrite, which resulted in the elimination of the effectiveness of zinc sulfate in depressing sphalerite floatability (Zn recovery: >90%). Despite the presence of much higher amounts of chalcopyrite and pyrite than sphalerite, the application of nitrogen (N_2) gas limiting the galvanic interaction between pyrite and chalcopyrite by reducing the dissolved oxygen (DO) concentration in the system effectively depressed the floatability of sphalerite (Zn recovery: <30%).

Keywords: flotation; porphyry copper deposits; chalcopyrite; sphalerite; pyrite; galvanic interaction

1. Introduction

Copper (Cu) has excellent electrical and thermal conductivities, high corrosion resistance, and outstanding malleability and ductility, which make it an essential component in various applications (e.g., building and construction, electrical and electronic products, industrial machinery and equipment, transportation, etc.) [1,2]. Moreover, copper is one of 17 critical metals used across a variety of clean energy generation and storage technologies, so its demand is projected to increase continuously to achieve a low-carbon future [3,4].

Porphyry copper deposits (PCDs) are the most important sources of copper worldwide, accounting for more than 60% of the annual production of copper and they constitute about 65% of known copper ore resources [5]. Ores excavated from these deposits are typically processed by flotation—a separation technique based on the difference in surface wettabilities of minerals—to recover copper sulfides (mostly chalcopyrite ($CuFeS_2$)) while removing associated gangue minerals such as pyrite (FeS_2) and quartz (SiO_2) [6,7]. After a multi-stage flotation of copper ores, the produced copper concentrates are processed via pyrometallurgical routes to produce copper metal [8].

To meet the high forecasted demand for copper, mining industries are finding it necessary to mine ores at increasingly deeper positions since copper reserves are finite and easy-to-mine ore deposits are dwindling [9–11]. In the case of PCDs, as mining areas become deeper laterally, it could be extended to the peripheral zone, where the amount of sphalerite (ZnS) increases compared to the ore zone [12]. When processing copper ores containing sphalerite via flotation, the depression of sphalerite floatability is an important issue for producing saleable copper concentrates because the impurities in copper concentrates have a negative impact on copper smelting and penalties are imposed for their treatment in the smelting process [13–15]. In general, Cu-Zn sulfide ores consisting of a Cu/Zn ratio of 0.4–1:1 (*w*/*w*) [16–20] are processed by a two-step flotation: (i) the selective flotation of Cu minerals while depressing sphalerite using zinc sulfate ($ZnSO_4$)—a sphalerite depressant—and other gangue minerals such as pyrite (FeS_2) and silicate minerals under alkaline conditions, and (ii) the recovery of sphalerite after its activation using copper sulfate ($CuSO_4$) [21,22]. However, it is unclear whether the conventional approach for Cu-Zn and Cu-Zn-Pb ores (i.e., depression of sphalerite using zinc sulfate) is effective in Cu/Zn separation in the processing of the copper ores excavated from PCDs with a much lower level of Zn (e.g., a Cu/Zn ratio of 7:1 (*w*/*w*)) compared to Cu-Zn or Cu-Zn-Pb sulfide ores [23]. Thus, this study investigated the flotation behaviors of chalcopyrite and sphalerite in ores with high Cu/Zn ratios such as PCDs. Specifically, flotation tests of a single mineral (chalcopyrite or sphalerite) and mixed minerals (chalcopyrite and sphalerite with and without pyrite—a major impurity of PCDs) were conducted. Finally, a method to enhance sphalerite depression in the flotation of the chalcopyrite–sphalerite–pyrite mixture was proposed.

2. Materials and Methods

2.1. Minerals and Reagents

Four types of minerals were used in this study: chalcopyrite ($CuFeS_2$, Copper Queen Mine, Cochise County, AZ, USA), sphalerite (ZnS, Kamioka Mine, Hida, Japan), pyrite (FeS_2, Huanzala Mine, Huanuco, Peru), and quartz (SiO_2, 99% purity, FUJIFILM Wako Pure Chemical Corporation, Osaka, Japan). They were characterized by X-ray fluorescence spectroscopy (XRF, EDXL300, Rigaku Corporation, Tokyo, Japan) and X-ray powder diffraction (XRD, MultiFlex, Rigaku Corporation, Tokyo, Japan). The chemical and mineralogical compositions of these samples are summarized in Table 1 and shown in Figure 1, respectively. The XRD pattern of chalcopyrite (Figure 1a) showed that it contains impurities such as ferro-actinolite ($Ca_2(Fe^{2+}, Mg)_5Si_8O_{22}(OH)_2$), sphalerite, and quartz, whereas those of sphalerite and pyrite (Figure 1b,c) showed that these samples are highly pure as indicated by the absence of peaks from common minerals such as quartz. The samples were ground using a vibratory disc mill (RS 200, Retsch Inc., Haan, Germany) and screened to obtain a size fraction of $-75 + 38$ μm. For the flotation experiments, potassium isopropyl xanthate (PIPX, Tokyo Chemical Industry Co., Ltd., Tokyo, Japan) as a collector, methyl isobutyl carbinol (MIBC, Tokyo Chemical Industry Co., Ltd., Tokyo, Japan) as a frother, and zinc sulfate ($ZnSO_4$, FUJIFILM Wako Pure Chemical Corporation, Osaka, Japan) as a depressant for sphalerite were used. Sodium hydroxide (NaOH, FUJIFILM Wako Pure Chemical Corporation, Osaka, Japan) and sulfuric acid (H_2SO_4, FUJIFILM Wako Pure Chemical Corporation, Osaka, Japan) were used as pH adjusters.

Table 1. The chemical composition of mineral samples based on XRF.

Sample	Mass Fraction (%)				
	Cu	Zn	Fe	S	Si
Chalcopyrite	28.3	0.5	32.8	17.7	11.0
Sphalerite	–	54.0	6.4	32.3	3.3
Pyrite	–	–	42.3	52.5	1.0

Note: "–" denotes below detection limit.

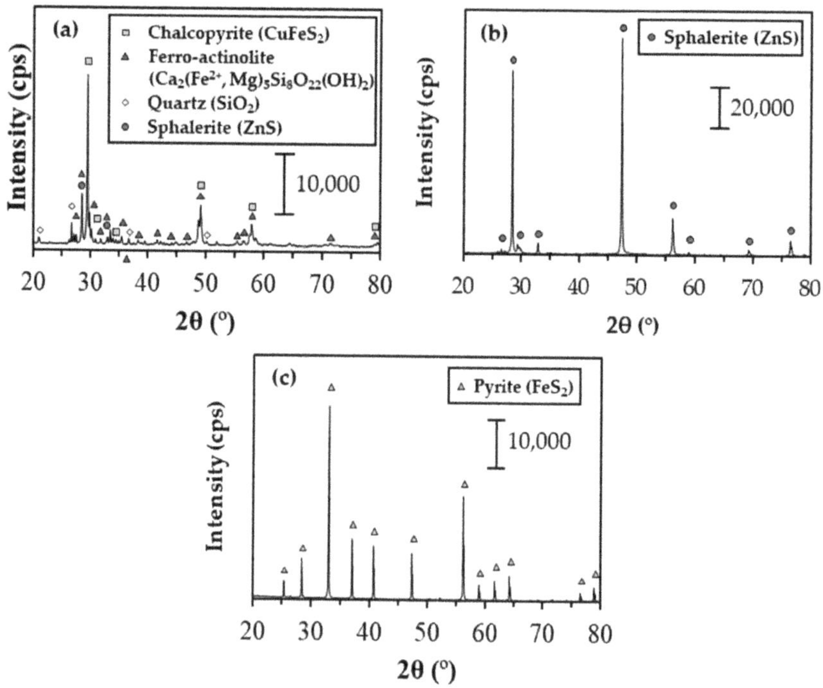

Figure 1. XRD patterns of mineral samples: (**a**) chalcopyrite ($CuFeS_2$), (**b**) sphalerite (ZnS), and (**c**) pyrite (FeS_2). Note differences in the scales of the y-axes.

2.2. Experimental Methods

2.2.1. Flotation

Flotation experiments were carried out using an agitator-type flotation machine (ASH-F30H, Kankyo-kanri Engineering, Akita, Japan) equipped with a 400 mL flotation cell under the following conditions: pH, 9; temperature, 25 °C; impeller speed, 1000 rpm; air flow rate, 1 L/min. The mixtures of ground sphalerite, chalcopyrite, pyrite, and quartz (−75 + 38 μm) were used as model samples. Three types of model samples were prepared: (i) 18 g of chalcopyrite or sphalerite and 2 g of quartz, (ii) 13 g of chalcopyrite, 1 g of sphalerite, and 6 g of quartz, (iii) 13 g of chalcopyrite, 1 g of sphalerite, 4 g of pyrite, and 2 g of quartz. The ratios of Cu and Zn in model samples (ii) and (iii) were based on the Cu ore sample obtained from South America. Meanwhile, quartz was added to not only adjust the amount of the model samples to 20 g but also to check the entrainment of minerals into the froth products.

Prior to the flotation experiments, each mineral sample was deslimed with 300 mL of distilled water, respectively [24]. After desliming, each mineral sample was mixed and repulped to 400 mL with distilled water containing various concentrations of zinc sulfate (0, 1, or 10 kg/t) in the flotation cell and conditioned with MIBC for 3 min, and then the flotation experiments were carried out based on the flowchart consisting of conditioning with PIPX for 3 min, followed by froth recovery for 3 min, which was repeated until the cumulative PIPX dosage reached 500 g/t (Figure 2). After flotation, froth and tailings were dried at 105 °C for 24 h and analyzed by XRF to determine the recovery of Cu, Zn, Fe, and Si. Some experiments were carried out in duplicate to ascertain that the differences observed were statistically significant.

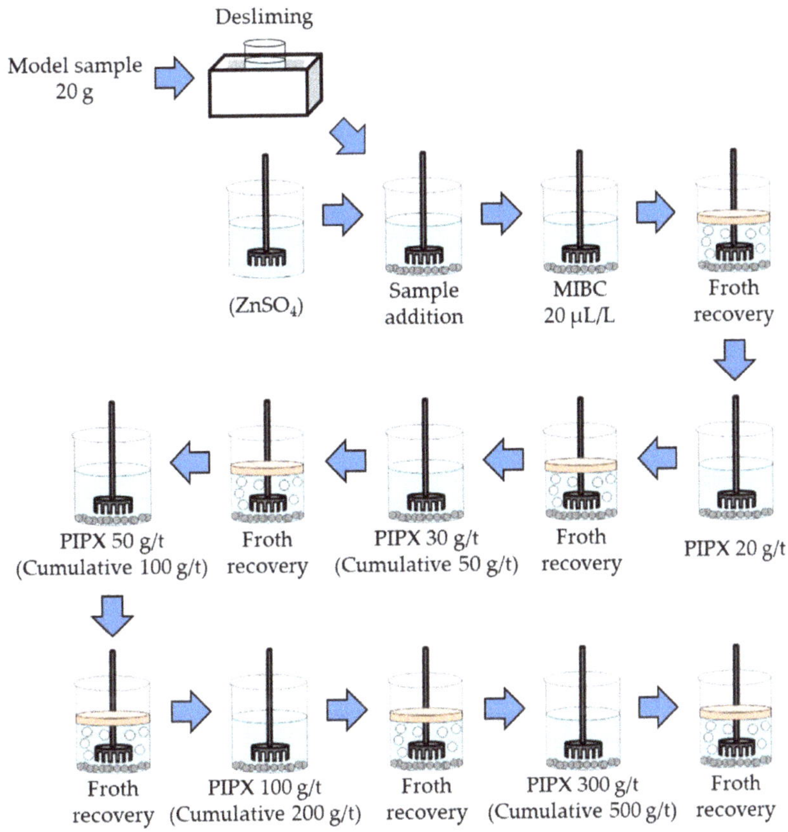

Figure 2. Flotation procedure with sequential addition of the collector (PIPX).

2.2.2. Contact Angle Measurements

Contact angle measurements were carried out in a solution containing 50 ppm zinc sulfate, 5 ppm PIPX, and 20 μL/L MIBC at pH 9 and 25 °C. A small cuboid crystal of sphalerite and pyrite (~10 mm (w) × 10 mm (d) × 10 mm (h)) was obtained by cutting the sphalerite and pyrite mineral specimens using a diamond cutter and fixed inside a plastic holder using Technovit® non-conductive resin (Kulzer GmbH, Wehrheim, Germany). The mineral surface was polished with a polishing machine (SAPHIR 250 M1, ATM GmbH, Mammelzen, Germany) against silicon carbide papers of increasing grit numbers (P180 → P320 → P600 → P1200), followed by polishing with diamond suspensions (3 and 1 μm). After this, the polished mineral sample was cleaned using an ultrasonicator (W-113 MK-II, Honda Electronics Co., Ltd., Toyohashi, Japan) for 5 min to remove any residually attached diamond particles, washed thoroughly with distilled water, and then immediately used for the experiments. The sphalerite and pyrite surfaces were reacted with a solution containing 50 ppm zinc sulfate in the flotation cell and conditioned for 3 min after adding the following reagents in sequence: 5 ppm PIPX and 20 μL/L MIBC. The concentrations of zinc sulfate, PIPX, and MIBC were the same as that used in the flotation experiments: 1 kg/t zinc sulfate, 100 g/t PIPX, and 20 μL/L MIBC, respectively. Bubble contact angles on the surfaces of sphalerite and pyrite were measured using the captive bubble method. A high-magnification digital microscope (VHX-1000, Keyence Corporation, Japan) and AutoCAD 2023 (Autodesk, Inc., San Rafael, CA, USA) were used to measure the bubble contact angles. The experiments were conducted in triplicates at various spots on the mineral surface to ascertain that the differences observed were statistically significant.

3. Results and Discussion

3.1. Single Mineral Flotation of Chalcopyrite or Sphalerite

To investigate the floatabilities of chalcopyrite and sphalerite, single mineral flotation experiments of chalcopyrite or sphalerite were carried out without Zn depressants. The Cu recovery in the absence of PIPX was 58% but significantly increased to 86% by adding 20 g/t PIPX because chalcopyrite has a high affinity with xanthate [25–28]. As the PIPX dosage increased from 20 to 200 g/t, the Cu recovery was mildly increased to 95% and remained constant after 200 g/t PIPX (Figure 3a). On the other hand, the Zn recovery was lower than 30% even at 500 g/t PIPX (Figure 3b) because sphalerite has a low affinity with xanthate, thus adding PIPX cannot improve the floatability of sphalerite [29].

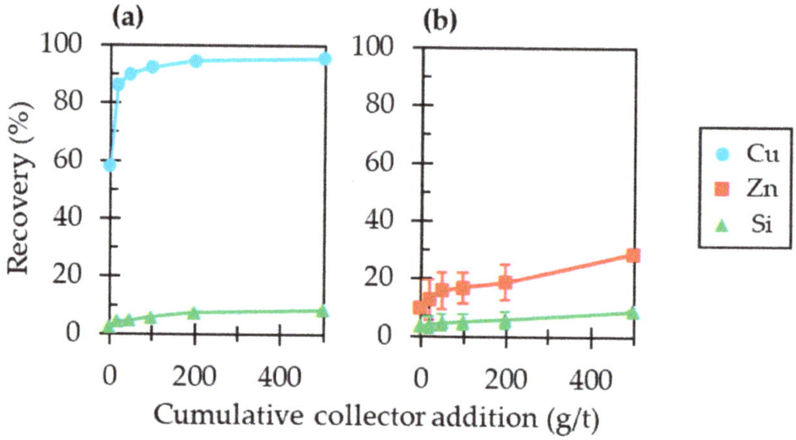

Figure 3. Results of single mineral flotation of (**a**) chalcopyrite or (**b**) sphalerite.

3.2. Effects of Chalcopyrite on the Floatability of Sphalerite

To investigate how the flotation behaviors of chalcopyrite and sphalerite are affected when they coexist in the absence of Zn depressants, flotation tests of mixed minerals (chalcopyrite and sphalerite) were conducted with various PIPX dosages. Compared to the results of the single mineral flotation of sphalerite (Figure 4a), the mixed mineral flotation results show that sphalerite floated together with chalcopyrite, indicating that the coexistence of chalcopyrite improved the floatability of sphalerite (Figure 4b). To elucidate why the floatability of sphalerite increased in the presence of chalcopyrite, the mixture of sphalerite and chalcopyrite was conditioned in the flotation cell following the flotation procedure, and the surface of sphalerite was analyzed by a scanning electron microscopy with energy-dispersive X-ray spectroscopy (SEM-EDS, JSM-IT200TM, JEOL Co., Ltd., Tokyo, Japan). Figure 5 shows the SEM photomicrograph of sphalerite after conditioning with chalcopyrite and the corresponding elemental maps of Zn, S, Cu, Fe, and O. The map of Cu was detected on the surface of sphalerite after conditioning with chalcopyrite and associated with the maps of S and O, whereas that of Cu did not correspond to that of Fe. These indicate that sphalerite was activated by Cu^{2+} dissolved from chalcopyrite, that is, a

CuS-like and/or $Cu(OH)_2$-like layer was formed on the surface of sphalerite as described in Equation (1) [30].

$$ZnS_{(s)} + Cu^{2+} = CuS_{(s)} + Zn^{2+} \quad (1)$$

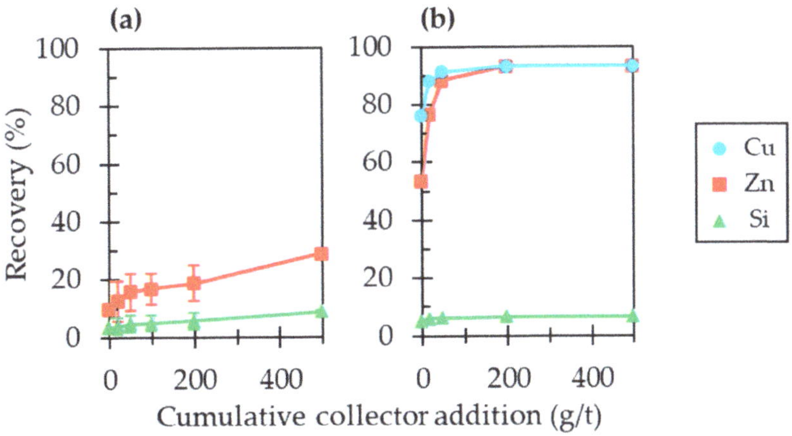

Figure 4. Flotation results of sphalerite in the absence of Zn depressants (**a**) without chalcopyrite and (**b**) with chalcopyrite. Note that the data at 100 g/t of collector in Figure 4b were absent due to the lack of froth amounts for the XRF.

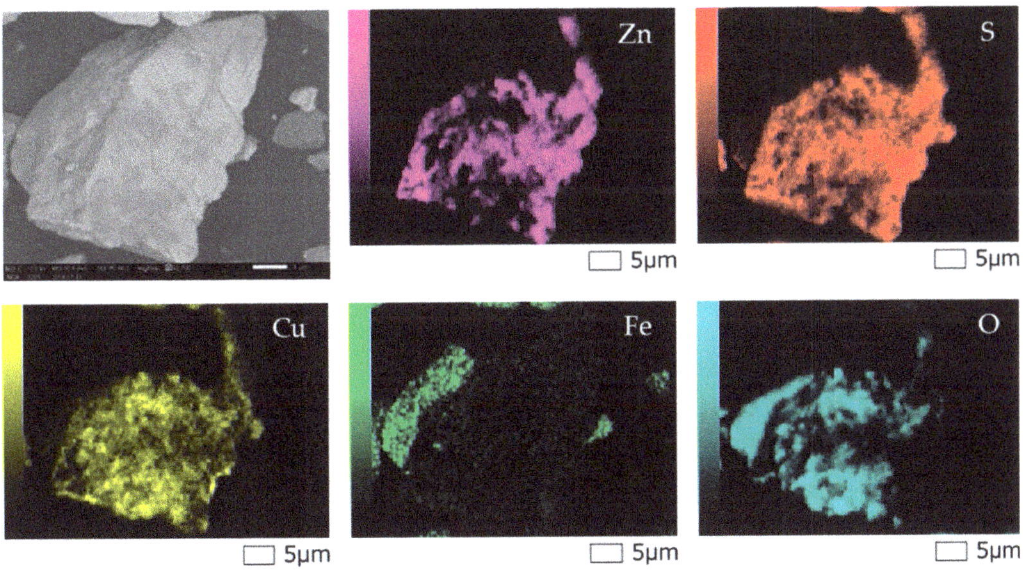

Figure 5. SEM photomicrograph of sphalerite after conditioning with chalcopyrite in the flotation cell with the corresponding elemental maps of Zn, S, Cu, Fe, and O.

Cu^{2+} was dissolved from copper oxides formed on the surface of chalcopyrite and caused unwanted activation of sphalerite [31,32]. The copper activation of sphalerite caused an increase in the floatability of sphalerite because xanthate was readily adsorbed on the Cu-compounds present on the sphalerite surface as illustrated in Figure 6. Thus, the increase in the floatability of sphalerite in the presence of chalcopyrite made the separation of sphalerite from chalcopyrite difficult without the aid of Zn depressants.

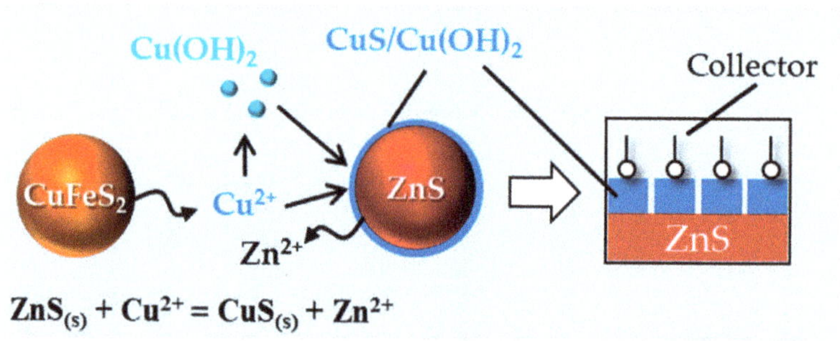

Figure 6. A schematic diagram of copper activation of sphalerite in the presence of chalcopyrite.

To depress the floatability of sphalerite in the presence of chalcopyrite, the effect of zinc sulfate, a conventional depressant, on sphalerite floatability was investigated. Figure 7 shows the flotation results of chalcopyrite and sphalerite with and without 1 kg/t zinc sulfate. As illustrated in Figure 7a, sphalerite was recovered as froth together with chalcopyrite in the absence of zinc sulfate; however, when zinc sulfate was added as a depressant for sphalerite, the Zn recovery was dramatically decreased to ~12% even at 500 g/t PIPX (Figure 7b). Meanwhile, the floatability of chalcopyrite was not affected by zinc sulfate; >90% of Cu recovery was achieved irrespective of the presence of zinc sulfate at >50 g/t PIPX (Figure 7). These results indicate that zinc sulfate is effective in depressing the floatability of sphalerite even in the presence of chalcopyrite. The possible mechanisms of how zinc sulfate could depress the floatability of sphalerite in the presence of chalcopyrite are as follows: (i) adsorption of zinc precipitates such as $Zn(OH)_2$ on the surface of sphalerite, rendering the surface of sphalerite hydrophilic where the adsorption of the collector is limited, (ii) inhibition of Cu activation by zinc precipitates, and/or (iii) deactivation of Cu-activated sphalerite via reverse reaction of Equation (1) [29,33].

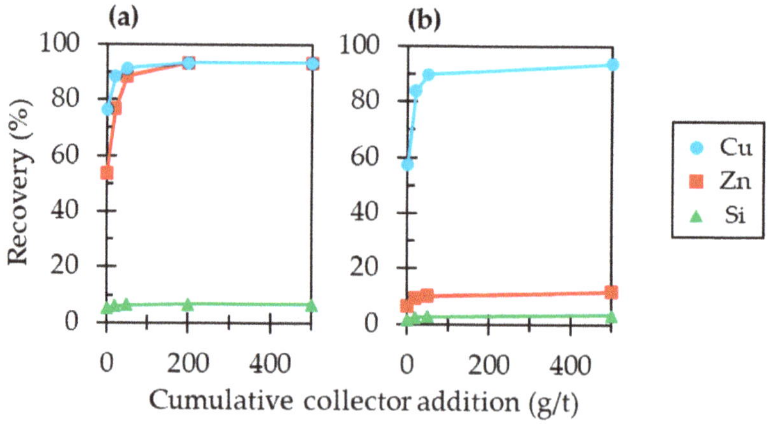

Figure 7. Flotation results of sphalerite and chalcopyrite with zinc sulfate (**a**) 0 kg/t and (**b**) 1 kg/t. Note that the data at 100 g/t of collector in Figure 7a and those at 100 and 200 g/t of collector in Figure 7b were absent due to the lack of froth amounts for the XRF.

3.3. Effects of Pyrite on Sphalerite Floatability in the Presence of Chalcopyrite

Copper ores contain not only valuable minerals such as chalcopyrite and sphalerite but also gangue minerals such as pyrite—a major gangue mineral of sulfide ores. The presence of pyrite has been reported to have a significant impact on the flotation behavior of minerals [34–39]. Although the addition of zinc sulfate improved the flotation separation of chalcopyrite from the chalcopyrite–sphalerite binary mixed minerals system, it is necessary to further investigate whether zinc sulfate would also be effective in the presence of pyrite. To clarify this, flotation tests of mixed minerals consisting of chalcopyrite, sphalerite, and pyrite were conducted with zinc sulfate (Figure 8). When pyrite coexisted with sphalerite and chalcopyrite, the recovery of Zn was dramatically increased (Figure 8b) compared to that in the absence of pyrite (Figure 8a). Though the amount of zinc sulfate increased from 1 to 10 kg/t, the floatability of sphalerite was high (>80%) (Figure 8c). These results indicate that pyrite inhibits the depressive effect of zinc sulfate toward sphalerite in the presence of chalcopyrite, which could be attributed to two possible mechanisms: (i) the reduced amount of zinc precipitates adsorbed onto the surface of sphalerite due to the presence of pyrite whose surface also attracts zinc precipitates, and (ii) the galvanic interaction between chalcopyrite and pyrite that promotes the anodic half-cell reaction of chalcopyrite, releasing more Cu^{2+} into the aqueous solution (Figure 9). These two possible mechanisms will be explored in more detail in the next sub-section.

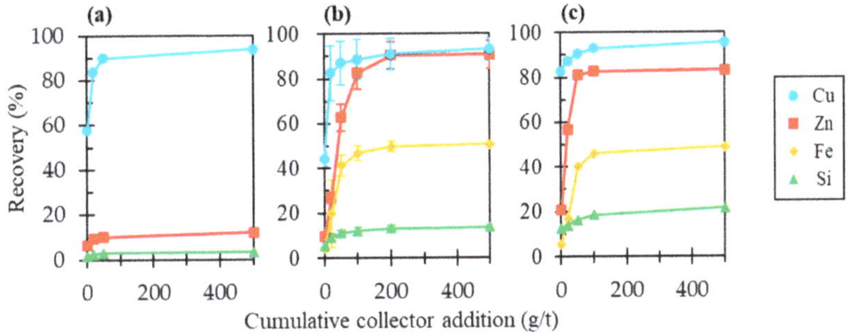

Figure 8. Flotation results of sphalerite and chalcopyrite (**a**) in the absence of pyrite and with zinc sulfate 1 kg/t, (**b**) in the presence of pyrite and zinc sulfate 1kg/t, and (**c**) in the presence of pyrite and zinc sulfate 10 kg/t. Note that the data at 100 and 200 g/t of collector in Figure 8a and those at 200 g/t of collector in Figure 8c were absent due to the lack of froth amounts for the XRF.

3.4. Mechanisms of Losing the Depressive Effect of Zinc Sulfate toward Sphalerite in the Presence of Both Pyrite and Chalcopyrite

3.4.1. Effects of Pyrite on the Hydrophilicity of Sphalerite

One of the considerable mechanisms of depressing sphalerite floatability with zinc sulfate may be due to the adsorption of zinc precipitates such as $Zn(OH)_2$ on the surface sphalerite, rendering its surface hydrophilic, thus limiting the adsorption of a collector [29,33]. If Zn precipitates are adsorbed onto the surface of not only sphalerite but also pyrite, it would result in reducing the amount of Zn precipitates adsorbed onto the surface of sphalerite, and thus the depression of sphalerite floatability by zinc sulfate may not be achieved. To investigate whether pyrite affects the adsorption of Zn precipitates on sphalerite, the contact angles of sphalerite in the absence and presence of pyrite were measured using a small cuboid crystal of sphalerite and pyrite under the same conditions as the flotation experiments; 50 ppm zinc sulfate, 5 ppm PIPX, and 20 μL/L MIBC at pH 9 and 25 °C. As shown in Figure 10, the difference in the contact angles between the sphalerite and air bubble in the absence and presence of pyrite was almost negligible, indicating that the reduced amount of Zn precipitates adsorbed onto the sphalerite surface by pyrite is not the reason for losing the depressive effect of zinc sulfate toward sphalerite.

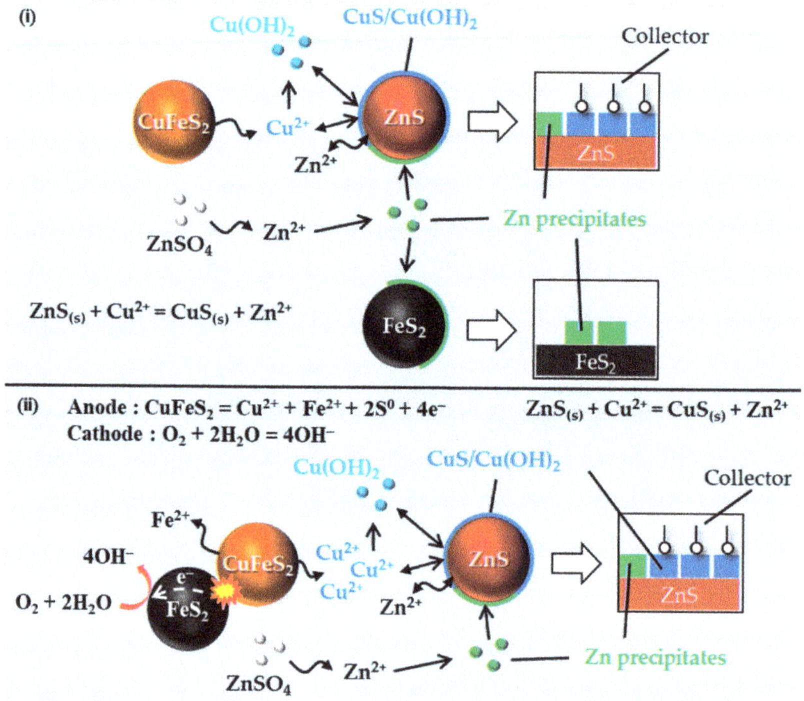

Figure 9. A schematic diagram of two possible mechanisms for why pyrite inhibits the depression of sphalerite with zinc sulfate in the presence of chalcopyrite: (**i**) the reduced amount of zinc precipitates adsorbed onto the surface of sphalerite due to the presence of pyrite whose surface also attracts zinc precipitates, (**ii**) galvanic interaction between chalcopyrite and pyrite that promotes the anodic half-cell reaction of chalcopyrite, releasing more Cu^{2+} into the aqueous solution.

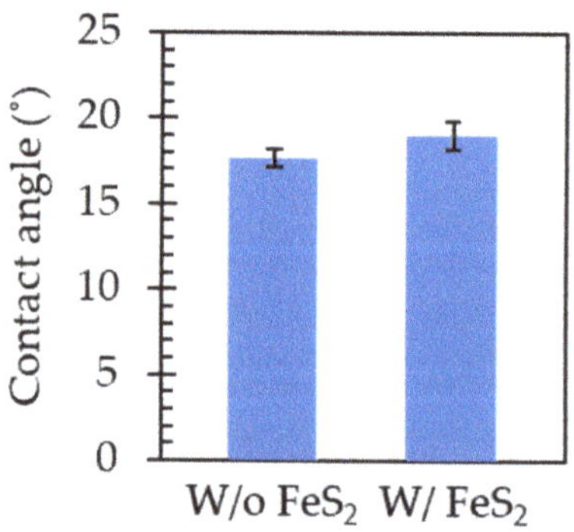

Figure 10. The contact angle of sphalerite in the absence and presence of pyrite.

3.4.2. Effects of the Galvanic Interaction between Chalcopyrite and Pyrite on the Floatability of Sphalerite

Pyrite is known to strongly affect the floatability of target minerals due to their electrochemical interactions with each other [40,41]. When pyrite and chalcopyrite are in contact, they form a galvanic cell where the former (i.e., pyrite) acts as a cathode due to a higher rest potential, whereas the latter (i.e., chalcopyrite) having a lower rest potential behaves as an anode [42]. The formation of a galvanic cell is well-known to enhance the dissolution of minerals or metals that act as anodes while it protects the cathode from corrosion [43–48]. That is, the dissolution of chalcopyrite as an anodic reaction (Equation (2)) is accelerated via the galvanic interaction, whereas on the surface of pyrite, the reduction reaction of dissolved oxygen (DO) as a cathodic reaction (Equation (3)) occurs [49].

$$CuFeS_2 = Cu^{2+} + Fe^{2+} + 2S^0 + 4e^- \quad (2)$$

$$O_2 + 2H_2O + 4e^- = 4OH^- \quad (3)$$

This indicates that the amount of dissolved Cu^{2+} in the flotation pulp would increase due to the galvanic interaction between chalcopyrite and pyrite, and the degree of sphalerite activation by Cu^{2+} would also be enhanced, resulting in the difficulty in depressing sphalerite floatability by zinc sulfate. Therefore, the enhanced dissolution of chalcopyrite due to the galvanic interaction with pyrite is one of the possible mechanisms that inhibits the ability of zinc sulfate to depress sphalerite floatability in the presence of both pyrite and chalcopyrite.

To investigate the effect of the galvanic interaction between chalcopyrite and pyrite on the floatability of sphalerite, mixed minerals flotation experiments of chalcopyrite, sphalerite, and pyrite with 1 kg/t zinc sulfate were carried out, where the amount of DO in the flotation pulp was limited. DO is an essential reactant for the galvanic interaction to take place, and thus the effect of the galvanic interaction between chalcopyrite and pyrite on the floatability of sphalerite can be clarified by conducting flotation experiments in the absence and presence of oxygen in the system. For the flotation under anaerobic conditions, ultrapure nitrogen (N_2) gas (99.99%) was introduced to distilled water for 15 min before conditioning to remove DO from the flotation pulp, and froth recovery was also conducted by injecting N_2 gas instead of air to limit the appearance of oxygen in the system. As shown in Figure 11, the recovery of Zn was dramatically decreased under anaerobic conditions; at 100 g/t PIPX, Zn recoveries under aerobic and anaerobic conditions were ~90% and ~20%, respectively. As confirmed in Figure 8c, when the galvanic interaction between pyrite and chalcopyrite occurred, it was impossible to depress the floatability of sphalerite even with the addition of 10 kg/t zinc sulfate. These results indicate that the effects of the galvanic interaction between pyrite and chalcopyrite on sphalerite floatability were minimized when the amount of DO in the flotation pulp was limited by employing N_2 gas, resulting in effective depression of sphalerite floatability in the presence of both chalcopyrite and pyrite. Therefore, the galvanic interaction between pyrite and chalcopyrite would be the most likely mechanism for pyrite inhibiting the depression of sphalerite floatability with zinc sulfate in the presence of chalcopyrite.

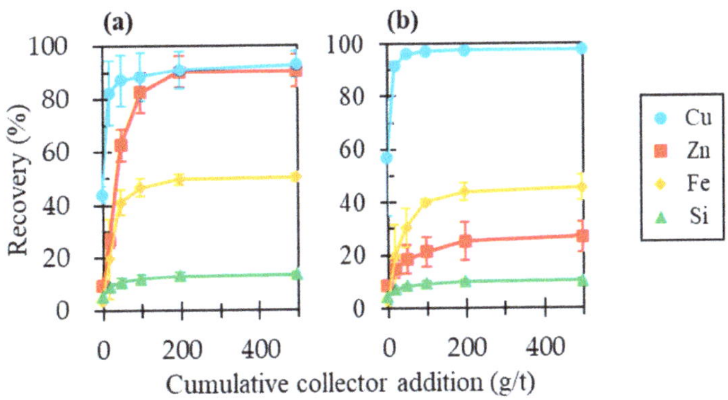

Figure 11. Flotation results of chalcopyrite, sphalerite, and pyrite in the presence of 1 kg/t zinc sulfate (**a**) with air and (**b**) with N_2 gas.

3.5. A Proposed Method to Efficiently Depress Sphalerite in Flotation of Cu Ores with a High Cu/Zn Ratio

Figure 12 shows the proposed method to efficiently depress sphalerite in the flotation of Cu ores with a high Cu/Zn ratio. When chalcopyrite and pyrite coexisted with sphalerite, zinc sulfate lost its ability to depress sphalerite floatability (Figure 8). This is most likely due to the galvanic interaction between chalcopyrite and pyrite that enhances the dissolution of chalcopyrite, and thus the amount of dissolved Cu^{2+} in the flotation pulp would be increased. As a result, Cu activation of sphalerite could be enhanced, so zinc sulfate becomes inefficient in depressing sphalerite (Figure 9ii). The cathodic reaction, i.e., the reduction of DO occurring on the pyrite surface (Equation (3)), is the first step of the galvanic interaction, so the application of N_2 gas minimizing the amount of DO in the system could limit the galvanic interaction, and as a result, zinc sulfate becomes effective in depressing sphalerite floatability. As the ratio of pyrite to chalcopyrite in copper ores increased, the galvanic interaction between them would become more intensive, which negatively would affect the floatability of chalcopyrite [50]. Copper ores contain not only chalcopyrite but also other copper sulfide minerals (e.g., chalcocite and bornite), which could also form a galvanic cell with pyrite due to their lower rest potential than pyrite [51]. Therefore, the application of N_2 gas to limit the galvanic interaction of copper sulfide minerals with pyrite could be a promising method to achieve Cu/Zn separation by effectively depressing sphalerite floatability in the actual processing of copper ores containing many mineralogical species such as several copper sulfide minerals and pyrite.

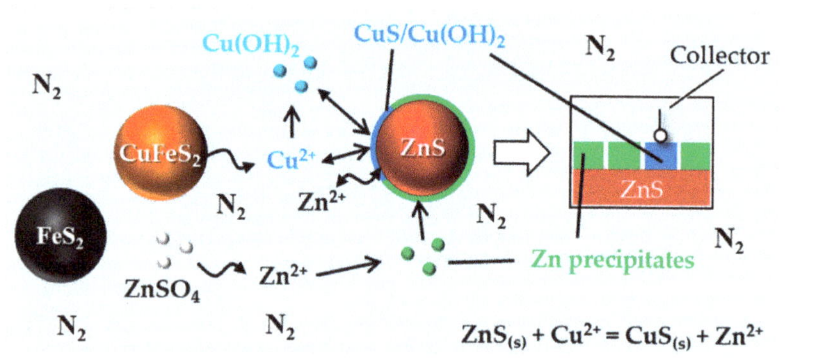

Figure 12. A schematic diagram of the proposed efficient method to depress sphalerite in flotation of Cu ores containing sphalerite and pyrite with zinc sulfate.

4. Conclusions

In this paper, the flotation behaviors of chalcopyrite and sphalerite in model samples mimicking Cu ores with high Cu/Zn ratios were investigated. The recovery of Zn was low in the single mineral flotation of sphalerite, but it dramatically increased when chalcopyrite coexisted due to the activation of sphalerite by Cu^{2+} dissolved from chalcopyrite. The addition of zinc sulfate successfully depressed the sphalerite floatability; however, the presence of pyrite inhibited its ability to depress sphalerite floatability most likely due to the galvanic interaction between pyrite and chalcopyrite. The application of N_2 gas that limits the galvanic interaction between pyrite and chalcopyrite by reducing the DO concentration in the system is an effective option to depress sphalerite with zinc sulfate, although the amounts of chalcopyrite and pyrite are much higher than sphalerite. Typically, copper ores contain not only chalcopyrite but also other copper sulfide minerals (e.g., chalcocite and bornite), which could form a galvanic cell with pyrite. Therefore, the application of N_2 gas to limit the galvanic interaction of copper sulfide minerals with pyrite could be a promising method to achieve Cu/Zn separation by effectively depressing sphalerite floatability in the actual processing of copper ores containing many mineralogical species such as several copper sulfide minerals and pyrite.

Author Contributions: Conceptualization, K.A., M.I. and K.H.; methodology, K.A., M.I. and I.P.; investigation, K.A. and N.O.; data curation, S.J., T.S., T.O. and R.S.M.; writing—original draft preparation, K.A.; writing—review and editing, M.I., N.O., S.J., I.P., K.H., T.K., T.T., T.S., T.O., R.S.M. and N.H.; visualization, K.A.; supervision, T.K. and N.H.; project administration, T.T. and K.S.; funding acquisition, M.I. All authors have read and agreed to the published version of the manuscript.

Funding: This research was funded by Japan Oil, Gas and Metals National Corporation (JOGMEC), and JSPS KAKENHI Grant Number JP21J20552.

Data Availability Statement: Not applicable.

Conflicts of Interest: The authors declare no conflict of interest.

References

1. Spatari, S.; Bertram, M.; Gordon, R.B.; Henderson, K.; Graedel, T.E. Twentieth Century Copper Stocks and Flows in North America: A Dynamic Analysis. *Ecol. Econ.* **2005**, *54*, 37–51. [CrossRef]
2. Bilal, M.; Park, I.; Hornn, V.; Ito, M.; Hassan, F.; Jeon, S.; Hiroyoshi, N. The Challenges and Prospects of Recovering Fine Copper Sulfides from Tailings Using Different Flotation Techniques: A Review. *Minerals* **2022**, *12*, 586. [CrossRef]
3. WBG Minerals for Climate Action: The Mineral Intensity of the Clean Energy Transition. Available online: https://pubdocs.worldbank.org/en/961711588875536384/Minerals-for-Climate-Action-The-Mineral-Intensity-of-the-Clean-Energy-Transition.pdf (accessed on 25 April 2022).
4. Park, I.; Kanazawa, Y.; Sato, N.; Galtchandmani, P.; Jha, M.K.; Tabelin, C.B.; Jeon, S.; Ito, M.; Hiroyoshi, N. Beneficiation of Low-Grade Rare Earth Ore from Khalzan Buregtei Deposit (Mongolia) by Magnetic Separation. *Minerals* **2021**, *11*, 1432. [CrossRef]
5. Park, I.; Hong, S.; Jeon, S.; Ito, M.; Hiroyoshi, N. Flotation Separation of Chalcopyrite and Molybdenite Assisted by Microencapsulation Using Ferrous and Phosphate Ions: Part I. Selective Coating Formation. *Metals* **2020**, *10*, 1667. [CrossRef]
6. Hornn, V.; Ito, M.; Shimada, H.; Tabelin, C.B.; Jeon, S.; Park, I.; Hiroyoshi, N. Agglomeration-Flotation of Finely Ground Chalcopyrite and Quartz: Effects of Agitation Strength during Agglomeration Using Emulsified Oil on Chalcopyrite. *Minerals* **2020**, *10*, 380. [CrossRef]
7. Bilal, M.; Ito, M.; Akishino, R.; Bu, X.; Ul Hassan, F.; Park, I.; Jeon, S.; Aikawa, K.; Hiroyoshi, N. Heterogenous Carrier Flotation Technique for Recovering Finely Ground Chalcopyrite Particles Using Coarse Pyrite Particles as a Carrier. *Miner. Eng.* **2022**, *180*, 107518. [CrossRef]
8. Park, I.; Hong, S.; Jeon, S.; Ito, M.; Hiroyoshi, N. A Review of Recent Advances in Depression Techniques for Flotation Separation of Cu–Mo Sulfides in Porphyry Copper Deposits. *Metals* **2020**, *10*, 1269. [CrossRef]
9. Park, I. Advances in Selective Flotation and Leaching Process in Metallurgy. *Metals* **2022**, *12*, 144. [CrossRef]
10. Hornn, V.; Park, I.; Ito, M.; Shimada, H.; Suto, T.; Tabelin, C.B.; Jeon, S.; Hiroyoshi, N. Agglomeration-Flotation of Finely Ground Chalcopyrite Using Surfactant-Stabilized Oil Emulsions: Effects of Co-Existing Minerals and Ions. *Miner. Eng.* **2021**, *171*, 107076. [CrossRef]
11. Tabelin, C.B.; Park, I.; Phengsaart, T.; Jeon, S.; Villacorte-Tabelin, M.; Alonzo, D.; Yoo, K.; Ito, M.; Hiroyoshi, N. Copper and Critical Metals Production from Porphyry Ores and E-Wastes: A Review of Resource Availability, Processing/Recycling Challenges, Socio-Environmental Aspects, and Sustainability Issues. *Resour. Conserv. Recycl.* **2021**, *170*, 105610. [CrossRef]
12. John, D.A.; Ayuso, R.A.; Barton, M.D.; Blakely, R.J.; Bodnar, R.J.; Dilles, J.H.; Gray, F.; Graybeal, F.T.; Mars, J.C.; McPhee, D.K.; et al. *Porphyry Copper Deposit Model, Chapter B of Mineral Deposit Models for Resource Assessment*; U.S. Geological Survey Scientific Investigations Report 2010-5070-B; U.S. Geological Survey: Menlo Park, CA, USA, 2010.
13. Tongamp, W.; Takasaki, Y.; Shibayama, A. Arsenic Removal from Copper Ores and Concentrates through Alkaline Leaching in NaHS Media. *Hydrometallurgy* **2009**, *98*, 213–218. [CrossRef]
14. Haga, K.; Tongamp, W.; Shibayama, A. Investigation of Flotation Parameters for Copper Recovery from Enargite and Chalcopyrite Mixed Ore. *Mater. Trans.* **2012**, *53*, 707–715. [CrossRef]
15. Haga, K.; Altansukh, B.; Shibayama, A. Volatilization of Arsenic and Antimony from Tennantite/Tetrahedrite Ore by a Roasting Process. *Mater. Trans.* **2018**, *59*, 1396–1403. [CrossRef]
16. Sinche-Gonzalez, M.; Fornasiero, D. Understanding the Effect of Sulphate in Mining-Process Water on Sulphide Flotation. *Miner. Eng.* **2021**, *165*, 106865. [CrossRef]
17. Bıçak, Ö.; Ekmekçi, Z.; Can, M.; Öztürk, Y. The Effect of Water Chemistry on Froth Stability and Surface Chemistry of the Flotation of a Cu–Zn Sulfide Ore. *Int. J. Miner. Process.* **2012**, *102–103*, 32–37. [CrossRef]
18. Natarajan, R.; Nirdosh, I.; Muthuswami, S.V. Flotation of a Copper-Zinc Ore Using p-Nonylcupferron as Collector. *Dev. Chem. Eng. Miner. Processing* **1997**, *5*, 183–193. [CrossRef]
19. Soltani, F.; Koleini, S.M.J.; Abdollahy, M. Optimization of Cu–Zn Massive Sulphide Flotation by Selective Reagents. *J. Inst. Eng. India Ser. D* **2014**, *95*, 125–134. [CrossRef]

20. Kant, C.; Rao, S.R.; Finch, J.A. Distribution of Surface Metal Ions among the Products of Chalcopyrite Flotation. *Miner. Eng.* **1994**, *7*, 905–916. [CrossRef]
21. Aikawa, K.; Ito, M.; Kusano, A.; Jeon, S.; Park, I.; Hiroyoshi, N. Development of a Sustainable Process for Complex Sulfide Ores Containing Anglesite: Effect of Anglesite on Sphalerite Floatability, Enhanced Depression of Sphalerite by Extracting Anglesite, and Recovery of Extracted Pb2+ as Zero-Valent Pb by Cementation Using Zero-Valent Fe. *Minerals* **2022**, *12*, 723. [CrossRef]
22. Woodcock, J.T.; Sparrow, G.J.; Bruckard, W.J.; Johnson, N.W.; Dunne, R. Plant Practice: Sulfide Minerals and Precious Metals. In *Froth Flotation a Century of Innovation*; Fuerstenau, M., Jameson, G., Yoon, R.H., Eds.; SME: Littleton, CO, USA, 2007; pp. 781–843; ISBN 978-0-873-35252-9.
23. Ito, M.; Orii, N.; Aikawa, K.; Jeon, S.; Park, I.; Hiroyoshi, N.; Magwaneng, R.; Kudo, K.; Sunada, K.; Takahashi, T. Effects of Several Depressants on the Floatability of Sphalerite in the Flotation of Copper-Zinc Sulfides. In Proceedings of the Mining and Materials Processing Institute of Japan (MMIJ) Fall Meeting 2020, online, 8–10 September 2020; Volume 7.
24. Aikawa, K.; Ito, M.; Kusano, A.; Park, I.; Oki, T.; Takahashi, T.; Furuya, H.; Hiroyoshi, N. Flotation of Seafloor Massive Sulfide Ores: Combination of Surface Cleaning and Deactivation of Lead-Activated Sphalerite to Improve the Separation Efficiency of Chalcopyrite and Sphalerite. *Metals* **2021**, *11*, 253. [CrossRef]
25. Bilal, M.; Ito, M.; Koike, K.; Hornn, V.; Ul Hassan, F.; Jeon, S.; Park, I.; Hiroyoshi, N. Effects of Coarse Chalcopyrite on Flotation Behavior of Fine Chalcopyrite. *Miner. Eng.* **2021**, *163*, 106776. [CrossRef]
26. Hornn, V.; Ito, M.; Shimada, H.; Tabelin, C.B.; Jeon, S.; Park, I.; Hiroyoshi, N. Agglomeration–Flotation of Finely Ground Chalcopyrite Using Emulsified Oil Stabilized by Emulsifiers: Implications for Porphyry Copper Ore Flotation. *Metals* **2020**, *10*, 912. [CrossRef]
27. Hornn, V.; Ito, M.; Yamazawa, R.; Shimada, H.; Tabelin, C.B.; Jeon, S.; Park, I.; Hiroyoshi, N. Kinetic Analysis for Agglomeration-Flotation of Finely Ground Chalcopyrite: Comparison of First Order Kinetic Model and Experimental Results. *Mater. Trans.* **2020**, *61*, 1940–1948. [CrossRef]
28. Park, I.; Hong, S.; Jeon, S.; Ito, M.; Hiroyoshi, N. Flotation Separation of Chalcopyrite and Molybdenite Assisted by Microencapsulation Using Ferrous and Phosphate Ions: Part II. Flotation. *Metals* **2021**, *11*, 439. [CrossRef]
29. Wills, B.A.; Finch, J.A. Froth flotation. In *Mineral Processing Technology: An Introduction to the Practical Aspects of Ore Treatment and Mineral Recovery*, 8th ed.; Elsevier: Oxford, UK, 2016; pp. 265–380; ISBN 9780080970530.
30. Chandra, A.P.; Gerson, A.R. A Review of the Fundamental Studies of the Copper Activation Mechanisms for Selective Flotation of the Sulfide Minerals, Sphalerite and Pyrite. *Adv. Colloid Interface Sci.* **2009**, *145*, 97–110. [CrossRef]
31. Lascelles, D.; Finch, J.A. Quantifying Accidental Activation. Part I. Cu Ion Production. *Miner. Eng.* **2002**, *15*, 567–571. [CrossRef]
32. Wei, Q.; Jiao, F.; Dong, L.; Liu, X.; Qin, W. Selective Depression of Copper-Activated Sphalerite by Polyaspartic Acid during Chalcopyrite Flotation. *Trans. Nonferrous Met. Soc. China* **2021**, *31*, 1784–1795. [CrossRef]
33. Cao, M.; Liu, Q. Reexamining the Functions of Zinc Sulfate as a Selective Depressant in Differential Sulfide Flotation—The Role of Coagulation. *J. Colloid Interface Sci.* **2006**, *301*, 523–531. [CrossRef]
34. Pecina-Treviño, E.T.; Uribe-Salas, A.; Nava-Alonso, F. Effect of Dissolved Oxygen and Galvanic Contact on the Floatability of Galena and Pyrite with Aerophine 3418A. *Miner. Eng.* **2003**, *16*, 359–367. [CrossRef]
35. Qin, W.; Wang, X.; Ma, L.; Jiao, F.; Liu, R.; Yang, C.; Gao, K. Electrochemical Characteristics and Collectorless Flotation Behavior of Galena: With and without the Presence of Pyrite. *Miner. Eng.* **2015**, *74*, 99–104. [CrossRef]
36. Barker, G.J.; Gerson, A.R.; Menuge, J.F. The Impact of Iron Sulfide on Lead Recovery at the Giant Navan Zn–Pb Orebody, Ireland. *Int. J. Miner. Process.* **2014**, *128*, 16–24. [CrossRef]
37. Qin, W.; Wang, X.; Ma, L.; Jiao, F.; Liu, R.; Gao, K. Effects of Galvanic Interaction between Galena and Pyrite on Their Flotation in the Presence of Butyl Xanthate. *Trans. Nonferrous Met. Soc. China* **2015**, *25*, 3111–3118. [CrossRef]
38. Wang, X.; Qin, W.; Jiao, F.; Wu, J. The Influence of Galvanic Interaction on the Dissolution and Surface Composition of Galena and Pyrite in Flotation System. *Miner. Eng.* **2020**, *156*, 106525. [CrossRef]
39. Aikawa, K.; Ito, M.; Segawa, T.; Jeon, S.; Park, I.; Tabelin, C.B.; Hiroyoshi, N. Depression of Lead-Activated Sphalerite by Pyrite via Galvanic Interactions: Implications to the Selective Flotation of Complex Sulfide Ores. *Miner. Eng.* **2020**, *152*, 106367. [CrossRef]
40. Owusu, C.; Brito e Abreu, S.; Skinner, W.; Addai-Mensah, J.; Zanin, M. The Influence of Pyrite Content on the Flotation of Chalcopyrite/Pyrite Mixtures. *Miner. Eng.* **2014**, *55*, 87–95. [CrossRef]
41. Mu, Y.; Peng, Y.; Lauten, R.A. The Galvanic Interaction between Chalcopyrite and Pyrite in the Presence of Lignosulfonate-Based Biopolymers and Its Effects on Flotation Performance. *Miner. Eng.* **2018**, *122*, 91–98. [CrossRef]
42. Nazari, G.; Dixon, D.G.; Dreisinger, D.B. Enhancing the Kinetics of Chalcopyrite Leaching in the Galvanox^{TM} Process. *Hydrometallurgy* **2011**, *105*, 251–258. [CrossRef]
43. Seng, S.; Tabelin, C.B.; Kojima, M.; Hiroyoshi, N.; Ito, M. Galvanic Microencapsulation (GME) Using Zero-Valent Aluminum and Zero-Valent Iron to Suppress Pyrite Oxidation. *Mater. Trans.* **2019**, *60*, 277–286. [CrossRef]
44. Seng, S.; Tabelin, C.B.; Makino, Y.; Chea, M.; Phengsaart, T.; Park, I.; Hiroyoshi, N.; Ito, M. Improvement of Flotation and Suppression of Pyrite Oxidation Using Phosphate-Enhanced Galvanic Microencapsulation (GME) in a Ball Mill with Steel Ball Media. *Miner. Eng.* **2019**, *143*, 105931. [CrossRef]
45. Jeon, S.; Bright, S.; Park, I.; Kuze, A.; Ito, M.; Hiroyoshi, N. A Kinetic Study on Enhanced Cementation of Gold Ions by Galvanic Interactions between Aluminum (Al) as an Electron Donor and Activated Carbon (AC) as an Electron Mediator in Ammonium Thiosulfate System. *Minerals* **2022**, *12*, 91. [CrossRef]

46. Jeon, S.; Bright, S.; Park, I.; Tabelin, C.B.; Ito, M.; Hiroyoshi, N. A Simple and Efficient Recovery Technique for Gold Ions from Ammonium Thiosulfate Medium by Galvanic Interactions of Zero-Valent Aluminum and Activated Carbon: A Parametric and Mechanistic Study of Cementation. *Hydrometallurgy* **2022**, *208*, 105815. [CrossRef]
47. Tabelin, C.B.; Igarashi, T.; Villacorte-Tabelin, M.; Park, I.; Opiso, E.M.; Ito, M.; Hiroyoshi, N. Arsenic, Selenium, Boron, Lead, Cadmium, Copper, and Zinc in Naturally Contaminated Rocks: A Review of Their Sources, Modes of Enrichment, Mechanisms of Release, and Mitigation Strategies. *Sci. Total Environ.* **2018**, *645*, 1522–1553. [CrossRef] [PubMed]
48. Park, I.; Tabelin, C.B.; Jeon, S.; Li, X.; Seno, K.; Ito, M.; Hiroyoshi, N. A Review of Recent Strategies for Acid Mine Drainage Prevention and Mine Tailings Recycling. *Chemosphere* **2019**, *219*, 588–606. [CrossRef] [PubMed]
49. Ekmekçi, Z.; Demirel, H. Effects of Galvanic Interaction on Collectorless Flotation Behaviour of Chalcopyrite and Pyrite. *Int. J. Miner. Process.* **1997**, *52*, 31–48. [CrossRef]
50. Lee, R.L.J.; Chen, X.; Peng, Y. Flotation Performance of Chalcopyrite in the Presence of an Elevated Pyrite Proportion. *Miner. Eng.* **2022**, *177*, 107387. [CrossRef]
51. Allison, S.A.; Goold, L.A.; Nicol, M.J.; Granville, A. A Determination of the Products of Reaction between Various Sulphide Minerals and Aqueous Xanthate. *Metall. Trans.* **1972**, *3*, 2613–2618. [CrossRef]

minerals

MDPI

Article

Development of a Sustainable Process for Complex Sulfide Ores Containing Anglesite: Effect of Anglesite on Sphalerite Floatability, Enhanced Depression of Sphalerite by Extracting Anglesite, and Recovery of Extracted Pb^{2+} as Zero-Valent Pb by Cementation Using Zero-Valent Fe

Kosei Aikawa [1,*], Mayumi Ito [2], Atsuhiro Kusano [1], Sanghee Jeon [2], Ilhwan Park [2] and Naoki Hiroyoshi [2]

[1] Division of Sustainable Resources Engineering, Graduate School of Engineering, Hokkaido University, Sapporo 060-8628, Japan; e531taikoteam@gmail.com

[2] Division of Sustainable Resources Engineering, Faculty of Engineering, Hokkaido University, Sapporo 060-8628, Japan; itomayu@eng.hokudai.ac.jp (M.I.); shjun1121@eng.hokudai.ac.jp (S.J.); i-park@eng.hokudai.ac.jp (I.P.); hiroyosi@eng.hokudai.ac.jp (N.H.)

* Correspondence: k-aikawa@frontier.hokudai.ac.jp; Tel.: +81-11-706-6315

check for updates

Citation: Aikawa, K.; Ito, M.; Kusano, A.; Jeon, S.; Park, I.; Hiroyoshi, N. Development of a Sustainable Process for Complex Sulfide Ores Containing Anglesite: Effect of Anglesite on Sphalerite Floatability, Enhanced Depression of Sphalerite by Extracting Anglesite, and Recovery of Extracted Pb^{2+} as Zero-Valent Pb by Cementation Using Zero-Valent Fe. *Minerals* **2022**, *12*, 723. https://doi.org/10.3390/min12060723

Academic Editor: Hyunjung Kim

Received: 26 April 2022
Accepted: 3 June 2022
Published: 6 June 2022

Abstract: The presence of anglesite ($PbSO_4$) in complex sulfide ores negatively affects the separation of Cu-Pb sulfides and sphalerite (ZnS) due to lead activation, and $PbSO_4$ rejected to tailings dams contaminates the surrounding environment with lead. To address these problems, this study investigated the application of ethylene diamine tetra acetic acid (EDTA) pretreatment extracting $PbSO_4$ to ZnS flotation and the recovery of the extracted Pb^{2+} as zero-valent Pb by cementation using zero-valent iron (ZVI). The application of EDTA pretreatment could extract ~99.8% of $PbSO_4$, thus depressing ZnS floatability from 82% to 30%. In addition, cementation using ZVI could recover ~99.7% of Pb^{2+} from the leachate of EDTA pretreatment.

Keywords: flotation; complex sulfide ores; sphalerite; anglesite; extraction; cementation

1. Introduction

In 2015, 193 governments adopted 17 sustainable development goals (SDGs) aiming to eradicate poverty, protect the planet, and ensure peaceful and prosperous lives for everyone by 2030 [1–3]. Number 13 of the SDGs—"Climate Action"—envisions drastic carbon dioxide (CO_2) emission reductions to combat climate change and achieve a carbon-neutral society by 2050 [4]. Copper (Cu), lead (Pb), and zinc (Zn) are classified as critical metals defined by the World Bank as essential in renewable energy and clean storage technologies (e.g., wind turbines, solar photovoltaic (PV) panels, batteries, etc.), and thus their demands are projected to significantly increase in the future to achieve a carbon-neutral society [5,6].

Complex sulfide ores, some of the most important sources of critical metals, are typically composed of several metal sulfide minerals such as chalcopyrite ($CuFeS_2$), galena (PbS), and sphalerite (ZnS). In the mineral processing of complex sulfide ores consisting of Cu-Pb-Zn sulfide minerals, flotation has been commonly adopted to produce concentrates of each mineral [7–16]. Some complex sulfide ores contain not only PbS but also anglesite ($PbSO_4$), the presence of which is problematic for the separation of Cu, Pb, and Zn by flotation due to the unwanted activation of ZnS by Pb^{2+} released from $PbSO_4$ [17–19]. In the general flotation circuit of Cu-Pb-Zn sulfide ores, Cu and Pb sulfides are first recovered as froth, and then ZnS is recovered with the assistance of activators (e.g., $CuSO_4$) [20,21]. When $PbSO_4$ is contained in complex sulfide ores, however, Pb^{2+} is readily released from $PbSO_4$ during conditioning and/or flotation due to its higher solubility compared to PbS—the solubility product (K_{sp}) of $PbSO_4$ and PbS is $10^{-7.79}$ and $10^{-26.77}$, respectively [22]—and activates the surface of ZnS via the formation of PbS-like compounds, as illustrated in Equation (1) [23,24];

$$ZnS(s) + Pb^{2+} = PbS(s) + Zn^{2+} \quad (1)$$

The activation of ZnS by Pb^{2+} is known to increase the ZnS floatability because of a higher affinity of PbS-like compounds for xanthate compared to ZnS [25–27]. As a result, ZnS is recovered as froth together with $CuFeS_2$ and PbS, making their separation difficult [17]. Therefore, a proper way of depressing ZnS floatability during flotation of complex sulfide ores containing $PbSO_4$ should be established.

In addition, $PbSO_4$ is typically distributed into flotation tailings because it has poor affinity for the collector (e.g., xanthate) [28,29]. The disposal of $PbSO_4$ into tailings storage facilities (TSFs) would cause lead pollution in the surrounding environment [30,31]. That is, the presence of $PbSO_4$ in complex sulfide ores would cause not only difficulties in the separation of ZnS from Cu and Pb sulfide minerals but also lead pollution in the surrounding environment of TSFs.

To address the above-mentioned problems, this study investigated the improvement of ZnS depression by extracting $PbSO_4$ and the recovery of extracted Pb^{2+} as zero-valent Pb by cementation using zero-valent iron (ZVI). Specifically, this study was devoted not only to optimizing the conditions for the extraction of $PbSO_4$ by ethylene diamine tetra acetic acid (EDTA) as well as the recovery of extracted Pb^{2+} as Pb^0 via cementation using ZVI but also to understanding how the absence/presence of $PbSO_4$ affected the floatability of ZnS through flotation experiments and surface characterizations of ZnS/$PbSO_4$ mixtures with and without EDTA pretreatment. Finally, a sustainable flowsheet for the processing of complex sulfide ores containing $PbSO_4$ with the combination of EDTA pretreatment, flotation, and cementation is proposed.

2. Materials and Methods

2.1. Minerals and Reagents

Five types of minerals were used in this study: sphalerite (ZnS, Kamioka Mine, Hida, Japan), lead sulfate ($PbSO_4$, 98% purity), lead carbonate ($PbCO_3$, 99% purity), lead chloride ($PbCl_2$, 99% purity), and quartz (SiO_2, 99% purity). The lead compounds and quartz were obtained from FUJIFILM Wako Pure Chemical Corporation (Osaka, Japan). A mineral specimen of ZnS was characterized using X-ray fluorescence spectroscopy (XRF, EDXL300, Rigaku Corporation, Tokyo, Japan) and X-ray powder diffraction (XRD, MultiFlex, Rigaku Corporation, Tokyo, Japan), and its chemical and mineralogical compositions ware shown in Table 1 and Figure 1, respectively. The XRD pattern of ZnS implied that ZnS was relatively pure due to the absence of the peaks of common minerals such as SiO_2. In preparation for the flotation experiments, ZnS was ground by a vibratory disc mill (RS 200, Retsch Inc., Haan, Germany) and screened to obtain a size fraction of −75 + 38 μm.

Table 1. Chemical composition of a mineral specimen of ZnS based on XRF.

Elements	Zn	Fe	Pb	S	Si
Mass fraction (%)	55.7	6.2	0.2	21.3	8.4

For the flotation experiments, potassium amyl xanthate (KAX, Tokyo Chemical Industry Co., Ltd., Tokyo, Japan) and methyl isobutyl carbinol (MIBC, FUJIFILM Wako Pure Chemical Corporation, Osaka, Japan) were used as collector and frother, respectively. Zinc sulfate ($ZnSO_4$) and sodium sulfite (Na_2SO_3), purchased from FUJIFILM Wako Pure Chemical Corporation (Osaka, Japan), were used as depressants for ZnS. Sulfuric acid (H_2SO_4, FUJIFILM Wako Pure Chemical Corporation, Osaka, Japan) and sodium hydroxide (NaOH, FUJIFILM Wako Pure Chemical Corporation, Osaka, Japan) were used as pH adjusters. For the extraction experiments, ethylene diamine tetra acetic acid (EDTA, FUJIFILM Wako Pure Chemical Corporation, Osaka, Japan) was used as an extractant. For the cementation experiments, zero-valent Fe (ZVI) powder (Fe^0, −45 μm, FUJIFILM Wako Pure Chemical Corporation, Osaka, Japan) was used as a reductant.

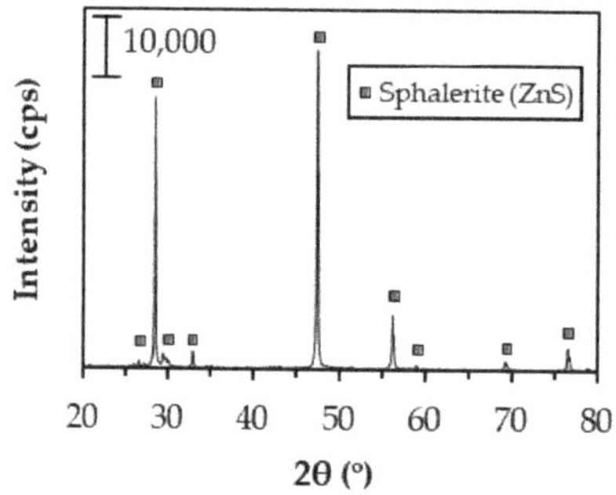

Figure 1. XRD pattern of a mineral specimen of ZnS.

2.2. Experimental Methods

2.2.1. Extraction Experiments of Lead Minerals Using EDTA

The extraction experiments of lead minerals were conducted in 50 mL Erlenmeyer flasks containing 1 g of the lead mineral and an EDTA solution (solid/liquid ratio: 1 g/10 mL). The solutions containing 100, 200, or 500 mM EDTA were used without pH adjustment (natural pH: 11.3). The flasks were shaken in a thermostat water bath shaker at a shaking speed of 120 rpm and an amplitude of 40 mm at 25 °C. After this, the leachates were collected by filtration using 0.2 μm syringe-driven membrane filters and immediately analyzed using an inductively coupled plasma atomic emission spectrometer (ICP-AES, ICPE 9820, Shimadzu Corporation, Kyoto, Japan) (margin of error = ±2%) to measure the concentration of Pb^{2+}. The experiments were performed in triplicate to ascertain that the differences observed were statistically significant, and the extraction efficiency of the lead mineral (E_{Pb}) was calculated using the following equation:

$$E_{Pb} = \frac{[Pb]_i}{[Pb]_{tot}} \times 100 \quad (2)$$

where $[Pb]_{tot}$ and $[Pb]_i$ are total Pb (mM) and extracted Pb (mM), respectively.

2.2.2. Cementation Experiments of the Extracted Pb^{2+}

The cementation experiments of Pb^{2+} extracted from $PbSO_4$ were conducted in 50 mL Erlenmeyer flasks. A volume of 10 mL of the leachate collected by filtration using 0.2 μm syringe-driven membrane filters after the extraction experiments ([EDTA], 500 mM; S/L ratio, 1 g/10 mL; time, 30 min) was added to the flask, and ultrapure nitrogen gas (N_2; 99.99%) was introduced for 15 min to remove the dissolved oxygen (DO) present in the leachate. Then, a known amount of ZVI powder was added to the flask, and N_2 was further introduced for 5 min. The flask was tightly capped with a silicon rubber plug and parafilm and shaken in a thermostat water bath shaker at a shaking speed of 120 rpm and an amplitude of 40 mm at 25 °C. After this, the leachate was filtered using a 0.2 μm syringe-driven membrane filter, and the filtrate was immediately analyzed using ICP-AES to measure the concentration of Pb^{2+}. Meanwhile, the residue was washed thoroughly with distilled water, dried in a vacuum-drying oven at 40 °C, and analyzed by scanning electron microscopy with energy-dispersive X-ray spectroscopy (SEM-EDS, JSM-IT200TM, JEOL Co., Ltd., Tokyo, Japan). The experiments were performed in triplicate to ascertain that the differences observed were statistically significant. Pb recovery (R_{Pb}) was calculated using the following equation:

$$R_{Pb} = \frac{[Pb]_I - [Pb]_F}{[Pb]_I} \times 100 \quad (3)$$

where $[Pb]_I$ and $[Pb]_F$ are the initial and the final concentrations of Pb (mM) in the leachate, respectively.

2.2.3. Flotation Experiments

The flotation experiments were carried out using an agitator-type flotation machine (ASH-F30H, Kankyo-kanri Engineering, Akita, Japan) equipped with a 400 mL flotation cell under the following conditions: pH, 6.5; temperature, 25 °C; pulp density, 5%; impeller speed, 1000 rpm; air flow rate, 1 L/min. In a 500 mL beaker, a model sample containing 15 g of ZnS and 5 g of $PbSO_4$ or SiO_2 was suspended in 300 mL of distilled water, and the supernatant was decanted to remove fine particles (<38 μm) [17]. After desliming, a model sample was repulped to 400 mL with distilled water in the flotation cell and conditioned for 3 min after adding the following reagents in sequence: ~5 kg/t of $ZnSO_4$ (100 ppm Zn^{2+}), 1 kg/t of Na_2SO_3, 20 g/t of KAX, and 20 μL/L of MIBC. Afterwards, air was introduced at a flowrate of 1 L/min, and froth was recovered for 3 min. The recovered froth/tailing products were dried at 105 °C for 24 h and analyzed by XRF to determine the recovery of Zn. The experiments were performed in duplicate to ascertain that the differences observed were statistically significant. An aliquot of the pulp of about 5 mL was collected 3 min after adding the depressant, filtered through a 0.2 μm syringe-driven membrane filter, and immediately analyzed using ICP-AES to check the extent of lead activation of ZnS. Meanwhile, the collected residues were washed thoroughly with distilled water, dried in a vacuum-drying oven at 40 °C, and analyzed by X-ray photoelectron spectroscopy (XPS, JPS-9200, JEOL Co., Ltd., Tokyo, Japan). The XPS analysis was conducted using an Al Kα X-ray source (1486.7 eV) operated at 100 W (voltage = 10 kV; current = 10 mA) under ultrahigh vacuum conditions (~10^{-7} Pa). The binding energies of photoelectrons were calibrated using C1s (285 eV) or $Zn2p_{3/2}$ (1022.0 eV) for charge correction. The XPS data were analyzed by Casa XPS, and deconvolutions of the spectra were carried out using an 80% Gaussian–20% Lorentzian peak model and a Shirley background [32–35].

In addition, flotation experiments of ZnS/$PbSO_4$ mixture after EDTA pretreatment were conducted to check how effective the removal of $PbSO_4$ was on the depression of ZnS floatability. A mixture of 15 g ZnS and 5 g $PbSO_4$ was mixed with 200 mL of a 500 mM EDTA solution (i.e., solid/liquid ratio: 20 g/200 mL) and shaken in a thermostat water bath shaker at a shaking speed of 120 rpm and an amplitude of 40 mm at 25 °C for 30 min. After this, the leachate was filtered using a 0.2 μm syringe-driven membrane filter, the filtrate was immediately analyzed using ICP-AES to measure the concentration of Pb^{2+}, and the cementation experiments to recover the extracted Pb^{2+} in the filtrate were carried out using 1 g/10 mL of ZVI for 24 h under the same conditions and with the same procedures mentioned in Section 2.2.2. Meanwhile, the residue obtained after EDTA pretreatment was washed thoroughly with distilled water and deslimed to remove fine particles [17]. After desliming, the residue was repulped to 400 mL with distilled water, and then flotation was conducted under the same conditions mentioned above. An aliquot of the pulp of about 5 mL was also collected 3 min after adding the depressant ($ZnSO_4$) and filtered through a 0.2 μm syringe-driven membrane filter, and the filtrate and residue were analyzed by ICP-AES and XPS, respectively.

3. Results and Discussion

3.1. Effects of $PbSO_4$ on ZnS Floatability

Figure 2 shows the flotation results of ZnS mixed with SiO_2 or $PbSO_4$. As can be seen, Zn recovery in the presence of SiO_2 was ~12% but significantly increased to >80% when $PbSO_4$ was present. It was reported that the affinity between ZnS and xanthate is poor, so the surface of ZnS needs to be modified using activators [21]. Since the solubility of $PbSO_4$ is higher than that of PbS, a substantial amount of Pb^{2+} might be released into the pulp during the conditioning process when the ore contains $PbSO_4$ [17]. As a result, lead activation of ZnS could occur (Equation (1)), thus increasing its floatability due to the formation of PbS-like compounds, having a higher affinity for xanthate than ZnS, on the mineral surface.

In short, the presence of $PbSO_4$ in the complex sulfide ores would negatively affect the separation of Cu-Pb from Zn. In addition, some portion of $PbSO_4$ will remain undissolved and be rejected as tailings due to the poor affinity of $PbSO_4$ for the collector (e.g., xanthate), which can cause lead contamination of the surrounding environment [28,29,36].

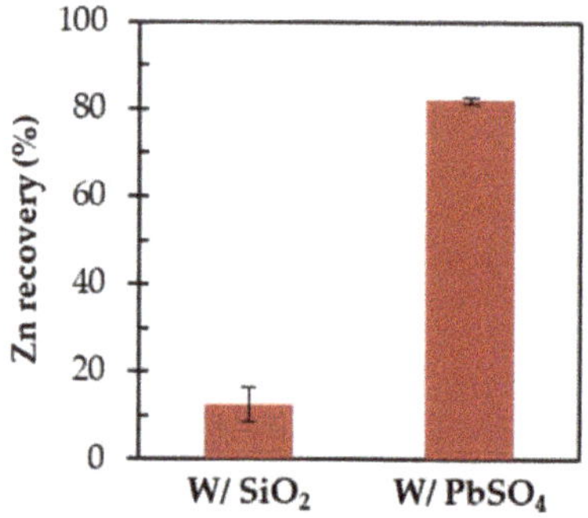

Figure 2. Flotation results of ZnS with SiO_2 or $PbSO_4$.

3.2. Extraction of $PbSO_4$ Using EDTA

To address the above-mentioned problems caused by the presence of $PbSO_4$ (i.e., unwanted activation of ZnS and lead contamination), its extraction prior to flotation was investigated. Among the extractants, EDTA was chosen in this study because of its ability to extract metal ions from sulfates like $PbSO_4$, without dissolving metal sulfides (e.g., PbS and ZnS) [37–40].

Figure 3a shows the extraction efficiency of $PbSO_4$ after 24 h with different concentrations of EDTA (100, 200, or 500 mM). When the stoichiometric ratio of EDTA/$[Pb]_{tot}$ was lower than 1, the amount of Pb^{2+} extracted was identical to the concentration of EDTA, which indicated that Pb^{2+} and EDTA formed a 1:1 complex (i.e., $[Pb(EDTA)]^{2-}$). When the stoichiometric ratio of EDTA/$[Pb]_{tot}$ was >1, almost all $PbSO_4$ (~98%) was extracted. The rate of $PbSO_4$ extraction was further studied with 500 mM EDTA, and the result showed that ~97% of $PbSO_4$ was extracted within 30 min (Figure 3b). EDTA could extract not only $PbSO_4$ but also other Pb minerals like $PbCO_3$ and $PbCl_2$ (Figure 4). These minerals are also unfloatable and will be rejected into the TSFs, indicating that their presence in the ores may result in the contamination of the TSFs with Pb. It was confirmed that EDTA pretreatment can extract almost all Pb minerals except for PbS, thus suppressing lead pollution to the surrounding environment of TSFs. It is noted that the concentration of EDTA should be chosen in the consideration of the amount of soluble minerals contained in ores, because EDTA would be consumed not only by Pb^{2+} but also by other metal ions from soluble minerals such as sulfates, oxides, hydroxides, and carbonates.

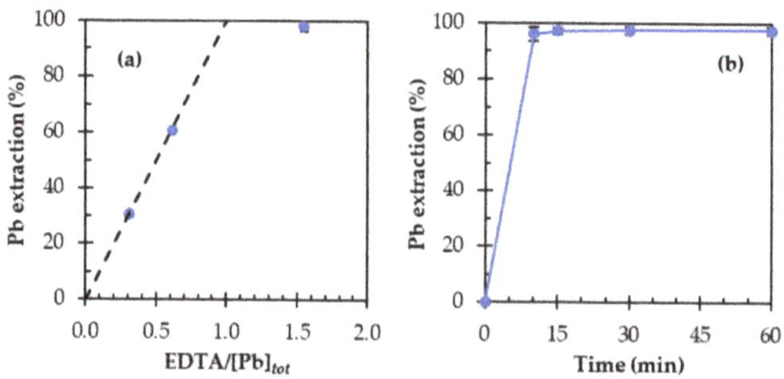

Figure 3. (**a**) Extraction efficiency of $PbSO_4$ after 24 h with different concentrations of EDTA (100, 200, or 500 mM) and (**b**) effect of time on $PbSO_4$ extraction with 500 mM EDTA. Note that the dotted line in Figure 3a corresponds to the ratio of EDTA/$[Pb]_{tot}$.

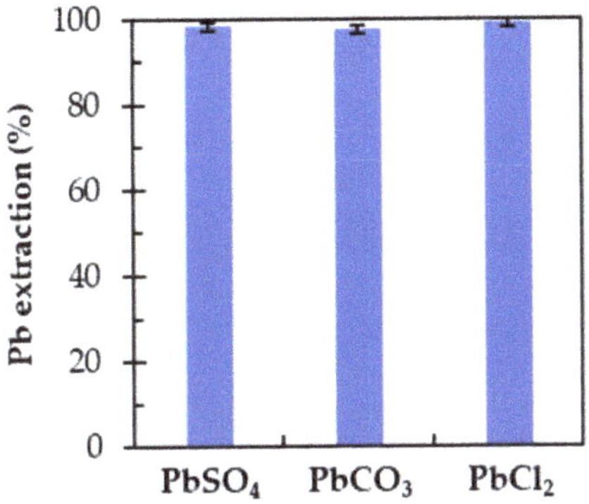

Figure 4. Extraction efficiency of lead minerals with 500 mM EDTA after 24 h.

3.3. *Recovery of Extracted Pb^{2+} as Zero-Valent Pb by Cementation*

After EDTA pretreatment, the recovery of Pb^{2+} from the leachate will not only add economic value but also protect the environment. Cementation is one of the effective methods to recover metal ions as zero-valent metals [41–45]. A general cementation reaction is illustrated in Equation (4): a metal (A^0) gives electrons to metal ion (B^{b+}), driven by the difference in standard redox potentials of the interacting metals and their ions, and as a result, B^{b+} is deposited on the surface of A^0 as B^0 [46–49]:

$$n\,A^0 + m\,B^{b+} \rightarrow n\,A^{a+} + m\,B^0\ (na = mb) \quad (4)$$

To recover the extracted Pb^{2+} as zero-valent Pb (Pb^0) from the EDTA leachate, cementation experiments were carried out using ZVI powder as a reductant after removing DO from the solution by purging with ultrapure nitrogen gas. Figure 5a shows the effects of the ZVI amount on the recovery of the extracted Pb^{2+} from the leachate after EDTA pretreatment using 500 mM EDTA for 30 min. When the amount of ZVI was 0.5 g/10 mL, ~40% of Pb^{2+} was recovered, although ~0.18 g/10 mL ZVI was stoichiometrically sufficient to recover all Pb^{2+} in the solution. As the cementation reaction progressed, the surface of ZVI was covered with cementation products (i.e., Pb^0), which hindered a further cementation reaction. This was probably the reason why 1 g/10 mL of ZVI was required to recover all Pb^{2+} from the EDTA leachate.

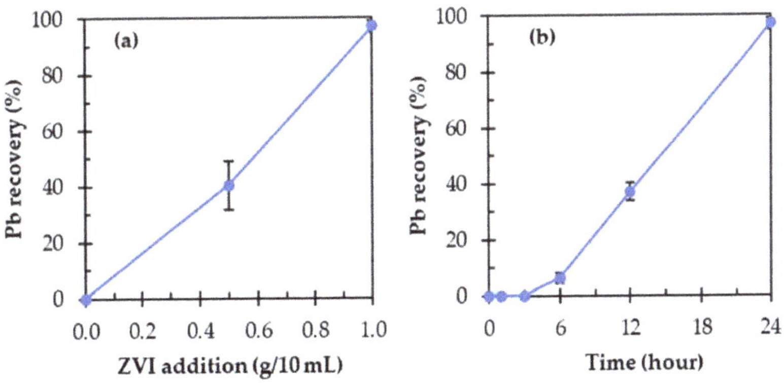

Figure 5. (**a**) Effects of the ZVI amount on the recovery of the extracted Pb after 24 h in the leachate after $PbSO_4$ extraction using 500 mM of EDTA for 30 min and (**b**) effects of time on the recovery of the extracted Pb^{2+} using 1 g/10 mL of ZVI in the leachate after EDTA pretreatment using 500 mM EDTA for 30 min.

Figure 5b shows the cementation results of Pb^{2+} by ZVI in time. As can be seen, almost no cementation occurred for 3 h, and then Pb^{2+} started being cemented gradually. The initial slow rate of Pb cementation can have two reasons: (i) the presence of Fe oxide films formed on the ZVI surface that limited the cementation of Pb^{2+} and (ii) an excess amount of free EDTA that redissolved the cemented Pb^0. After 3 h, the concentration of free EDTA ligands was most likely decreased due to its consumption in dissolving Fe oxide films and/or in forming complexes with Fe species released by the cementation reaction. It is important to note that after 24 h, ~97% cementation was achieved. To confirm the recovery of Pb^{2+} on ZVI, the residue obtained from the cementation experiment using 1 g/10 mL of ZVI for 24 h was analyzed using SEM-EDS. As shown in Figure 6, the deposition of Pb compounds on ZVI powder was observed, suggesting that the extracted Pb^{2+} was reductively deposited via the cementation reaction:

$$Pb(EDTA)^{2-} + Fe^0 = Pb^0 + EDTA + Fe^{2+} \quad (5)$$

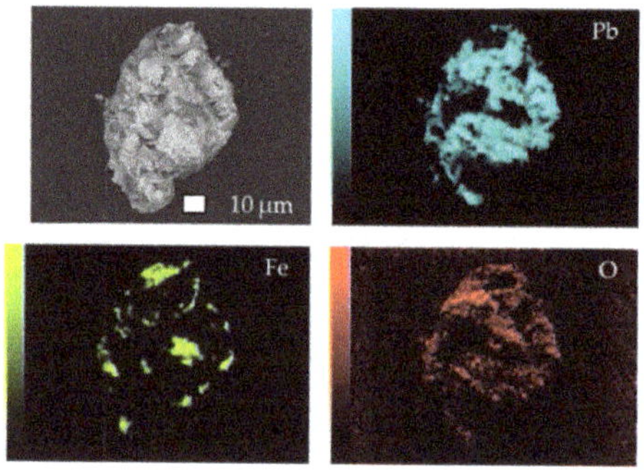

Figure 6. SEM photomicrograph of the residue of the cementation experiment using 1 g/10 mL of ZVI for 24 h with the corresponding elemental maps of Fe, Pb, and O.

Equation (5) consists of two half-cell reactions [50]:

$$Pb(EDTA)^{2-} + 2e^- = Pb^0 + EDTA^{4-} \quad (6)$$

$$Fe^0 = Fe^{2+} + 2e^- \quad (7)$$

The above-mentioned residue was also analyzed by XRD to further characterize Pb compounds formed on ZVI (Figure 7). As shown in the XRD pattern, the peaks of Fe^0, Fe_2O_3, Pb^0, and PbO were detected, indicating that Pb^{2+} was reductively deposited as Pb^0 on the surface of ZVI, as illustrated in Equation (5). The presence of PbO and Fe_2O_3 was due most likely to the oxidation of Pb^0 and ZVI surface during drying and storing of the sample [51–53].

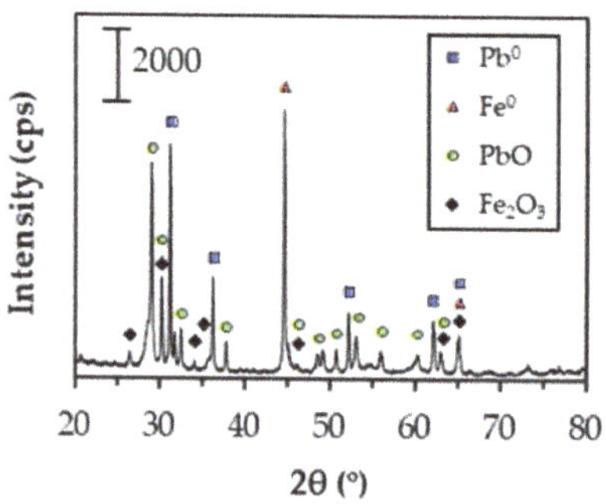

Figure 7. XRD pattern of the residue of the cementation experiment using 1 g/10 mL of ZVI for 24 h.

3.4. Effects of EDTA Pretreatment on ZnS Depression in Flotation

As described in Section 3.2, almost all $PbSO_4$ was extracted with EDTA, so the effects of EDTA pretreatment on ZnS depression in flotation were investigated. Flotation experiments using model samples ($ZnS/PbSO_4$ mixture) with and without EDTA pretreatment were conducted (Figure 8). Prior to the flotation experiments, the mixture of 15 g ZnS and 5 g $PbSO_4$ was treated with 500 mM of EDTA for 30 min (solid/liquid ratio: 20 g/200 mL), and ~99.8% of $PbSO_4$ was extracted. The cementation experiments using 1 g/10 mL of ZVI for 24 h were also carried out to recover the extracted Pb^{2+} in the filtrate obtained after EDTA pretreatment, and ~99.7% of the extracted Pb^{2+} was recovered. Meanwhile, the residue obtained after EDTA pretreatment was deslimed and repulped to 400 mL with distilled water in the flotation cell, and then the flotation experiments were conducted with ~5 kg/t of $ZnSO_4$ (100 ppm Zn^{2+}) and 1 kg/t of Na_2SO_3 as depressants at pH 6.5. As shown in Figure 8, the floatability of Zn was clearly depressed from ~82% to ~30% when EDTA pretreatment was applied.

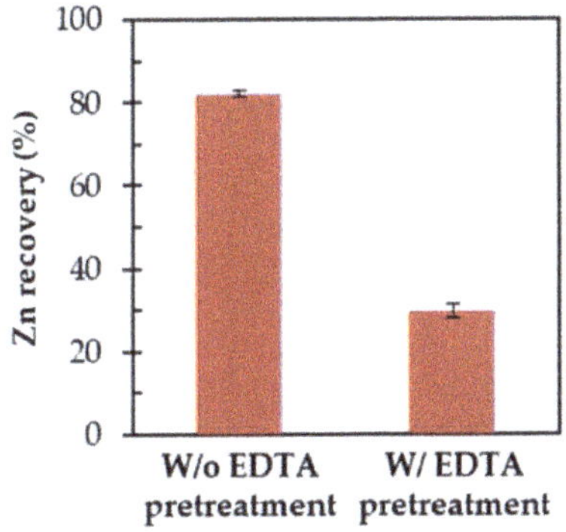

Figure 8. Effects of EDTA pretreatment on the floatability of ZnS in the presence of $PbSO_4$.

To confirm the depression of ZnS floatability after EDTA pretreatment, a pretreated sample reacted with zinc sulfate for 3 min, a ~5 mL aliquot of the pulp, was collected, and the filtrate and the residue were analyzed by ICP-AES and XPS, respectively. The concentration of Pb^{2+} in the filtrate without EDTA pretreatment was 5.3 ppm; it decreased to 0.2 ppm with EDTA pretreatment. On the other hand, the concentrations of Zn^{2+} in the filtrates with and without EDTA pretreatment were 101 and 99.4 ppm, respectively. Based on thermodynamics calculations using Equation (8), the possibility of lead activation of ZnS (Equation (1)) could be evaluated using the measured values of Pb^{2+} and Zn^{2+} [17,54,55]. In this study, the changes in free energy were calculated using measured values of Pb^{2+} and Zn^{2+} concentrations.

Table 2 shows the calculated results of the changes in free energy based on the equilibrium constants of lead activation of ZnS (Equation (1)). The change in free energy without EDTA pretreatment was negative, while that with EDTA pretreatment was positive. When the change in free energy is positive, the reverse reaction of lead activation (Equation (1)) would occur spontaneously, indicating that lead activation would be limited by EDTA pretreatment. These results support the flotation results that the depression of ZnS floatability was achieved by the pretreatment, which decreased Pb^{2+} concentration during flotation by extracting $PbSO_4$ in advance.

$$\Delta G = -RT \ln K + \ln\left(\mathrm{Zn}^{2+}/\mathrm{Pb}^{2+}\right) \tag{8}$$

$$K = \frac{K_{sp}^{ZnS}}{K_{sp}^{PbS}} \tag{9}$$

Table 2. Calculation results of the change in free energy with and without EDTA pretreatment.

K	ΔG (kJ/mol) Without EDTA Pretreatment	ΔG (kJ/mol) With EDTA Pretreatment
1000 [a]	−6.9	1.8
704 [b]	−6.1	2.7
1059 [c]	−7.1	1.7
1127 [d]	−7.2	1.5

[a] The equilibrium constant was obtained from a previous study [54]. [b, c, d] The equilibrium constants were calculated using Equation (9) by the reported K_{sp} values of ZnS and PbS, respectively [56–58].

To confirm whether the lead activation of sphalerite was limited by the extraction of $PbSO_4$ using EDTA, untreated ZnS as well as $ZnSO_4$-pretreated ZnS/$PbSO_4$ mixtures with and without EDTA pretreatment were analyzed by XPS. The $Pb4f_{7/2}$ core-level spectra of the samples are shown in Figure 9, and the corresponding curve-fitting parameters are summarized in Table 3. Since the Zn3s peak overlapped with the $Pb4f_{7/2}$ signals, the spectrum was resolved by curve fitting to subtract the area due to the Zn3s peak [55,59]. As illustrated in Figure 9b and c, the deconvoluted XPS spectra of the residues with and without $PbSO_4$ extraction showed two types of Pb species: (1) Pb^{2+}—S of PbS (143.6 and 138.6 eV) and (2) Pb^{2+}—SO_4 of $PbSO_4$ (144.7 and 139.8 eV), while the deconvoluted XPS spectrum of untreated ZnS showed no Pb species (Figure 9a) [26,60]. The decrease in the intensity of Pb^{2+}—SO_4 was in a good agreement with the extraction efficiency of $PbSO_4$ using EDTA for the ZnS/$PbSO_4$ mixture (~99.8%). When EDTA pretreatment was applied prior to the flotation experiments, the peak intensity ratio of PbS/ZnS decreased from 2.3 to 0.2, indicating that EDTA pretreatment was effective in limiting lead activation of ZnS. This is in line with the calculated results of the change in free energy based on thermodynamics (Table 2).

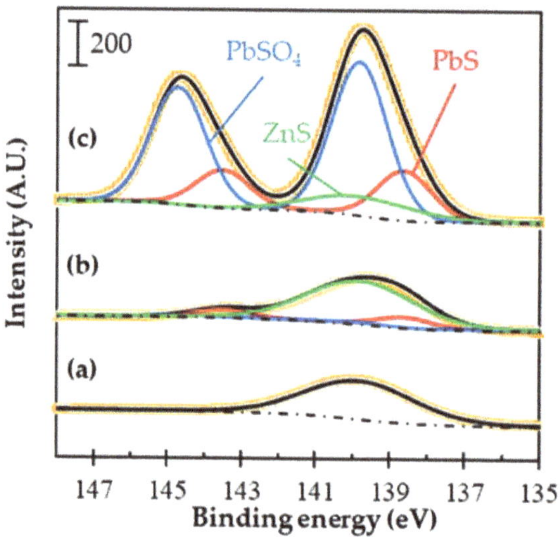

Figure 9. XPS Pb4f7/2 spectra of (**a**) raw ZnS, (**b**) ZnS/$PbSO_4$ with EDTA pretreatment, and (**c**) ZnS/$PbSO_4$ without EDTA pretreatment.

Table 3. XPS peak parameters for $Pb4f_{7/2}$ spectra and relative abundances of Pb species.

Binding Energy (eV)	FWHM	Assignments	Contents (at.%)		
			Untreated [a]	W/ EDTA [b] Pretreatment	W/o EDTA [a] Pretreatment
138.6 ± 0.05 143.5 ± 0.05	1.7	PbS	0	17.1	21.5
139.8 ± 0.05 144.7 ± 0.05	1.7	$PbSO_4$	0	8.8	48.4
139.8 ± 0.05	3.4	ZnS	100	80.5	9.5

[a] The binding energies of photoelectrons were calibrated using C1s (285 eV) for charge correction. [b] The binding energies of photoelectrons were calibrated using $Zn2p_{3/2}$ (1022.0 eV) for charge correction.

3.5. Implication of This Study

The results of this study have significant implications not only for understanding how the presence of $PbSO_4$ in complex sulfide ores affects the floatability of ZnS but also for proposing a sustainable process flowsheet covering both the improved flotation separation of complex sulfide ores and the detoxification of solid/solution wastes contaminated with Pb species. Firstly, the presence of $PbSO_4$ might have a detrimental impact on the flotation separation of complex sulfide ores due to unwanted lead activation of ZnS that improves its floatability, which could not be depressed with $ZnSO_4$—the conventional depressant for ZnS. Secondly, EDTA washing could extract almost all $PbSO_4$ from the ZnS/$PbSO_4$ mixture, and consequently, the floatability of ZnS decreased due to the limited amount of PbS-like compounds formed on the ZnS surface by lead activation. Finally, cementation using ZVI could recover Pb^{2+} extracted from $PbSO_4$ during EDTA pretreatment as Pb^0, which will not only add economic value but also protect the environment.

Based on the findings of this study, a sustainable process flowsheet for complex sulfide ores is proposed (Figure 10). In this proposed flowsheet, $PbSO_4$ is first extracted by EDTA pretreatment before flotation. The residue obtained after EDTA pretreatment is fed to a flotation stage where ZnS floatability is effectively depressed by the conventional depressant for ZnS due to the decrease in the ratio of Pb^{2+}/Zn^{2+} of the flotation pulp, and then gangue minerals like SiO_2 and FeS_2 would be disposed of into a tailings dam after the recovery of ZnS with the assistance of activators (e.g., $CuSO_4$). Meanwhile, the leachate obtained from EDTA pretreatment is rich in Pb^{2+} that can be recovered via cementation using ZVI as Pb^0. To keep up with the high demand for critical metals following the SDGs, mine developments should be in harmony with the environment. This proposed flowsheet can achieve enhanced selective flotation of complex sulfide ores (depression of ZnS floatability) while preventing lead pollution to the surrounding environment of tailings dams (removal of toxic $PbSO_4$ before flotation) and maximizing the recovery of critical elements by cementation (recovery of Pb^0 using ZVI).

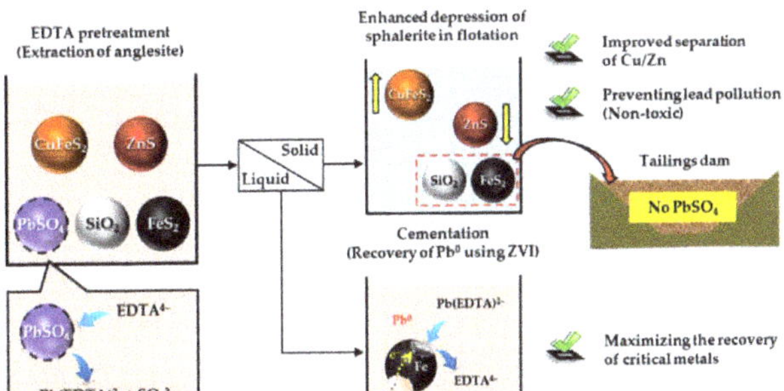

Figure 10. Proposed enhanced depression of sphalerite (ZnS) by extracting anglesite ($PbSO_4$) using ethylene diamine tetra acetic acid (EDTA) and recovery of the extracted Pb^{2+} as zero-valent Pb (Pb^0) by cementation using zero-valent Fe (ZVI).

4. Conclusions

In this paper, a pretreatment of flotation to extract $PbSO_4$ for the prevention of lead pollution to the surrounding environment of tailings dams and the depression of ZnS floatability combined with the recovery of extracted Pb^{2+} by cementation was investigated, and the findings of this study are summarized as follows:

1. ZnS floatability increased in the presence of $PbSO_4$.
2. The conventional depressants for ZnS, zinc sulfate and sodium sulfite, were not effective in depressing the floatability of ZnS when $PbSO_4$ was present.
3. Almost of all $PbSO_4$ (>97%) was extracted using EDTA, and >97% of the extracted Pb^{2+} could be recovered as Pb^0 by cementation using ZVI.
4. A pretreatment of flotation extracting $PbSO_4$ using EDTA was effective in depressing ZnS floatability.
5. The proposed method for complex sulfide ores containing $PbSO_4$, a combination of extraction of $PbSO_4$ using EDTA and recovery of extracted Pb^{2+} as zero-valent Pb by cementation using ZVI, could achieve enhanced selective flotation of complex sulfide ores (depression of ZnS floatability) while preventing lead pollution to the surrounding environment of tailings dams (removal of toxic $PbSO_4$ before flotation) and maximizing the recovery of critical elements by cementation (recovery of Pb^0 using ZVI).

Author Contributions: Conceptualization, K.A., M.I. and N.H.; methodology, K.A.; investigation, A.K.; data curation, A.K. and S.J.; writing—original draft preparation, K.A.; writing—review and editing, M.I., S.J., I.P. and N.H.; visualization, K.A.; supervision, N.H.; project administration, K.A.; funding acquisition, K.A. All authors have read and agreed to the published version of the manuscript.

Funding: This research was funded by JSPS KAKENHI Grant Number JP21J20552.

Data Availability Statement: Not applicable.

Conflicts of Interest: The authors declare no conflict of interest.

References

1. Hák, T.; Janoušková, S.; Moldan, B. Sustainable Development Goals: A Need for Relevant Indicators. *Ecol. Indic.* **2016**, *60*, 565–573. [CrossRef]
2. Tabelin, C.B.; Park, I.; Phengsaart, T.; Jeon, S.; Villacorte-Tabelin, M.; Alonzo, D.; Yoo, K.; Ito, M.; Hiroyoshi, N. Copper and Critical Metals Production from Porphyry Ores and E-Wastes: A Review of Resource Availability, Processing/Recycling Challenges, Socio-Environmental Aspects, and Sustainability Issues. *Resour. Conserv. Recycl.* **2021**, *170*, 105610. [CrossRef]
3. Park, I.; Kanazawa, Y.; Sato, N.; Galtchandmani, P.; Jha, M.K.; Tabelin, C.B.; Jeon, S.; Ito, M.; Hiroyoshi, N. Beneficiation of Low-grade Rare Earth Ore from Khalzan Buregtei Deposit (Mongolia) by Magnetic Separation. *Minerals* **2021**, *11*, 1432. [CrossRef]
4. Goal 13 | Department of Economic and Social Affairs. Available online: https://sdgs.un.org/goals/goal13 (accessed on 26 April 2022).
5. Nansai, K.; Nakajima, K.; Kagawa, S.; Kondo, Y.; Suh, S.; Shigetomi, Y.; Oshita, Y. Global Flows of Critical Metals Necessary for Low-Carbon Technologies: The Case of Neodymium, Cobalt, and Platinum. *Environ. Sci. Technol.* **2014**, *48*, 1391–1400. [CrossRef] [PubMed]
6. WBG Minerals for Climate Action: The Mineral Intensity of the Clean Energy Transition. Available online: https://pubdocs.worldbank.org/en/961711588875536384/Minerals-for-Climate-Action-The-Mineral-Intensity-of-the-Clean-Energy-Transition.pdf (accessed on 25 April 2022).
7. Bilal, M.; Ito, M.; Koike, K.; Hornn, V.; Ul Hassan, F.; Jeon, S.; Park, I.; Hiroyoshi, N. Effects of Coarse Chalcopyrite on Flotation Behavior of Fine Chalcopyrite. *Miner. Eng.* **2021**, *163*, 106776. [CrossRef]
8. Bilal, M.; Ito, M.; Akishino, R.; Bu, X.; Ul Hassan, F.; Park, I.; Jeon, S.; Aikawa, K.; Hiroyoshi, N. Heterogenous Carrier Flotation Technique for Recovering Finely Ground Chalcopyrite Particles Using Coarse Pyrite Particles as a Carrier. *Miner. Eng.* **2022**, *180*, 107518. [CrossRef]
9. Hornn, V.; Ito, M.; Shimada, H.; Tabelin, C.B.; Jeon, S.; Park, I.; Hiroyoshi, N. Agglomeration–Flotation of Finely Ground Chalcopyrite Using Emulsified Oil Stabilized by Emulsifiers: Implications for Porphyry Copper Ore Flotation. *Metals* **2020**, *10*, 912. [CrossRef]
10. Hornn, V.; Ito, M.; Shimada, H.; Tabelin, C.B.; Jeon, S.; Park, I.; Hiroyoshi, N. Agglomeration-Flotation of Finely Ground Chalcopyrite and Quartz: Effects of Agitation Strength during Agglomeration Using Emulsified Oil on Chalcopyrite. *Minerals* **2020**, *10*, 380. [CrossRef]
11. Hornn, V.; Ito, M.; Yamazawa, R.; Shimada, H.; Tabelin, C.B.; Jeon, S.; Park, I.; Hiroyoshi, N. Kinetic Analysis for Agglomeration-Flotation of Finely Ground Chalcopyrite: Comparison of First Order Kinetic Model and Experimental Results. *Mater. Trans.* **2020**, *61*, 1940–1948. [CrossRef]

12. Hornn, V.; Park, I.; Ito, M.; Shimada, H.; Suto, T.; Tabelin, C.B.; Jeon, S.; Hiroyoshi, N. Agglomeration-Flotation of Finely Ground Chalcopyrite Using Surfactant-Stabilized Oil Emulsions: Effects of Co-Existing Minerals and Ions. *Miner. Eng.* **2021**, *171*, 107076. [CrossRef]
13. Park, I.; Hong, S.; Jeon, S.; Ito, M.; Hiroyoshi, N. A Review of Recent Advances in Depression Techniques for Flotation Separation of Cu–Mo Sulfides in Porphyry Copper Deposits. *Metals* **2020**, *10*, 1269. [CrossRef]
14. Park, I.; Hong, S.; Jeon, S.; Ito, M.; Hiroyoshi, N. Flotation Separation of Chalcopyrite and Molybdenite Assisted by Microencapsulation Using Ferrous and Phosphate Ions: Part I. Selective Coating Formation. *Metals* **2020**, *10*, 1667. [CrossRef]
15. Park, I.; Hong, S.; Jeon, S.; Ito, M.; Hiroyoshi, N. Flotation Separation of Chalcopyrite and Molybdenite Assisted by Microencapsulation Using Ferrous and Phosphate Ions: Part II. Flotation. *Metals* **2021**, *11*, 439. [CrossRef]
16. Seng, S.; Tabelin, C.B.; Makino, Y.; Chea, M.; Phengsaart, T.; Park, I.; Hiroyoshi, N.; Ito, M. Improvement of Flotation and Suppression of Pyrite Oxidation Using Phosphate-Enhanced Galvanic Microencapsulation (GME) in a Ball Mill with Steel Ball Media. *Miner. Eng.* **2019**, *143*, 105931. [CrossRef]
17. Aikawa, K.; Ito, M.; Kusano, A.; Park, I.; Oki, T.; Takahashi, T.; Furuya, H.; Hiroyoshi, N. Flotation of Seafloor Massive Sulfide Ores: Combination of Surface Cleaning and Deactivation of Lead-Activated Sphalerite to Improve the Separation Efficiency of Chalcopyrite and Sphalerite. *Metals* **2021**, *11*, 253. [CrossRef]
18. Ministry of Economic, Trade and Industry (METI) and Japan Oil, Gas and Metals National Corporation (JOGMEC). Summary Report on Comprehensive Evaluations of Development Plan of Seafloor Massive Sulfide Deposits. Available online: http://www.jogmec.go.jp/content/300359550.pdf (accessed on 26 February 2021).
19. Zeng, Z.; Chen, Z.; Qi, H. Two Processes of Anglesite Formation and a Model of Secondary Supergene Enrichment of Bi and Ag in Seafloor Hydrothermal Sulfide Deposits. *J. Mar. Sci. Eng.* **2021**, *10*, 35. [CrossRef]
20. Woodcock, J.T.; Sparrow, G.J.; Bruckard, W.J.; Johnson, N.W.; Dunne, R. *Plant Practice: Sulfide Minerals and Precious Metals. InFroth Flotation a Century of Innovation*; Fuerstenau, M., Jameson, G., Yoon, R.H., Eds.; SME: Littleton, CO, USA, 2007; pp. 781–843, ISBN 978-0-873-35252-9.
21. Wills, B.A.; Finch, J.A. Froth flotation. In *Mineral Processing Technology: An Introduction to the Practical Aspects of Ore Treatment and Mineral Recovery*, 8th ed.; Elsevier: Oxford, UK, 2016; pp. 265–380, ISBN 9780080970530.
22. Ball, J.W.; Nordstrom, D.K. *User's Manual for WATEQ4F, with Revised Thermodynamic Data Base and Text Cases for Calculating Speciation of Major, Trace, and Redox Elements in Natural Waters*; USGS Numbered Series Open-File Report 91-183; U.S. Geological Survey: Reston, VA, USA, 1991.
23. Rashchi, F.; Sui, C.; Finch, J.A. Sphalerite Activation and Surface Pb Ion Concentration. *Int. J. Miner. Process.* **2002**, *67*, 43–58. [CrossRef]
24. Laskowski, J.S.; Liu, Q.; Zhan, Y. Sphalerite Activation: Flotation and Electrokinetic Studies. *Miner. Eng.* **1997**, *10*, 787–802. [CrossRef]
25. Houot, R.; Raveneau, P. Activation of Sphalerite Flotation in the Presence of Lead Ions. *Int. J. Miner. Process.* **1992**, *35*, 253–271. [CrossRef]
26. Aikawa, K.; Ito, M.; Segawa, T.; Jeon, S.; Park, I.; Tabelin, C.B.; Hiroyoshi, N. Depression of Lead-Activated Sphalerite by Pyrite via Galvanic Interactions: Implications to the Selective Flotation of Complex Sulfide Ores. *Miner. Eng.* **2020**, *152*, 106367. [CrossRef]
27. Trahar, W.J.; Senior, G.D.; Heyes, G.W.; Creed, M.D. The Activation of Sphalerite by Lead—A Flotation Perspective. *Int. J. Miner. Process.* **1997**, *49*, 121–148. [CrossRef]
28. Rashchi, F.; Dashti, A.; Arabpour-Yazdi, M.; Abdizadeh, H. Anglesite Flotation: A Study for Lead Recovery from Zinc Leach Residue. *Miner. Eng.* **2005**, *18*, 205–212. [CrossRef]
29. Fuerstenau, M.C.; Olivas, S.A.; Herrera-Urbina, R.; Han, K.N. The Surface Characteristics and Flotation Behavior of Anglesite and Cerussite. *Int. J. Miner. Process.* **1987**, *20*, 73–85. [CrossRef]
30. Silwamba, M.; Ito, M.; Hiroyoshi, N.; Tabelin, C.B.; Hashizume, R.; Fukushima, T.; Park, I.; Jeon, S.; Igarashi, T.; Sato, T.; et al. Alkaline Leaching and Concurrent Cementation of Dissolved Pb and Zn from Zinc Plant Leach Residues. *Minerals* **2022**, *12*, 393. [CrossRef]
31. Schindler, M.; Santosh, M.; Dotto, G.; Silva, L.F.O.; Hochella, M.F. A Review on Pb-Bearing Nanoparticles, Particulate Matter and Colloids Released from Mining and Smelting Activities. *Gondwana Res.* **2021**. [CrossRef]
32. Park, I.; Tabelin, C.B.; Seno, K.; Jeon, S.; Inano, H.; Ito, M.; Hiroyoshi, N. Carrier-Microencapsulation of Arsenopyrite Using Al-Catecholate Complex: Nature of Oxidation Products, Effects on Anodic and Cathodic Reactions, and Coating Stability under Simulated Weathering Conditions. *Heliyon* **2020**, *6*, e03189. [CrossRef] [PubMed]
33. Shirley, D.A. High-Resolution X-ray Photoemission Spectrum of the Valence Bands of Gold. *Phys. Rev. B* **1972**, *5*, 4709–4714. [CrossRef]
34. Tabelin, C.B.; Corpuz, R.D.; Igarashi, T.; Villacorte-Tabelin, M.; Ito, M.; Hiroyoshi, N. Hematite-Catalysed Scorodite Formation as a Novel Arsenic Immobilisation Strategy under Ambient Conditions. *Chemosphere* **2019**, *233*, 946–953. [CrossRef] [PubMed]
35. Zoleta, J.B.; Itao, G.B.; Resabal, V.J.T.; Lubguban, A.A.; Corpuz, R.D.; Ito, M.; Hiroyoshi, N.; Tabelin, C.B. Improved Pyrolysis Behavior of Ammonium Polyphosphate-Melamine-Expandable (APP-MEL-EG) Intumescent Fire Retardant Coating System Using Ceria and Dolomite as Additives for I-Beam Steel Application. *Heliyon* **2020**, *6*, e03119. [CrossRef]

36. Silwamba, M.; Ito, M.; Hiroyoshi, N.; Tabelin, C.B.; Hashizume, R.; Fukushima, T.; Park, I.; Jeon, S.; Igarashi, T.; Sato, T.; et al. Recovery of Lead and Zinc from Zinc Plant Leach Residues by Concurrent Dissolution-Cementation Using Zero-Valent Aluminum in Chloride Medium. *Metals* **2020**, *10*, 531. [CrossRef]
37. Bicak, O. A Technique to Determine Ore Variability in a Sulphide Ore. *Miner. Eng.* **2019**, *142*, 105927. [CrossRef]
38. Grano, S.R.; Ralston, J.; Johnson, N.W. Characterization and Treatment of Heavy Medium Slimes in the Mt. Isa Mines Lead-Zinc Concentrator. *Miner. Eng.* **1988**, *1*, 137–150. [CrossRef]
39. Kant, C.; Rao, S.R.; Finch, J.A. Distribution of Surface Metal Ions among the Products of Chalcopyrite Flotation. *Miner. Eng.* **1994**, *7*, 905–916. [CrossRef]
40. Rumball, J.A.; Richmond, G.D. Measurement of Oxidation in a Base Metal Flotation Circuit by Selective Leaching with EDTA. *Int. J. Miner. Process.* **1996**, *48*, 1–20. [CrossRef]
41. Choi, S.; Jeon, S.; Park, I.; Ito, M.; Hiroyoshi, N. Addition of Fe_3O_4 as Electron Mediator for Enhanced Cementation of Cd^{2+} and Zn^{2+} on Aluminum Powder from Sulfate Solutions and Magnetic Separation to Concentrate Cemented Metals from Cementation Products. *J. Environ. Chem. Eng.* **2021**, *9*, 106699. [CrossRef]
42. Jeon, S.; Bright, S.; Park, I.; Tabelin, C.B.; Ito, M.; Hiroyoshi, N. The Effects of Coexisting Copper, Iron, Cobalt, Nickel, and Zinc Ions on Gold Recovery by Enhanced Cementation via Galvanic Interactions between Zero-Valent Aluminum and Activated Carbon in Ammonium Thiosulfate Systems. *Metals* **2021**, *11*, 1352. [CrossRef]
43. Choi, S.; Jeon, S.; Park, I.; Ito, M.; Hiroyoshi, N. Enhanced Cementation of Co^{2+} and Ni^{2+} from Sulfate and Chloride Solutions Using Aluminum as an Electron Donor and Conductive Particles as an Electron Pathway. *Metals* **2021**, *11*, 248. [CrossRef]
44. Jeon, S.; Tabelin, C.B.; Park, I.; Nagata, Y.; Ito, M.; Hiroyoshi, N. Ammonium Thiosulfate Extraction of Gold from Printed Circuit Boards (PCBs) of End-of-Life Mobile Phones and Its Recovery from Pregnant Leach Solution by Cementation. *Hydrometallurgy* **2020**, *191*, 105214. [CrossRef]
45. Jeon, S.; Tabelin, C.B.; Takahashi, H.; Park, I.; Ito, M.; Hiroyoshi, N. Enhanced Cementation of Gold via Galvanic Interactions Using Activated Carbon and Zero-Valent Aluminum: A Novel Approach to Recover Gold Ions from Ammonium Thiosulfate Medium. *Hydrometallurgy* **2020**, *191*, 105165. [CrossRef]
46. Jeon, S.; Bright, S.; Park, I.; Kuze, A.; Ito, M.; Hiroyoshi, N. A Kinetic Study on Enhanced Cementation of Gold Ions by Galvanic Interactions between Aluminum (Al) as an Electron Donor and Activated Carbon (AC) as an Electron Mediator in Ammonium Thiosulfate System. *Minerals* **2022**, *12*, 91. [CrossRef]
47. Jeon, S.; Bright, S.; Park, I.; Tabelin, C.B.; Ito, M.; Hiroyoshi, N. A Simple and Efficient Recovery Technique for Gold Ions from Ammonium Thiosulfate Medium by Galvanic Interactions of Zero-Valent Aluminum and Activated Carbon: A Parametric and Mechanistic Study of Cementation. *Hydrometallurgy* **2022**, *208*, 105815. [CrossRef]
48. Choi, S.; Yoo, K.; Alorro, R.D.; Tabelin, C.B. Cementation of Co Ion in Leach Solution Using Zn Powder Followed by Magnetic Separation of Cementation-Precipitate for Recovery of Unreacted Zn Powder. *Miner. Eng.* **2020**, *145*, 106061. [CrossRef]
49. Choi, S.; Jeon, S.; Park, I.; Tabelin, C.B.; Ito, M.; Hiroyoshi, N. Enhanced Cementation of Cd^{2+}, Co^{2+}, Ni^{2+}, and Zn^{2+} on Al from Sulfate Solutions by Activated Carbon Addition. *Hydrometallurgy* **2021**, *201*, 105580. [CrossRef]
50. Silwamba, M.; Ito, M.; Tabelin, C.B.; Park, I.; Jeon, S.; Takada, M.; Kubo, Y.; Hokari, N.; Tsunekawa, M.; Hiroyoshi, N. Simultaneous Extraction and Recovery of Lead Using Citrate and Micro-Scale Zero-Valent Iron for Decontamination of Polluted Shooting Range Soils. *Environ. Adv.* **2021**, *5*, 100115. [CrossRef]
51. Xi, Y.; Mallavarapu, M.; Naidu, R. Reduction and Adsorption of Pb^{2+} in Aqueous Solution by Nano-Zero-Valent Iron—A SEM, TEM and XPS Study. *Mater. Res. Bull.* **2010**, *45*, 1361–1367. [CrossRef]
52. Taylor, J.A.; Perry, D.L. An X-ray Photoelectron and Electron Energy Loss Study of the Oxidation of Lead. *J. Vac. Sci. Technol. A Vac. Surf. Film.* **1998**, *2*, 771. [CrossRef]
53. Silwamba, M.; Ito, M.; Hiroyoshi, N.; Tabelin, C.B.; Fukushima, T.; Park, I.; Jeon, S.; Igarashi, T.; Sato, T.; Nyambe, I.; et al. Detoxification of Lead-Bearing Zinc Plant Leach Residues from Kabwe, Zambia by Coupled Extraction-Cementation Method. *J. Environ. Chem. Eng.* **2020**, *8*, 104197. [CrossRef]
54. El-Shall, H.E.; Elgillani, D.A.; Abdel-Khalek, N.A. Role of Zinc Sulfate in Depression of Lead-Activated Sphalerite. *Int. J. Miner. Process.* **2000**, *58*, 67–75. [CrossRef]
55. Basilio, C.I.; Kartio, I.J.; Yoon, R.-H. Lead Activation of Sphalerite during Galena Flotation. *Miner. Eng.* **1996**, *9*, 869–879. [CrossRef]
56. Helgeson, H.C. Thermodynamics of Hydrothermal Systems at Elevated Temperatures and Pressures. *Am. J. Sci.* **1969**, *267*, 729–804. [CrossRef]
57. Latimer, W.M. *The Oxidation States of the Elements and Their Potentials in Aqueous Solutions*, 2nd ed.; Prentice-Hall, Inc.: Upper Saddle River, NJ, USA, 1952; pp. 72, 152, 169, ISBN 978-0-758-15876-5.
58. Leckie, J.O.; James, R.O. *Aqueous Environmental Chemistry of Metals*; Rubin, A.J., Ed.; Ann Arbor Science Publishers: Ann Arbor, MI, USA, 1974; pp. 1–76, ISBN 978-0-250-40060-7.
59. Liu, J.; Ejtemaei, M.; Nguyen, A.V.; Wen, S.; Zeng, Y. Surface Chemistry of Pb-Activated Sphalerite. *Miner. Eng.* **2020**, *145*, 106058. [CrossRef]
60. Hernan, L.; Morales, J.; Sanchez, L.; Tirado, J.L.; Espinos, J.P.; Gonzalez Elipe, A.R. Diffraction and XPS Studies of Misfit Layer Chalcogenides Intercalated with Cobaltocene. *Chem. Mater.* **1995**, *7*, 1576–1582. [CrossRef]

minerals

MDPI

Article

Diagnosis and Optimization of Gold Ore Flotation Circuit via Linear Circuit Analysis and Mass Balance Simulation

Seongsoo Han [1,†], Minuk Jung [2,3,†], Wonjae Lee [2], Seongmin Kim [3], Kyoungmun Lee [1], Geun-tae Lim [4], Ho-Seok Jeon [3,5], Siyoung Q. Choi [1,*] and Yosep Han [3,5,*]

1 Department of Chemical and Biomolecular Engineering, Korea Advanced Institute of Science and Technology (KAIST), 291 Daehak-ro, Daejeon 34141, Korea; sshan12@kaist.ac.kr (S.H.); lkm2387@kaist.ac.kr (K.L.)
2 Department of Earth Resources and Environmental Engineering, Hanyang University, 222 Wangsimni-ro, Seoul 04763, Korea; mujung@kigam.re.kr (M.J.); lwj9004@hanyang.ac.kr (W.L.)
3 Resources Recovery Research Center, Mineral Resources Division, Korea Institute of Geoscience & Mineral Resources (KIGAM), 124 Gwahak-ro, Daejeon 34132, Korea; smkim@kigam.re.kr (S.K.); hsjeon@kigam.re.kr (H.-S.J.)
4 Gold Mining Division, Suncement Co., Ltd., Haenam-gun 59006, Korea; gtlim@suncement.co.kr
5 Department of Resources Recycling, University of Science and Technology (UST), 217 Gajeong-ro, Daejeon 34113, Korea
* Correspondence: sqchoi@kaist.ac.kr (S.Q.C.); yosep@kigam.re.kr (Y.H.); Tel.: +82-42-350-3914 (S.Q.C.); +82-42-868-3181 (Y.H.)
† These authors contributed equally to this work.

Citation: Han, S.; Jung, M.; Lee, W.; Kim, S.; Lee, K.; Lim, G.-t.; Jeon, H.-S.; Choi, S.Q.; Han, Y. Diagnosis and Optimization of Gold Ore Flotation Circuit via Linear Circuit Analysis and Mass Balance Simulation. *Minerals* **2021**, *11*, 1065. https://doi.org/10.3390/min11101065

Academic Editors: Marthias Silwamba and Ilhwan Park

Received: 2 September 2021
Accepted: 27 September 2021
Published: 29 September 2021

Abstract: The aim of this study is to diagnose and optimize a closed multistage gold ore flotation circuit in an operational industrial plant. Linear circuit analysis (LCA), a partition-based model, and a mass balance model using flotation first-order kinetics are employed to diagnose the current process. The result shows that the current circuit operates with high recovery but the gold grade of the final concentrate is low owing to the low buoyancy ratio. Hence, several alternative circuits with different streamlines and cell arrangements are proposed and simulated using LCA and a mass balance model. The result suggests that if the current process is changed to an alternative circuit in which the floated product stream of the rougher bank is changed, then the gold grade of the concentrate can be improved by 128%. Finally, the current circuit is optimized by changing it to an alternative circuit. This study provides a methodology for adapting the simulation of optimization for the flotation circuit of an industrial plant via LCA and mass balance simulation.

Keywords: flotation circuit; gold ore; diagnosis; optimization; mass balance; process simulation

1. Introduction

Efficient metal production is in demand owing to an increase in metal demand worldwide and the depletion of resources. Accordingly, methods to improve the economic efficiency of metal production have garnered attention [1–7]. The production of metal from minerals involves the beneficiation process, i.e., mineral processing, which physically separates the target mineral (i.e., concentrate) and gangue (i.e., tailing). It comprises several unit processes, such as comminution, classification, and separation [8,9]. In the separation process, flotation is the final step that determines the efficient recovery of the concentrate, which is the final product of beneficiation [10–12]. In general, the flotation process comprises a complex multistage closed circuit, which improves the grade and recovery of concentrates such as roughers, scavengers, and cleaner circuits [13–15]. Hence, the installation and operation of a plant through an optimal design for a highly efficient separation process is crucial, particularly in the froth flotation step [16–25]. Therefore, it is believed that the overall energy-saving effect should be achieved through an optimized flotation process owing to the improvement in productivity. However, because most large-scale plant flotation processes in operation have been preinstalled and operate continuously, process optimization for an efficient concentrate production is difficult to control [1,4,15]. To optimize flotation, the floatability of unit cells was evaluated primarily via laboratory-scale experiments [3]. However, this method is prone to errors when scaled up to accommodate the plant. Furthermore, it is expensive to configure and test multiple plant-scale flotation circuits [14,19,26,27].

Hence, the diagnosis and optimization of circuits via computer simulations have been proposed [12,14,16–19]. A general diagnostic and optimization model of the flotation process includes a mass balance model based on flotation kinetics [2,28]. Furthermore, Noble and Luttrel [26] proposed a linear circuit analysis (LCA) as a process diagnosis and optimization model. This methodology is a partition-based separation model for estimating the arrangement and interconnection of unit processes in a circuit. This evaluation method does not require many experiments or knowledge associated with process variables, and it can represent the separation efficiency of the circuit as one indicator [26]. Therefore, the diagnosis and optimization of the flotation circuit can be performed easily and more accurately if a mass balance model and LCA are applied.

In this study, the flotation process of a beneficiation plant during operation was diagnosed and optimized using LCA and mass balance simulations. The target circuit was that of a flotation process in an industrial gold ore beneficiation plant in Jeollanam-do province, Korea. This beneficiation plant was first introduced by Lee et al. [29]. The plant was not optimized and diagnosed when it was initially installed. In addition, the flotation process was primarily operated based on the operator's experience. Hence, it is difficult to predict the recovery and grade of the concentrate based on changes in the quality and throughput of the ore, and process optimizations such as circuit changes and flotation cell addition are futile. Therefore, the abovementioned target circuit was selected as the test bed for this study. The flotation circuit information, such as the mineral composition of ores, throughput, and cell size, was used in a mass balance simulation to calculate the gold grade of the concentrate and the recovery. The simulation results were validated by comparing them with industrial data. Subsequently, several alternative circuits for the flotation process were proposed for process optimization. Finally, the separation efficiency of the alternative circuits and the current process were compared and analyzed using LCA and mass balance simulations. The circuit was optimized by adopting the process that afforded the best separation efficiency.

2. Materials and Methods

2.1. Materials and Reagents

Gold ores were obtained from the Gasado gold mine in Jeollanam-do, Korea. It is known that the gold ore consists of pyrite, muscovite, and quartz [30]. The gold ores were initially crushed using a jaw crusher and a cone crusher, and then ground using a ball mill in the gold ore beneficiation plant. Water was obtained from the plants. The particle sizes of 50% passing (d_{50}) and 80% passing (d_{80}) were 35 and 94 μm, respectively. Potassium amyl xanthate (KAX, $C_5H_{11}KOS_2$, TCI Co., Ltd., Tokyo, Japan) and Aeropromoter 3477 (AP3477, isobutyl dithiophosphates, Cytec Industries Inc., Woodland Park, NJ, USA) were used as collectors. Poly-propylene glycol (PPG; average molecular weight ~425, Sigma–Aldrich, St. Louis, MO, USA) was used as a frother.

2.2. Characterization of Gold Ores

The gold ores processed by the target beneficiation plant were Gasado gold ores [29]. To evaluate the mineralogical properties of these gold ores, X-ray fluorescence spectrometry (XRF, PW2404, Phillips, Tokyo, Japan), X-ray diffraction (XRD, SmartLab, Rigaku, Tokyo, Japan), and density analysis (gas pycnometer, AccuPyc II, Micromeritics, Norcross, GA, USA) were performed. Gold in all ores was analyzed using the fire-assay technique. To obtain photographs of gold ore, gold ore specimens were prepared by cold mounting with an epoxy resin (KEM90 Resin, ATM GmbH, Mammelzen, Germany) and a hardener

(KEM90 Hardener, ATM, Berlin, Germany). The specimens were polished using a polishing machine (SAPHIR 520m, ATM GmbH, Mammelzen, Germany) equipped with cloth pads and diamond suspensions. Subsequently, the gold ore specimens were directly observed using a digital optical microscope (DVM2500, Leica, Wetzlar, Germany), where a white LED was used as the light source.

2.3. Flotation Circuit Description

The target flotation circuit is a beneficiation plant located in Haenam-gun, Jeollanam-do, Korea. The grinding and classification process of this plant was introduced by Lee et al. [29]. Currently, the throughput of this process is 170 t/d (7.08 t/h). All of the flotation machines are of the Denver sub-A type. Chemicals (collector: KAX, AP3477, frother: PPG) were simultaneously added to the conditioner. Modifiers such as activators and depressors were not used in this process. Groundwater was used as the process water, and its pH was 6–6.5.

The flowsheet is shown in Figure 1, and Table 1 lists the number of cells, cell size, and effective volume of this circuit. This flotation circuit comprised two rougher banks, a cleaner, and a scavenger. Each rougher bank contained four cells. The scavenger and cleaner were composed of four cells. The floated product of cleaner 1 bank was the final concentrate, and the sunk product of scavenger #4 cell was the final tailings. This process was designed to be cleaned three times by transferring the rougher concentrate to the cleaner #3 bank. Furthermore, it was designed as a closed circuit, in which the cleaner tailing and scavenger concentrate were fed back to the conditioner.

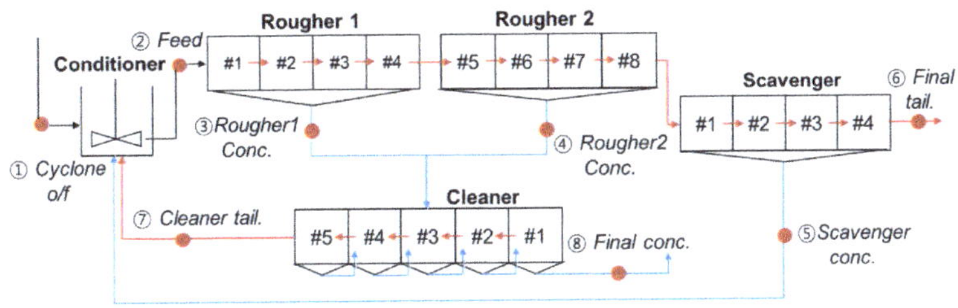

Figure 1. Flowsheet of gold ore flotation circuit. Blue line represents stream of floated products, and red line represents stream of sinked products after flotation. Red point represents sampling point: ① cyclone overflow (o/f); ② feed; ③ rougher 1 concentrate; ④ rougher 2 concentrate; ⑤ scavenger concentrate ⑥ final tailing; ⑦ cleaner tailing; ⑧ final concentrate.

Table 1. Detailed information regarding flotation circuit in target beneficiation plant.

	Number of Banks	Number of Cells	Cell Size, m^3 *	Effective Volume, m^3
Conditioner	-	1	2.4 × 2.4	9.8
Rougher	2	4	1.5 × 1.3 × 1.2	2.1
Scavenger	1	4	1.5 × 1.4 × 1.2	2.1
Cleaner	5	1	1.1 × 1.4 × 0.9	1.2

* Conditioner: diameter × higher; other cells: width × length × height.

The final concentrate was delivered to a refinery (LS-Nikko Copper Inc., Ulsan, Korea), and gold was produced through pyrometallurgy. If the Au grade of the final concentrate is less than 100 g/t, then an economical gold production is impossible at the refinery owing to the high cost of refining gold. Therefore, the Au grade of the final concentrate should exceed 100 g/t. Gold is an extremely expensive precious metal that can maximize profits by improving the recovery of concentrates.

2.4. Measurement of Flotation Rate Constant

Flotation experiments were performed to calculate the floatation rate constant for minerals in the gold ore (pyrite, quartz, muscovite, and gold). The experimental conditions were matched with the industrial conditions of a plant during the flotation process. For the flotation tests, a gold ore slurry was prepared with 35 wt.% solid pulp density. The prepared pulps were mixed with 70 g/t of KAX, 35 g/t of AP3477, and 40 g/t of PPG for 10 min at 1200 rpm. The tests were performed in the first cell of the rougher 1 bank (#1 Rougher 1). The flotation cell was a Denver sub-A type (Metso, Finland). Air was injected into the cells. The flotation gas rate was 5 L/min, which was determined by the suction of air into the pulp through the impeller at 1200 rpm. The total flotation time was 8 min. The water used was maintained at room temperature (25 °C), and its pH was 6.5 after conditioning. Five concentrates were skimmed off after 0.5, 1, 2, 4, and 8 min. All solid samples obtained from the flotation tests were dried at 40 °C and then weighed. The gold grade of all the products was determined using the fire-assay method. The contents of other elements were analyzed using XRD and XRF. The first-order rate constant model was used in this study because the flotation process is generally regarded as a first-order process. The rate constant was calculated using Equation (1), as follows [9]:

$$R = \mathrm{R}_{\infty}(1 - \exp(-\mathrm{k}\tau)) \tag{1}$$

Here, R is the recovery of the target minerals, R_{∞} the maximum recovery, k the rate constant, and τ the flotation time. The recovery of the target concentrate is expressed as shown in Equation (2) [9,31,32]:

$$R = \frac{Cc}{Ff} \times 100 \tag{2}$$

C and c are the weight and grade of the concentrate, respectively; F and f are the weight and grade of the feed, respectively. The rate constant was calculated via nonlinear regression analysis using MATLAB 2021a (MATLAB, Natick, MA, USA).

2.5. Optimization of Process: Alternative Process Proposal

Herein, alternative circuits were proposed to optimize the current gold ore flotation process. The proposed alternative processes are illustrated in Figure 2. Considering the space problem and throughput of the plant, four alternative processes were proposed.

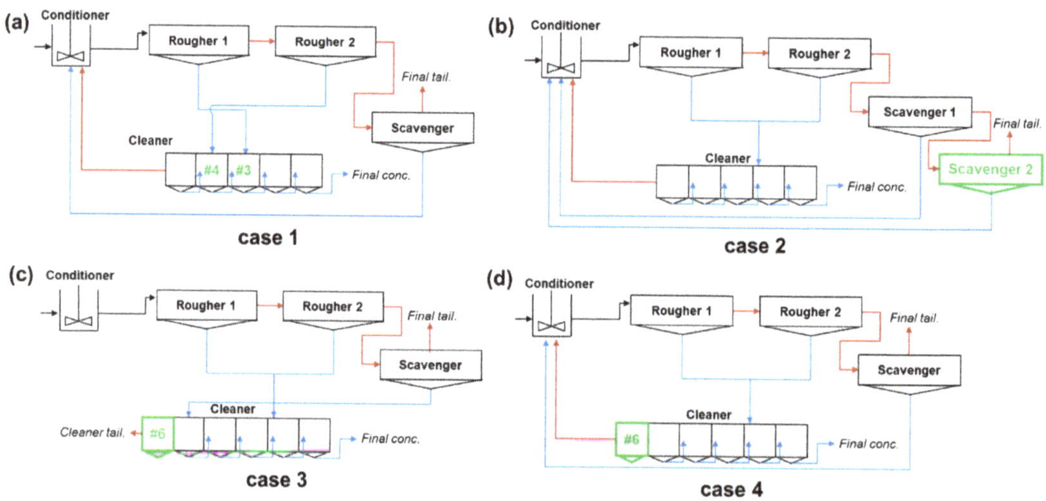

Figure 2. Flow diagram of flotation circuit of alternative circuits. (**a**) Case 1: transfer float from rougher 1 bank to cleaner 3,

transfer float from rougher 2 bank to cleaner 4; (**b**) Case 2: scavenger bank is added; (**c**) Case 3: open circuit with addition of cleaner bank; (**d**) Case 4: closed circuit with addition of cleaner bank. Blue line represents stream of floated products, and red line represents stream of sinked products after flotation. Green text represents changes in the existing process.

Case 1 involves a circuit that transfers the floated products from the rougher 1 bank to cleaner #3, and the floated products from the rougher 2 bank to cleaner #4. In the current circuit, the floated products of all rougher banks were transferred to cleaner #3. Therefore, the outstreams of the floated product of the rougher banks in Case 1 differed from that of the current process. Case 1 was proposed to increase the gold grade of the concentrate. In Case 2, a scavenger bank was added to increase the recovery rate of gold. Case 3 involves an open circuit with an additional cleaner bank, and it was proposed to significantly improve the gold grade of the final concentrate. Finally, Case 4 was designed as the closed-circuit version of Case 3.

2.6. LCA

LCA, which is based on partition-based separation, is effective for analyzing the connection and configuration of each unit [26,33,34]. In addition, because the circuit configuration allows the separation efficiency (SE) to be determined, the performance of circuits with different configurations can be evaluated easily [26]. The evaluation was performed by calculating the overall circuit concentrate-to-feed ratio (C/F) using the partition probability (P) of each unit process. The recovery of a unit is expressed as the probability of the unit process as a function of the experimental variable (Z). The recovery of the entire circuit can be expressed in the form of an algebraic expression using this P [26,33,34]. For example, as shown in Figure 3, if the recoveries of each process are P_1 and P_2, then the recovery (C/F) of the entire process becomes $P_1 \times P_2$. In this process, if the recovery as a function of the experimental variable of the unit process is known, then a partition curve of the overall recovery with respect to the experimental variable can be constructed. Assuming that P is the same for each process, the SE of the entire process can be compared with the efficiency of the unit process based on the slope value ($d(C/F)/dP$, $P = 0.5$) at the point where P is 50% [26,35]. As shown in Figure 3, C/F is P^2 when $P_1 = P_2 = P$; therefore, when $P = 0.5$, the slope value is 1, which is the same as the efficiency of the unit process. In the case of a complex process such as a multistage flotation circuit, it is difficult to calculate the algebraic expression directly. Therefore, the recovery and SE of the entire circuit were calculated using the matrix reduction algorithm proposed by Noble and Luttrell [26].

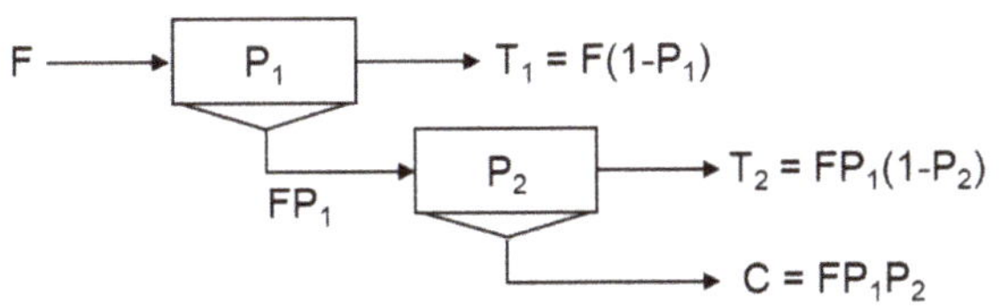

Figure 3. Example of linear circuit analysis. F is the feed rate, T_1 the sunk product rate, P_1 and P_2 the recovery for each unit bank, and C the concentrate rate.

2.7. Simulation of Mass Balance

Using LCA, the recovery of the total circuit can be estimated as a function of the recovery of the unit bank. We simulated the mass balance to compare the performances of the current and alternative flotation circuits.

We assumed that the flotation process was in a steady state to simulate the mass balance, grade of concentrate, and recovery. This assumption is reasonable as the target flotation process has been in progress for a significant amount time, while the reagents and operating conditions were maintained. Therefore, in this study, the steady state was assumed for the diagnosis and optimization of the flotation circuit. The mass balance model based on flotation kinetics is expressed in Equation (3) [36].

$$Q_{ci} = Q_{fi}\left[\frac{k_i\tau}{1+k_i\tau}\right] \quad (3)$$

In Equation (3), Q_{ci} is the mass flowrate of the concentrate (t/d), Q_{fi} the mass flowrate of the feed (t/d), τ the flotation time (min), and k_i the flotation kinetic constant (1/min) for each i component (k_{Au}, k_{Py}, k_{Qtz}, and k_{Mu}). To calculate the mass balance of each node, Equations (4)–(6) were applied [35].

$$Q_f = Q_{Au} + Q_{Py} + Q_{Mu} + Q_{Qtz} + Q_w \quad (4)$$

$$Q_c = Q_{ci} + Q_{cw} = \sum_i Q_{fi}\left[\frac{k_i\tau}{1+k_i\tau}\right] + Q_{cw} \quad (5)$$

$$Q_t = Q_f - Q_c \quad (6)$$

Here, Q_f is the feed mass rate; Q_w is the water mass rate; Q_c is the concentrate mass rate; Q_{cw} is the water mass rate in the concentrate, which can be obtained from the pulp density; Q_{Au}, Q_{Py}, Q_{Qtz}, and Q_{Mu} are the throughputs of gold, pyrite, quartz, and muscovite, respectively, which can be obtained from each mineral grade of the ore; Q_t is the flow rate of the tailings. The flotation residence time (τ) can be calculated using the volume of the cell occupied by the slurry (V_p) and the feed mass rate.

$$\tau = \frac{V_p}{Q_f} \quad (7)$$

To estimate the recovery and grade of the concentrate, the simulation was performed using the USIM-PAC 3.2 software (©Caspeo/BRGM, Orléans, France). Among the many available mineral processing software, USIM-PAC was selected because it enables the mass balance of the circuit to be predicted conveniently, and it contains all the considered flotation models [21]. An iterative calculation algorithm was applied to calculate the mass balance of the closed circuit, in which the in-stream and out-stream were repeated. In this algorithm, the convergence criterion was set to 10^{-4}. Convergence is achieved when the sum of all the least-square differences between the calculated and previous iteration flow rates becomes less than the convergence criterion.

Based on this simulation, the mass balance of the current circuit was calculated to determine the performance of the current circuit. Subsequently, it was used to evaluate the performance of the proposed alternative process.

2.8. Validation of Simulation

In the industrial plant, samples were obtained from each node. Subsequently, the mineral composition of the collected samples was analyzed, and the results were compared with the simulated results to validate the recovery and grade estimated from the simulation. The sampling points are indicated by red points in Figure 1. Mineral composition analyses, as described in Section 2.2, were performed.

3. Results and Discussion

3.1. Characterization of Gold Ores

According to a geological survey, the Gasado gold ore used in this study is a hydrothermal deposit formed by volcanic activity at the end of the Cretaceous period, and that gold is a pyritic gold ore coexisting with pyrite [30,37–40]. Figure 4 shows the results of XRD analysis of the gold ore and micrographs of the gold ore specimens. Table 2 shows the mineral compositions measured via XRF analysis and the gold grade from fire-assay analysis. Mineral analysis revealed that the gold ore contained primarily quartz, muscovite, and pyrite. Therefore, gold, pyrite, quartz, and muscovite were selected as representative minerals of this gold ore, and the flotation circuit was diagnosed and evaluated based on the quality and recovery of these minerals.

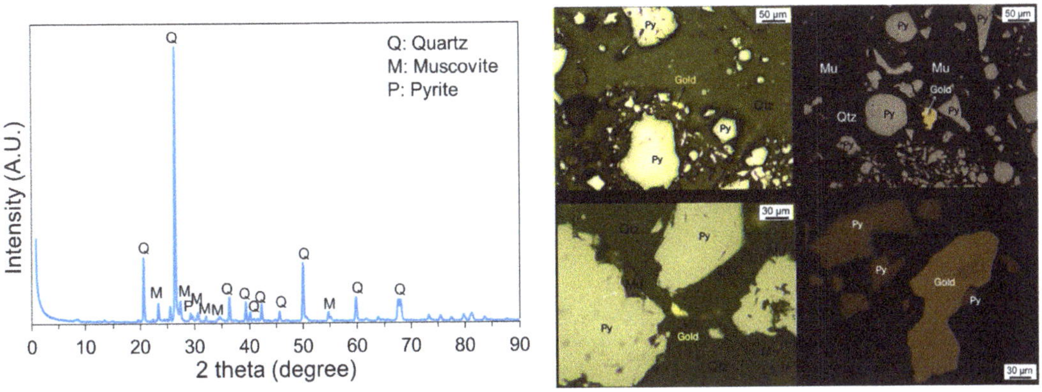

(**a**) (**b**)

Figure 4. Mineralogical properties of gold ore: (**a**) XRD patterns and (**b**) photographs of gold ore (Py: pyrite, Qtz: quartz, Mu: muscovite).

Table 2. Mineral composition of gold ore.

	Au (g/t)	Pyrite (wt.%)	Quartz (wt.%)	Muscovite (wt.%)	Others (wt.%)
gold ore	6.8 ± 1.2	9.6 ± 0.6	63.1 ± 1.1	28.0 ± 1.3	0.5 ± 0.2

3.2. Flotation Kinetics

Figure 5 shows the flotation recovery of gold, pyrite, quartz, and muscovite as a function of the flotation time. The maximum recovery (R_∞) and flotation kinetic rate constant (k) of the minerals were obtained by performing a nonlinear regression analysis of the experimental results using Equation (1). Because the coefficient of determination (R^2) of all minerals was calculated to be 0.96 or more, the experimental results can be assumed to have correlated significantly with the first flotation kinetic model (Equation (1)) [27,41].

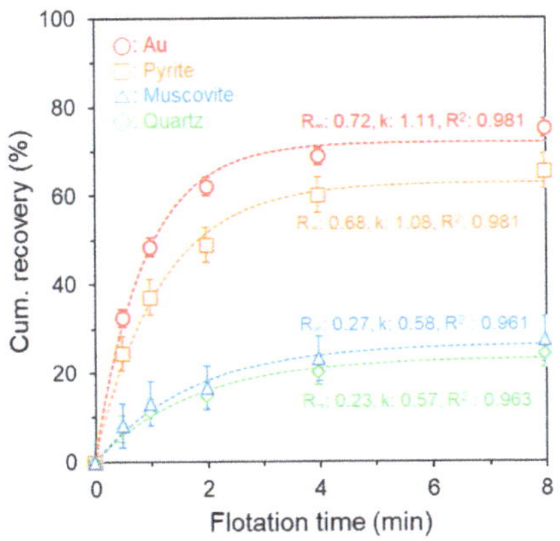

Figure 5. Cumulative flotation recovery of Gasado gold ore for gold (red circle), pyrite (orange rectangular), muscovite (blue triangle), and quartz (green diamond) as a function of flotation time. Dotted line represents flotation recovery calculated using Equation (1). R_∞ is the maximum recovery, k the flotation rate constant, and R^2 the coefficient of determination.

In general, the R_∞ and k values of gold and pyrite were higher than those of quartz and muscovite. The flotation rate constant was determined based on the bubble–particle interaction, which was dominated by the hydrophobicity of the mineral surface. Therefore, the high flotation rate constants of pyrite and gold were calculated because the hydrophobicity of their surfaces was increased by the thiol-type collectors (KAX, AP3477). In addition, hydrophilic minerals such as quartz and muscovite have low flotation rate constants. It is noteworthy that R_∞ is governed by the degree of liberation. If the degree of liberation is 100%, then the R_∞ of gold and pyrite, which are hydrophobic minerals, is 1. However, the photographs of the specimens presented in Figure 4b shows that the grain size of gold is less than 30 μm and that of pyrite is less than 50 μm. Because the d_{80} of the feed was 94 μm, the R_∞ of all minerals was between 0 and 1.

The k and R_∞ values obtained from this batch flotation experiment were used to simulate the mass balance of the flotation circuit.

3.3. LCAs

Figure 6 shows the LCA results of the current process and the alternative processes, where their partition functions and sharpness parameters are included. In this regard, the sharpness parameter is the SE, which is a frequently used indicator to technically evaluate separation processes and can be useful when the recovery and grade of two products are different [26]. The graph presented in Figure 6 can be demarcated by a diagonal line, where the upper area of the partition curve is associated with the scavenging effect, which increases the recovery, whereas the lower area is associated with the cleaning effect, which increases the grade [26]. In this context, assuming that the gold recovery of the unit bank is ≥0.5, and that of silica or mica, which is a gangue, is lower than 0.5, if the flotation circuit exhibits the scavenging effect when the recovery rate of the unit bank exceeds 0.5, and the cleaning effect when it is lower than 0.5, then the SE value should be high.

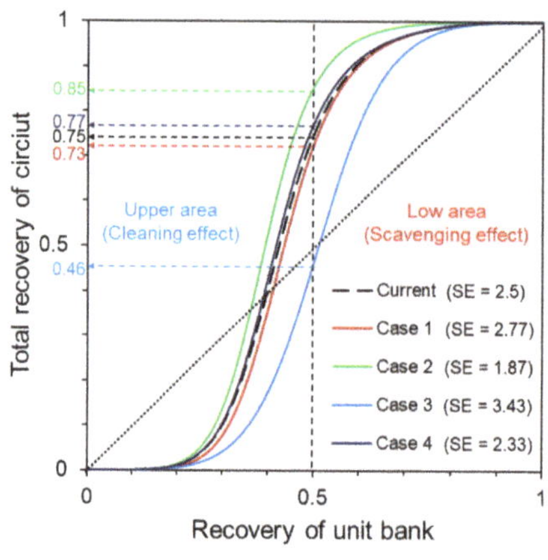

Figure 6. Circuit partition functions of current process (black line) and alternative process (cases 1–4). Dotted line represents diagonal line. SE is separation efficiency (d(C/F)/dP, P = 0.5).

Comparing the current process with the other proposed processes, its SE (= 2.5) ranked third, and the flotation process efficiency can be improved by changing the circuit to Case 1 using a simple circuit arrangement and changing it to Case 3 using an open circuit. However, considering that the recovery of gold, an expensive metal, is critical, the process of Case 3 is difficult to apply because the recovery of the entire process (= 0.46) is extremely low. Finally, the circuit for Case 1 can be considered as a circuit that can increase the enrichment ratio while maintaining a high gold recovery. The enrichment ratio is the ratio of the concentrate grade to the feed grade. Comparing the other cases with the current process, the gold recovery did not differ significantly, and the SE value was low. Therefore, it was difficult to improve the quality of the concentrate.

3.4. Mass Balance of Current Flotation Circuit

The recovery of each unit bank and the concentrate grade of the current circuit were simulated by applying the throughput, pulp density, flotation kinetic constant, and maximum recovery of the minerals to the mass balance model. The simulated distributions were compared with the industrial results for each stream to validate the simulation results. Figure 7 shows the grade of the minerals calculated via simulation and the content analysis results of minerals obtained from actual plants. Regarding the grade distribution shown in Figure 7, the simulated results agreed well with the industrial results. Therefore, the mass balance of the industrial plants was predicted via simulations [29].

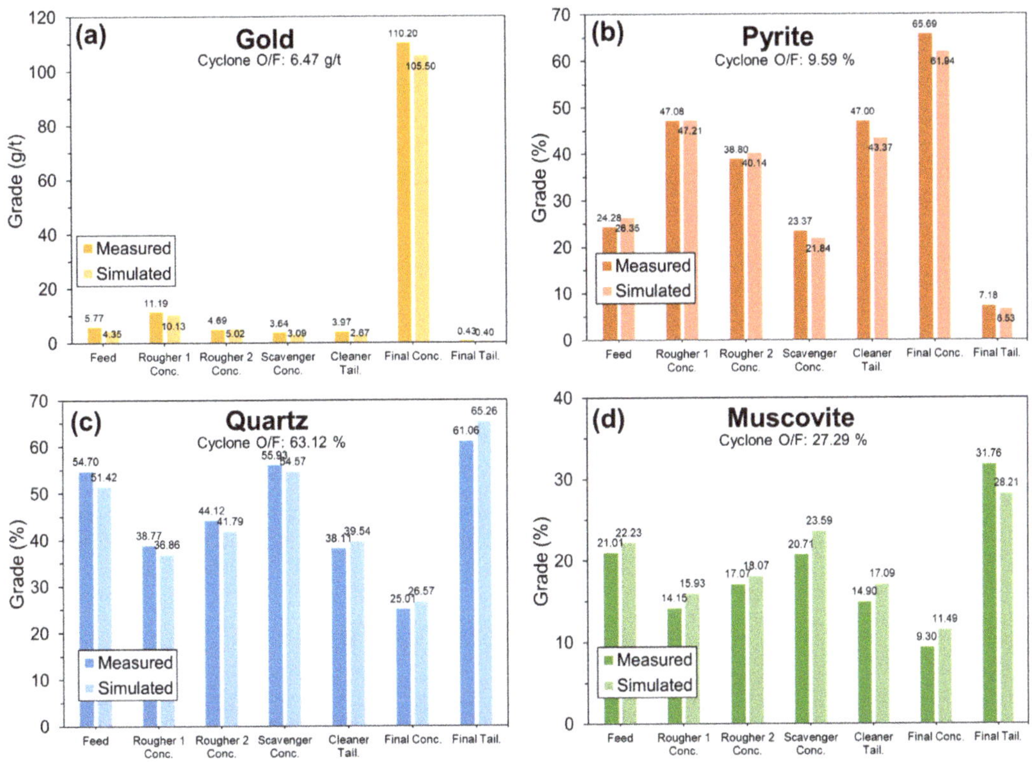

Figure 7. Comparison between experimental and simulated grade distributions of each stream in industrial plant: (**a**) Gold, (**b**) pyrite, (**c**) quartz, and (**d**) muscovite.

Figure 8 shows the mass balance results for the current circuit. The concentrate product rate was estimated to be 0.4 t/h (9.6 t/d), and the gold grade was 105.5 g/t for this closed flotation circuit at a throughput of 7.08 t/h (= 170 t/d). The total recovery of the circuit was 94.92%, and the enrichment ratio was 16.3 (= 105.5/6.47). Hence, it was confirmed that the current process operated at an extremely high recovery rate. Mass balance analysis revealed that the current process is highly effective for recovering gold from pyrite, quartz, and muscovite. These results are similar to those estimated using LCA. The overall recovery is important because gold is a precious and expensive metal. However, the enrichment ratio of the process was low. If the gold grade of the ore decreases significantly as compared with that of the current ore, then the final concentrate grade may be less than 100 g/t. As mentioned in Section 2.3, when the gold grade of the concentrate is less than 100 g/t, the smelting cost is high. In this case, the process must be modified to increase the gold grade of the concentrate.

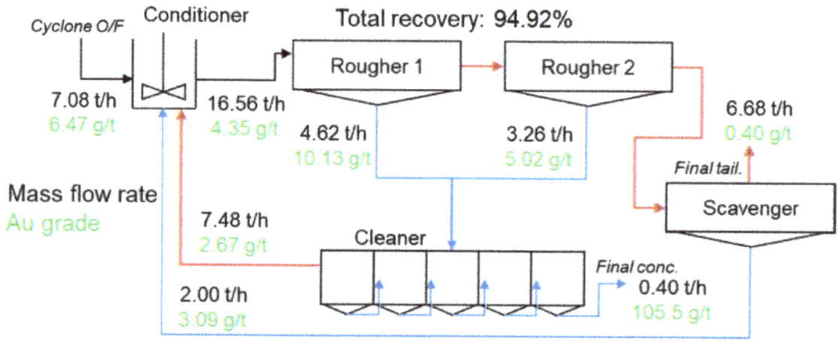

Figure 8. Simulated mass balance of current circuit of gold ore flotation. Blue line represents stream of floated products, and red line represents stream of sinked products after flotation. Black text is the mass flow rate. Green text is the gold grade at each node.

3.5. Comparison between Current Circuit and Alternative Circuits

Figure 9 shows a comparison of the recovery and concentrate grades for all alternative circuits. Case 1 does not involve the addition of a flotation bank to the existing process and does not incur additional equipment costs because it pertains to a circuit where the floated product stream of the existing rougher banks is changed. However, because of this change, the gold grade of the final concentrate improved from 105.5 to 134.7 g/t. The recovery was expected to be 90.92%. In Case 2, an alternative circuit with a scavenger bank added, and the overall recovery of 96.2%, which is slightly higher than 94.4% of the existing process. Because of the addition of scavenger trails, additional cost is incurred to change the process. In Case 3, which involves an alternative circuit designed as an open circuit, the gold grade of the final concentrate was 203.42 g/t, which yielded the highest-quality concentrate. However, the overall recovery was calculated to be as low as 62%. Meanwhile, Case 4, which involved the addition of only a cleaner unit bank to the existing process, did not differ significantly from the current process. Although the cleaner was added (which incurred additional cost), the economic efficiency did not increase as compared with that of the current process.

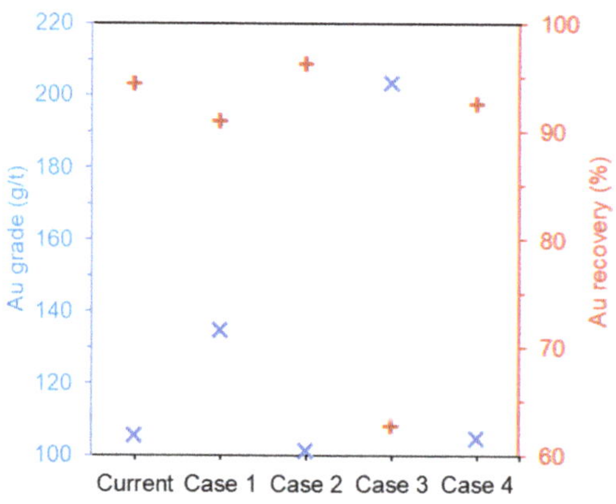

Figure 9. Simulated mass balance of current process for processing gold ore.

In summary, Cases 2 and 4, which are alternative processes involving additional unit banks, did not differ significantly from the current process. If the process is designed with an open circuit, as in Case 3, then the recovery will be reduced significantly. Hence, the open circuit is not suitable for the gold flotation process, where recovery is important. In Case 1, where the stream was changed without adding equipment to the current process, the grade of the concentrate improved by 128% (=134.7/105.5) compared to the gold grade of concentrate of current process., while a high recovery of 90.92% was maintained.

Figure 10 shows the simulated mass balance of the alternative process of Case 1. Compared with Figure 8, which shows the mass balance of the current process, the mass flow rate differed slightly, but it was confirmed that the final gold grade of the concentrate improved significantly. The enrichment ratio of the process of Case 1 was 20.8 (=134.73/6.47), which was a significant improvement compared with that of the current process. In particular, this process is more advantageous than the existing process when the gold quality of the ore is reduced.

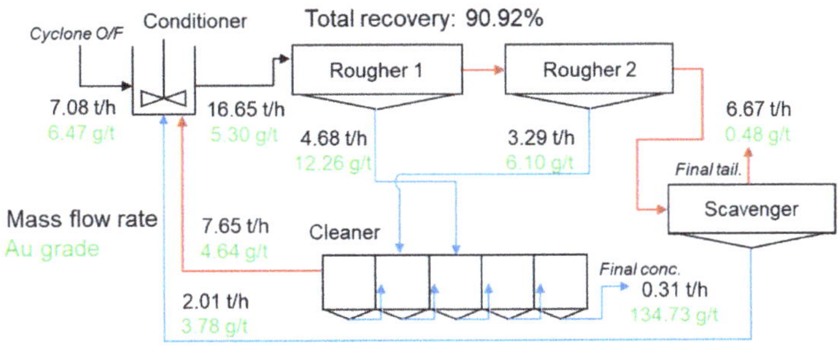

Figure 10. Simulated mass balance of current process using alternative process of Case 1. Blue line represents stream of floated products, and red line represents stream of sinked products after flotation. Black text is the mass flow rate. Green text is the gold grade at each node.

Therefore, the alternative process of Case 1 is expected to yield a higher SE than the current process. This expectation is consistent with the LCA results. Therefore, based on the LCA and mass balance analysis, the flotation process should be changed to the alternative process of Case 1. Currently, the plant has adopted the Case 1 process.

4. Conclusions

In this study, the diagnosis and optimization of the flotation circuit of gold ore in an industrial plant were performed. Optimization was proposed to design an alternative process with an SE that is higher than that of the current process. The SE was evaluated via LCA based on partition-based separation and mass balance simulation using a flotation first-order rate constant model. Because the simulated grade distributions of each of the streams were consistent with the industrial results, the simulation results were considered reasonable. After diagnosing the current process, an operation with high gold recovery was achieved. However, the operation was unsuitable for processing low-grade ore because of the low enrichment ratio. Therefore, LCA and mass balance simulations were performed for various alternative processes. Based on the results obtained, it was discovered that the alternative process of Case 1 should be adopted instead of the current process, as the former maintained high recovery while improving the enrichment ratio. Therefore, the process of Case 1 was adopted in the plant. Finally, we demonstrated that an operating plant can be diagnosed and optimized using LCA and mass balance simulations. In conclusion, our study provides a methodology for adapting the simulation of flotation circuit optimization for industrial plants, and based on these results, future research to be applied to other plants with various minerals compositions will be needed.

Author Contributions: Conceptualization: S.H., M.J., G.-t.L. and Y.H.; methodology: S.H., M.J., W.L., S.K., H.-S.J., S.Q.C. and Y.H.; validation: S.H., M.J., W.L., S.K., K.L., S.Q.C. and Y.H.; formal analysis: S.H., M.J., W.L.; investigation: S.H., M.J., G.-t.L., K.L., H.-S.J. and Y.H.; writing original draft: S.H., M.J., W.L., S.K. and Y.H.; writing review and editing: S.H., M.J., S.K., S.Q.C. and Y.H.; resources: G.-t.L.; supervision: S.Q.C., Y.H. All authors have read and agreed to the published version of the manuscript.

Funding: This research was supported by the Basic Research Project (GP2020-013, 21-3212-1) of the Korea Institute of Geoscience and Mineral Resources (KIGAM) funded by the Ministry of Science, ICT, and Future Planning of Korea, as well as by the Basic Science Research Program through the National Research Foundation of Korea (NRF) funded by the Ministry of Education (2021R1A6A3A01087783) and this work was supported by the Korea Institute of Energy Technology Evaluation and Planning (KETEP) and the Ministry of Trade, Industry & Energy (MOTIE) of the Republic of Korea (No. 20216110100040).

Conflicts of Interest: The authors declare no conflict of interest.

References

1. Hodouin, D. Methods for automatic control, observation, and optimization in mineral processing plants. *J. Process Control* **2011**, *21*, 211–225. [CrossRef]
2. Mendez, A.; Galvez, E.; Cisternas, L. Modeling of grinding and classification circuits as applied to the design of flotation processes. *Comput. Chem. Eng.* **2009**, *33*, 97–111. [CrossRef]
3. Irannajad, M.; Nuri, O.; Allahkarami, E. A new approach in separation process evaluation. Efficiency ratio and upgrading curves. *Physicochem. Probl. Miner. Process.* **2018**, *54*, 847–857.
4. Hodouin, D.; Jamsa-Jounela, S.-L.; Carvalho, M.T.; Bergh, L. State of the art and challenges in mineral processing control. *Control Eng. Pract.* **2001**, *9*, 995–1005. [CrossRef]
5. Han, Y.; Kim, H.; Park, J. Millimeter-sized spherical ion-sieve foams with hierarchical pore structure for recovery of lithium from seawater. *Chem. Eng. J.* **2012**, *210*, 482–489. [CrossRef]
6. Kim, S.; Park, S.; Han, S.; Han, Y.; Park, J. Silanol-rich ordered mesoporous silica modified thiol group for enhanced recovery performance of Au(III) in acidic leachate solution. *Chem. Eng. J.* **2018**, *351*, 1027–1037. [CrossRef]
7. Kim, Y.; Han, Y.; Kim, S.; Jeon, H. Green extraction of lithium from waste LAS glass ceramics using a water leaching processes. *Process Saf. Environ. Protect.* **2021**, *148*, 765–774. [CrossRef]
8. Wills, B.A.; Finch, J. *Wills' Mineral Processing Technology: An Introduction to the Practical Aspects of Ore Treatment and Mineral Recovery*; Butterworth-Heinemann: Oxford, UK, 2015.
9. Rajamani, R.; Herbst, J. Optimal control of a ball mill grinding circuit—I. Grinding circuit modeling and dynamic simulation. *Chem. Eng. Sci.* **1991**, *46*, 861–870. [CrossRef]
10. Han, Y.; Kim, S.; Go, B.; Lee, S.; Park, S.; Jeon, H. Optimized magnetic separation for efficient recovery of V and Ti enriched concentrates from vanadium-titanium magnetite (VTM) ore: Effect of grinding and magnetic Intensity. *Powder Tech.* **2021**, *391*, 282–291. [CrossRef]
11. Bouchard, J.; Desbiens, A.; Villar, R.; Nunez, E. Column flotation simulation and control: An overview. *Miner. Eng.* **2009**, *22*, 519–529. [CrossRef]
12. Mehrotra, S.P. Design of optimal flotation circuits—A review. *Miner. Metall. Process.* **1988**, *5*, 142–152. [CrossRef]
13. Welsby, S.D.D.; Vianna, S.M.; Franzidis, J.-P. A continuous pilot-scale flotation rig for the systematic study of flotation variables. *Miner. Eng.* **2010**, *23*, 826–829. [CrossRef]
14. Calisaya, D.; Lopez-Valdivieso, A. A strategy for the identification of optimal flotation circuits. *Miner. Eng.* **2016**, *96–97*, 157–167. [CrossRef]
15. Liu, J.; MacGregor, J. Froth-based modeling and control of flotation processes. *Miner. Eng.* **2008**, *21*, 642–651. [CrossRef]
16. Cisternas, L.; Lucay, F.; Acosta-Flore, R.; Galvez, E. A quasi-review of conceptual flotation design methods based on computational optimization. *Miner. Eng.* **2018**, *117*, 24–33. [CrossRef]
17. Ferreira, J.P.; Loveday, B.K. An improved model for simulation of flotation circuits. *Miner. Eng.* **2000**, *13*, 1441–1453. [CrossRef]
18. Bergh, L.; Yianatos, J. Control of rougher flotation circuits aided by industrial simulator. *J. Process Control.* **2013**, *23*, 140–147. [CrossRef]
19. Vasebi, A.; Poulin, E.; Hodouin, D. Determining a dynamic model for flotation circuits using plant data to implement a Kalman filter for data reconciliation. *Miner. Eng.* **2015**, *83*, 192–200. [CrossRef]
20. Vallejos, P.; Yianatos, J.; Grau, R.; Yanez, A. Evaluation of flotation circuits design using a novel approach. *Miner. Eng.* **2020**, *158*, 106591. [CrossRef]
21. Villeneuve, J.; Guillaneau, J.-C.; Durance, M.-V. Flotation modelling: A wide range of solutions for solving industrial problems. *Miner. Eng.* **1995**, *8*, 409–420. [CrossRef]
22. Yianatos, J.; Carrasco, C.; Bergh, L.; Vinnett, L.; Toress, C. Modelling and simulation of rougher flotation circuits. *Int. J. Miner. Process.* **2012**, *112–113*, 63–70. [CrossRef]
23. Maldonado, M.; Araya, R.; Finch, J. Optimization flotation bank performance by recovery profiling. *Miner. Eng.* **2011**, *24*, 939–943. [CrossRef]

24. Radmehr, V.; Shafaei, S.; Noaparast, M.; Abdollahi, H. Optimizing flotation circuit recovery by effective stage arrangements: A case study. *Minerals* **2018**, *8*, 417. [CrossRef]
25. Gorain, B.K.; Franzidis, J.-P.; Manlapig, E.V. The empirical prediction of bubble surface area flux in mechanical flotation cells from cell design and operating data. *Miner. Eng.* **1999**, *12*, 309–322. [CrossRef]
26. Noble, A.; Luttrell, G. The matrix reduction algorithm for solving separation circuits. *Miner. Eng.* **2014**, *64*, 97–108. [CrossRef]
27. Seppälä, P.; Sorsa, A.; Paavola, M.; Ruuska, J.; Remes, A.; Kumar, H.; Leiviskä, K. Development and calibration of a dynamic flotation circuit model. *Miner. Eng.* **2016**, *96*, 168–176. [CrossRef]
28. Yalcin, E.; Kelebek, S. Flotation kinetics of a pyritic gold ore. *Int. J. Miner. Process.* **2011**, *98*, 48–54. [CrossRef]
29. Lee, W.; Jung, M.; Han, S.; Park, S.; Park, S. Simulation of layout rearrangement in the grinding/classification process for increasing throughput of industrial gold ore plant. *Miner. Eng.* **2020**, *157*, 106545. [CrossRef]
30. Kim, C.; Choi, S.; Choi, S.; Lee, I. Hydrothermal alteration and its genetic implication in the Gasado volcanic-hosted epithermal gold-silver deposit: Use in exploration. *J. Miner. Soc. Korea* **2002**, *15*, 205–220.
31. Kim, S.; Baek, S.; Han, Y.; Jeon, H. Laboratory testing of Scheelite flotation from raw ore in Sangdong mine for process development. *Minerals* **2020**, *10*, 971. [CrossRef]
32. Han, Y.; Han, S.; Kim, B.; Yang, J.; Choi, J.; Kim, H.; You, K.; Kim, H. Flotation separation of quartz from apatite and surface forces in bubble-particle interactions: Role of pH and cationic amine collector contents. *J. Ind. Eng. Chem.* **2019**, *70*, 107–115. [CrossRef]
33. Meloy, T. Analysis and optimization of mineral processing and coal-cleaning circuits—Circuit analysis. *Int. J. Miner. Process.* **1983**, *10*, 61–80. [CrossRef]
34. Meloy, T. Optimizing for grade or profit in mineral processing circuits—Circuit analysis. *Int. J. Miner. Process.* **1983**, *11*, 89–99. [CrossRef]
35. Noble, A.; Luttrrell, G.H.; Amini, S.H. Linear circuit analysis: A tool for addressing challenges and identifying opportunities in process circuit design. *Mining. Metall. Explor.* **2019**, *36*, 159–171. [CrossRef]
36. Yianatos, J.; Vallejos, P.; Grau, R.; Yanez, A. New approach for flotation process modelling and simulation. *Miner. Eng.* **2020**, *156*, 106482. [CrossRef]
37. Moon, D.; Koh, S.; Lee, K. Geochemistry of the Moisan epithermal gold-silver deposit in Haenam area. *Econ. Environ. Geol.* **2010**, *43*, 491–503.
38. Han, I.; Ahn, Y.; Chi, S.; Bae, K.; Kim, H. Froth flotation of Au-Ag ore from epithermal deposit in Haenam. *J. Korean Soc. Miner. Energy Resour. Eng.* **2011**, *48*, 701–712.
39. Lee, W.; Jung, M.; Han, S.; Park, S.; Lim, G.; Park, J. Mathematical modeling and computational simulation on ball milling in domestic goldsilver mineral processing plant. J. Korean Soc. *Miner. Energy Resour. Eng.* **2017**, *54*, 502–511.
40. Han, S.; Jeong, M.; Lee, W.; Lim, G.; Park, J. Simulation of grinding/classification circuit in domestic gold ore processing plant using energy-based grinding model and mathematical classification model. *J. Korean Soc. Miner. Energy Resour. Eng.* **2018**, *55*, 8–19. [CrossRef]
41. Han, S.; You, K.; Kim, K.; Park, J. Measurement of the attachment force between an air bubble and a mineral surface: Relationship between the attachment force and flotation kinetics. *Langmuir* **2019**, *35*, 9364–9373. [CrossRef]

minerals

MDPI

Article

Beneficiation of Low-Grade Rare Earth Ore from Khalzan Buregtei Deposit (Mongolia) by Magnetic Separation

Ilhwan Park [1,*], Yuki Kanazawa [2], Naoya Sato [2], Purevdelger Galtchandmani [2], Manis Kumar Jha [3], Carlito Baltazar Tabelin [4], Sanghee Jeon [1], Mayumi Ito [1] and Naoki Hiroyoshi [1]

1 Division of Sustainable Resources Engineering, Faculty of Engineering, Hokkaido University, Sapporo 060-8628, Japan; shjun1121@eng.hokudai.ac.jp (S.J.); itomayu@eng.hokudai.ac.jp (M.I.); hiroyosi@eng.hokudai.ac.jp (N.H.)
2 Division of Sustainable Resources Engineering, Graduate School of Engineering, Hokkaido University, Sapporo 060-8628, Japan; kanazawa.yuki.c2@elms.hokudai.ac.jp (Y.K.); satonaoy@mmc.co.jp (N.S.); galtchandmani.p@bayanairag.com (P.G.)
3 Metal Extraction and Recycling Division, CSIR-National Metallurgical Laboratory, Jamshedpur 831007, India; mkjha@nmlindia.org
4 School of Minerals and Energy Resources Engineering, The University of New South Wales, Sydney, NSW 2052, Australia; c.tabelin@unsw.edu.au
* Correspondence: i-park@eng.hokudai.ac.jp

Abstract: The global demand for rare earth elements (REEs) is expected to increase significantly because of their importance in renewable energy and clean storage technologies, which are critical for drastic carbon dioxide emission reduction to achieve a carbon-neutral society. REE ore deposits around the world are scarce and those that have been identified but remain unexploited need to be developed to supply future demands. In this study, the Khalzan Buregtei deposit located in western Mongolia was studied with the aim of upgrading low-grade REE ore via magnetic separation techniques. The total REE content in this ore was ~6720 ppm (~3540 ppm light REE (LREE) + ~3180 ppm heavy REE (HREE)) with bastnaesite, pyrochlore, synchysite, and columbite-(Fe) identified as the main REE-bearing minerals. As the particle size fraction decreased from −4.0 + 2.0 mm to −0.5 + 0.1 mm, the recovery by dry high-intensity magnetic separation (DHIMS) increased from 20% to 70% of total rare earth oxide (TREO) while the enrichment ratio reached 2.8 from 1.3. Although effective, gangue minerals such as quartz and aluminosilicates were recovered (~22%) due most likely to insufficient liberation. Meanwhile, the wet high-intensity magnetic separation (WHIMS) could produce a magnetic concentrate with TREO recovery of ~80% and enrichment ratio of 5.5 under the following conditions: particle size fraction, −106 + 75 μm; feed flow rate, 3.2 L/min; magnetic induction, 0.8 T. These results indicate that combining DHIMS and WHIMS to upgrade the low-grade REE ore from the Khalzan Buregtei deposit is an effective approach.

Keywords: rare earth elements; magnetic separation; upgradation; Khalzan Buregtei deposit

Citation: Park, I.; Kanazawa, Y.; Sato, N.; Galtchandmani, P.; Jha, M.K.; Tabelin, C.B.; Jeon, S.; Ito, M.; Hiroyoshi, N. Beneficiation of Low-Grade Rare Earth Ore from Khalzan Buregtei Deposit (Mongolia) by Magnetic Separation. *Minerals* **2021**, *11*, 1432. https://doi.org/10.3390/min11121432

Academic Editor: William Skinner

Received: 19 November 2021
Accepted: 15 December 2021
Published: 18 December 2021

1. Introduction

Climate change is one of the most important and urgent issues faced by humanity this century that needs a solution [1–3]. The global mean temperature for 2020 was approximately 1.2 ± 0.1 °C warmer than pre-industrial levels (i.e., the period 1850–1900) due to rising anthropogenic greenhouse gases (GHGs) in the atmosphere, in particular carbon dioxide (CO_2) [4]. In 2015, the Paris Agreement was adopted by the 196 parties with the aim of limiting the global temperature rise to "well below 2 °C", preferably to 1.5 °C, compared to pre-industrial levels [5]. In accordance with the Paris Agreement, 193 governments adopted the 17 sustainable development goals (SDGs), Goal 13 of which is "Climate Action" aiming at reducing CO_2 emission to combat climate change and achieve a carbon-neutral society by 2050 [6,7]. To become carbon-neutral, conventional fossil-fuel-based electricity generation and transport technologies need to be replaced with renewable energy technologies as well as electric vehicles (EVs) [7–9].

However, low-carbon technologies are more intensive towards material, mineral and metal rather than conventional fossil-fuel-based technologies [7,8]. Among the materials for low-carbon technologies, rare earth elements (REEs)—a group of 17 elements consisting of lanthanides (lanthanum (La) to lutetium (Lu)) as well as scandium (Sc) and yttrium (Y)—are of great importance because significant quantities of REEs are utilized for manufacturing strong permanent magnet, a critical component used in generators for wind turbines and traction motors for EVs [8,10–12]. For example, 2 tons of REEs are required to build a 3 MW wind turbine, and the mass of REEs in a hybrid electric vehicle (HEV) with nickel-metal hydride (NiMH) batteries is approximately 3.5–4.5 kg [7,13–15]. Therefore, the stable supply of REEs for the next 20–30 years is important in the success of transitioning into a carbon-neutral society.

Although REEs are not scarce—the crustal abundance of REEs is about 7×10^{-5}–4×10^{-3} wt.%, more abundant than gold (5×10^{-7} wt.%) and silver (1×10^{-5} wt.%)—they are rarely found in deposits concentrated enough to be exploited economically [10,16]. The global production of REEs is dominated by only four countries: China (57.6%), USA (15.6%), Burma (12.3%), and Australia (7.0%) [17]. Due to their geographic production concentration, the REE supply chain is vulnerable to disruptions, and thus diversifying their supply sources is critical in maintaining a robust and stable supply chain.

According to the report by Bundesanstalt für Geowissenschaften und Rohstoffe (BGR) [18], Mongolia has a great potential to benefit from REEs as the country has four major REE deposits, named Khalzan Buregtei, Mushgai Khudag, Khotgor and Luginin Gol deposits (Figure 1a), with combined reserves of approximately 1.25 Mt REO. Among them, the Khalzan Buregtei deposit located in western Mongolia, approximately 50 km north of the city of Khovd, is considered the most important deposit because it is enriched in heavy REE (HREE), which is more expensive than light REE (LREE), as well as high field-strength elements (HFSEs) such as zirconium (Zr), niobium (Nb), tantalum (Ta), and yttrium (Y) [19–21]. Moreover, according to a report by the U.S. Geological Survey (USGS) [22], the global production of HREEs is lower than that of LREEs, which makes the development of appropriate beneficiation strategies for the Khalzan Buregtei deposit increasingly important.

The production of REE concentrates from ores has been conducted via the combination of gravity separation, magnetic separation, and flotation [16,23]. For example, shaking tables—one of the gravity separation techniques—either alone or in combination with flotation or magnetic separation were employed by a few hard rock REE beneficiation plants in Sichuan and Shandong in China, while Bayan Obo plant and some of the Sichuan operations use low-intensity magnetic separation (LIMS) to eliminate highly magnetic gangue minerals (e.g., magnetite (Fe_3O_4)) from the ground ore, two stages of wet high-intensity magnetic separation (WHIMS) to first recover hematite (Fe_2O_3) and REE minerals from gangue minerals at higher magnetic field intensity and then separate hematite and REE minerals at lower magnetic field intensity, followed by flotation to further improve the grade of REE concentrates [24,25]. In summary, gravity and magnetic separation techniques are commonly adopted to preconcentrate REEs, while flotation is employed at the end of the process to produce saleable REE concentrates.

Beneficiation of the Khalzan Buregtei deposit has not been reported to date, so the appropriate preconcentration process should be established prior to investigating flotation. Between gravity and magnetic separation techniques, the former is less attractive for processing low-grade and fine-grained ores [16]; thus, this study aimed to investigate the applicability of dry/wet magnetic separation techniques for upgrading low-grade rare earth ore from the Khalzan Buregtei deposit. Specifically, the effects of particle size, feed flow rate and magnetic induction, the main factors affecting magnetic separation, on total rare earth oxide (TREO) recovery/grade were examined.

2. Materials and Methods

2.1. Rare Earth Ore Sample

The rare earth ore sample used in this study was obtained from the Khalzan Buregtei deposit, Khovd, Mongolia. As shown in Figure 1b, this deposit has multiple geological units consisting of seven intrusive phases, and among them, the 5th (orange area) and 7th (blue-gray area) phases are mineralized rocks forming the rare-metal peralkaline granitic ore [26]. The mineral compositions of the 5th and 7th phases are very similar, but the former is recognized as having coarse- and medium-grained rare-metal peralkaline granite while the latter consists of fine- and medium-grained miarolitic rare-metal peralkaline granite [26]. In this study, the 5th phase ore was used. The ore sample was crushed by a jaw crusher (BB 51, Retsch Inc. Haan, Germany), ground with a ball mill, and screened to obtain the predetermined size fractions. For the chemical and mineralogical analyses of the ore, the sample was further ground to <50 μm using an agate mortar and pestle and then analyzed by X-ray fluorescence spectroscopy (XRF, EDXL300, Rigaku Corporation, Tokyo, Japan) and X-ray powder diffraction (XRD, MultiFlex, Rigaku Corporation, Tokyo, Japan). The XRD pattern was analyzed using the Match!® software (Crystal Impact GbR, Bonn, Germany). Thin sections of the ore were also prepared and analyzed by scanning electron microscopy with energy-dispersive X-ray spectroscopy (SEM-EDS, JSM-IT200, JEOL Ltd., Tokyo, Japan).

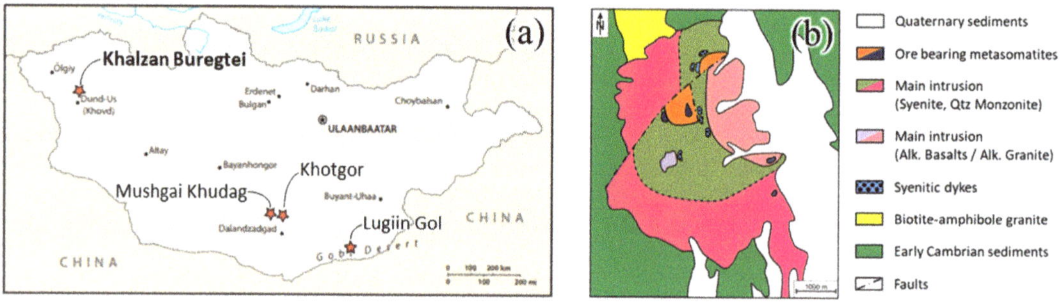

Figure 1. (**a**) A schematic location map of major Mongolian REE deposits (reprinted with permission from Muff and Tamiraa [18]), and (**b**) a geological sketch map of the Khalzan Buregtei deposit (reprinted with permission from Gronen et al. [20]).

2.2. Dry High-Intensity Magnetic Separation

Dry magnetic separation was conducted to recover rare earth minerals from coarse size fractions (−4 + 2, −2 + 0.5, and −0.5 + 0.1 mm) of the crushed ore sample. A dry high-intensity rare earth roll separator (RE6D06W-1, Eriez Magnetics Co. Ltd., Chiba, Japan), consisting of a rare earth roll (75.4 mm diameter × 150 mm width) and an idler roll enveloped with a 0.25 mm thickness Kevlar belt, was used (Figure 2a). The sample was fed to the separator through the vibratory feeder, and magnetic separation was conducted with a belt speed of 8.7 m/min and magnetic field intensity of 1.2 T. The recovered non-magnetic fractions were re-circulated back to the separator to improve REE recovery, which was repeated twice (total pass = 3). The final magnetic and non-magnetic fractions were ground to <50 μm and analyzed by XRF.

2.3. Wet High-Intensity Magnetic Separation

A dry magnetic separator is typically inefficient in processing fine-grained particles due to the influence of airflow, particle-particle adhesion and particle-rotor adhesion, so a wet high-intensity magnetic separator (WHIMS, L-4, Eriez Co. Ltd., Chiba, Japan) (Figure 2b) was adopted for recovering REEs from fine fractions (−212 + 106, −106 + 75, and −75 μm) [27]. In the separation cell, 16 pieces of fine, expanded metal matrix (opening size: 2 mm × 5 mm) were placed to increase the number of high field gradient sites where magnetic particles could be collected. For each test, 30 g of sample was suspended in

600 mL of deionized (DI) water with 250 mg/L of sodium hexametaphosphate (SHMP, $(NaPO_3)_6$, Wako Pure Chemical Industries, Ltd., Osaka, Japan) as a dispersant, and then dispersed for 3 min with an ultrasonicator (W-113, Honda Electronics Co., Ltd., Toyohashi, Japan). Afterwards, a predetermined magnetic induction (0.2–1.0 T) was applied to the separation cell, and then the pulp was fed into the funnel placed on the top of WHIMS at a flow rate of 0–5.8 L/min, which was adjusted by controlling the discharge valve. A flow rate of 0 L/min means the feed was retained in the separation cell for 3 min by fully closing the discharge valve. When the slurry was entirely passed through the WHIMS, 1 L of DI water was introduced to wash the collected magnetic particles, which minimize the entrainment of non-magnetic particles in the magnetic concentrate. The recovered non-magnetic fraction was re-circulated back to the WHIMS to improve REE recovery, which was repeated twice (total pass = 3). The final magnetic and non-magnetic fractions were dried at 105 °C for 24 h, ground to <50 μm, and analyzed by XRF. Some of the experiments were done in triplicate, and the relative standard deviations were <5%.

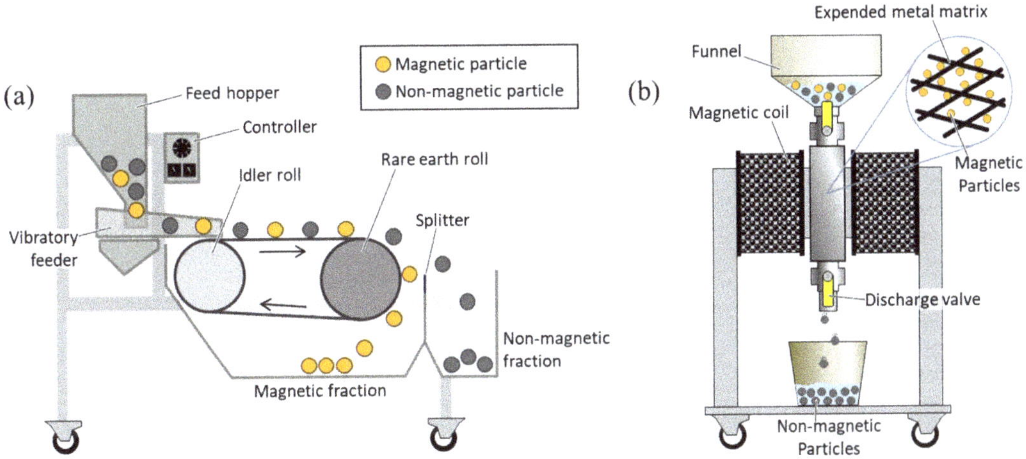

Figure 2. Schematic diagrams of (**a**) dry high-intensity magnetic separator and (**b**) wet high-intensity magnetic separator.

3. Results and Discussion

3.1. Characterization of REE-Bearing Ore of Khalzan Buregtei Deposit, Mongolia

Table 1 shows the chemical composition of the ore sample of the Khalzan Buregtei deposit. As can be seen, the sample is rich in Si (69.0%) and Al (16.1%), while Fe, K, Zr, Ca, Nb, Ti, and Na are contained in the sample as minor non-REE constituents in the ranges of 0.05–4.1%. The TREO content in the sample is 6720 ppm, and LREE (La to Eu) and HREE (Gd to Lu plus Y) contents are 3540 and 3180 ppm, respectively.

Table 1. Elemental composition of REE-bearing sample from the Khalzan Buregtei deposit, Mongolia.

Non-REE	Content (%)	REE	Content (ppm)
SiO_2	69.0 ± 2.2	La_2O_3	548 ± 184
Al_2O_3	16.1 ± 0.8	CeO_2	1376 ± 460
Fe_2O_3	4.1 ± 0.79	Eu_2O_3	1004 ± 493
K_2O	3.5 ± 0.42	Gd_2O_3	391 ± 63
ZrO_2	3.6 ± 0.63	Y_2O_3	1809 ± 339
CaO	1.3 ± 0.12	Dy_2O_3	525 ± 117
Nb_2O_5	0.4 ± 0.11	Er_2O_3	73 ± 21
TiO_2	0.4 ± 0.07	Yb_2O_3	181 ± 36
Na_2O	0.05 ± 0.12	Other REEs	813 ± 282
Others	0.8 ± 0.18	TREO	6720 ± 1257

As shown in Figure 3, mineral particles were sorted based on the color to further identify the distribution of REEs; that is, mineral particles were divided into five groups (e.g., black, brown, brownish yellow, yellow, and transparent), and the chemical composition of each group was given in Table 2. The TREO contents were found to decrease in the following order: black (8640 ppm) > brown (7930 ppm) > brownish yellow (3030 ppm) > yellow (1690 ppm) > transparent (560 ppm). In the case of non-REE components, the contents of Fe, Zr, and Nb were high in black and brown groups where TREO contents were highest, which implies that REEs are most likely incorporated in Fe-, Zr-, and/or Nb-bearing minerals. Meanwhile, the major constituent, Si, was inversely proportional to TREO contents; that is, Si content increased from 66.2% in the black group to 93.9% in the transparent group, indicating that particles in the transparent group are most likely quartz (SiO_2).

Figure 3. Photographs of samples sorted based on color: (**a**) black, (**b**) brown, (**c**) brownish yellow, (**d**) yellow, and (**e**) transparent particles.

Table 2. Elemental composition of samples sorted based on color.

Elements		(a)	(b)	(c)	(d)	(e)
Non-REE (%)	SiO_2	66.2 ± 0.1	68.6 ± 0.4	71.2 ± 0.1	75.6 ± 0.2	93.9 ± 0.0
	Al_2O_3	14.4 ± 0.1	15.4 ± 0.5	18.6 ± 0.1	14.3 ± 0.1	4.0 ± 0.01
	Fe_2O_3	8.2 ± 0.03	4.6 ± 0.03	1.3 ± 0.01	2.0 ± 0.00	0.4 ± 0.00
	K_2O	2.8 ± 0.12	3.1 ± 0.04	6.8 ± 0.03	2.7 ± 0.10	0.8 ± 0.01
	ZrO_2	3.1 ± 0.03	5.9 ± 0.09	0.9 ± 0.01	0.7 ± 0.01	0.6 ± 0.00
	CaO	1.6 ± 0.01	2.3 ± 0.01	0.1 ± 0.01	1.1 ± 0.01	0.3 ± 0.01
	Nb_2O_5	0.4 ± 0.00	0.3 ± 0.00	0.2 ± 0.00	0.2 ± 0.00	n.d.*
REE (ppm)	La_2O_3	150 ± 9	240 ± 8	140 ± 6	160 ± 8	n.d.*
	CeO_2	490 ± 38	660 ± 17	270 ± 15	460 ± 7	70 ± 8
	Nd_2O_3	n.d.*	420 ± 0	n.d.*	n.d.*	n.d.*
	Sm_2O_3	n.d.*	280 ± 9	510 ± 79	n.d.*	50 ± 9
	Eu_2O_3	3310 ± 650	1360 ± 910	420 ± 148	n.d.*	130 ± 57
	Gd_2O_3	1290 ± 50	n.d.*	120 ± 20	570 ± 30	n.d.*
	Y_2O_3	1960 ± 10	2960 ± 40	1020 ± 10	210 ± 2	160 ± 1
	Dy_2O_3	1090 ± 92	1410 ± 70	200 ± 17	230 ± 28	60 ± 7
	Er_2O_3	n.d.*	40 ± 47	100 ± 13	n.d.*	20 ± 8
	Tm_2O_3	230 ± 93	460 ± 30	150 ± 13	60 ± 20	40 ± 9
	Yb_2O_3	120 ± 50	100 ± 18	100 ± 4	n.d.*	30 ± 6
	TREO	8640 ± 526	7930 ± 183	3030 ± 213	1690 ± 26	560 ± 33

* Abbreviation: n.d., not detected.

The mineralogical composition of the sample was determined by XRD (Figure 4). As shown in the XRD pattern, quartz and aluminosilicates (e.g., albite ($NaAlSi_3O_8$) and microcline ($KAlSi_3O_8$)) were identified as the main gangue minerals in the sample, which is consistent with the XRF result (Table 1). Meanwhile, hematite (Fe_2O_3), aegirine ($NaFe^{3+}Si_2O_6$), arfvedsonite ($NaNa_2(Fe^{2+}{}_4Fe^{3+})Si_8O_{22}(OH)_2$), zircon ($ZrSiO_4$), and columbite-(Fe) ($FeNb_2O_6$) were minerals containing minor non-REE constituents (e.g., Fe, Zr, and Nb). Due to the low grade of REEs in the sample, REE-bearing minerals were not detected in the XRD pattern, thus, SEM-EDS analysis of thin sections was conducted to clarify the types of REE-bearing minerals in the sample. As illustrated in Figure 5, the major REE-bearing minerals were identified to be bastnaesite-(Ce) ($Ce(CO_3)F$) and pyrochlore ($(Na,Ca,REE)_2Nb_2O_6(OH,F)$), which are locked within the quartz/albite matrix. Moreover, synchysite ($Ca(REE)(CO_3)_2F$) and columbite -(Fe) were found to be REE-bearing minerals (Figure S1). Zircon has been reported to incorporate REEs into its structure [20,28,29], but REE signals were not detected on zircon (Figure S1). The grain sizes of REE-carbonates (e.g, bastnaesite and synchysite) and pyrochlore varied from 10 to 154 μm and 10 to 243 μm, respectively (Figures 4 and S1).

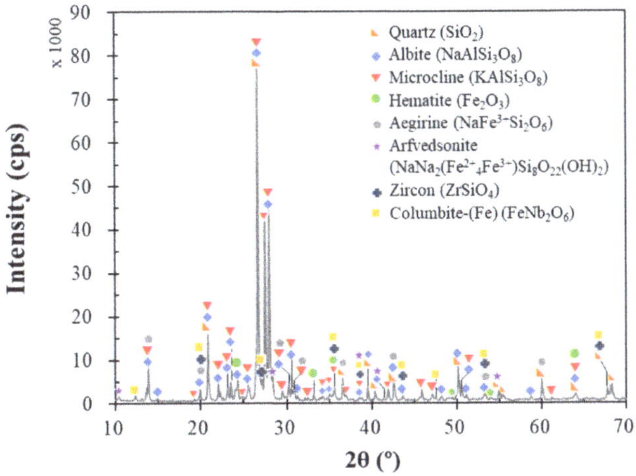

Figure 4. XRD pattern of the REE-bearing sample from the Khalzan Buregtei deposit, Mongolia.

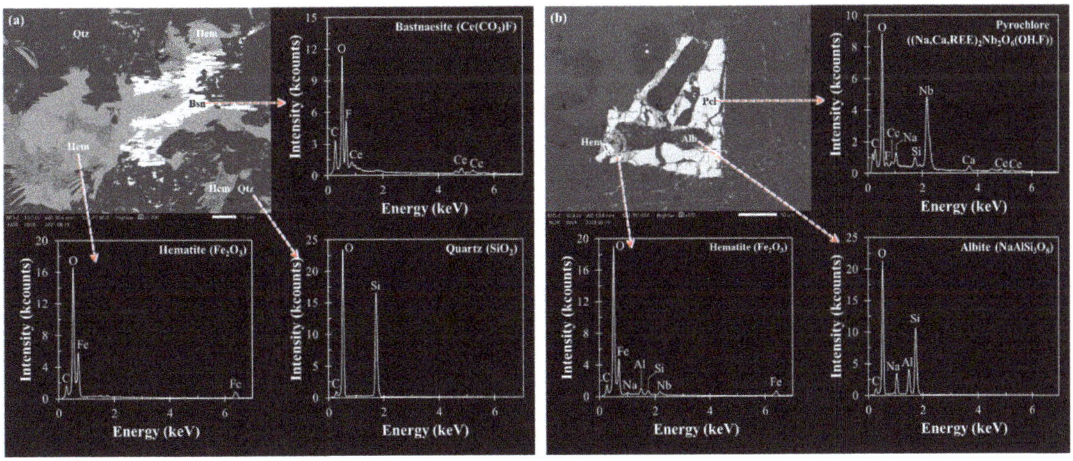

Figure 5. Back-scattered electron (BSE) images and EDS point analyses of (**a**) bastnaesite-hematite-quartz and (**b**) pyrochlore-hematite-albite ores in the sample. Note: Bsn, bastnaesite; Hem, hematite; Qtz, quartz; Pcl, pyrochlore; Alb, albite.

3.2. Dry High-Intensity Magnetic Separation

Upgrading REO from coarse size fractions (−4 + 2 mm, −2 + 0.5 mm, and −0.5 + 0.1 mm) was carried out using dry high-intensity magnetic separation. Figure 6 shows the recovery of TREO, Fe and Si, their enrichment ratios, and Newton's efficiency (η), which were calculated according to the following equations:

$$\text{Recovery } (\%) = \frac{\text{Cc}}{\text{Ff}} \times 100\%, \tag{1}$$

$$\text{Enrichment ratio} = \frac{\text{c}}{\text{f}}, \tag{2}$$

$$\eta(\%) = \text{R}_\text{c} - (100 - \text{R}_\text{t}), \tag{3}$$

where F and C are the weights of feed and concentrate, f and c are the grades of elements (REEs, Fe, and Si) in feed and concentrate, and R_c and R_t are the recoveries of REEs in the concentrate and gangue minerals in the tailing.

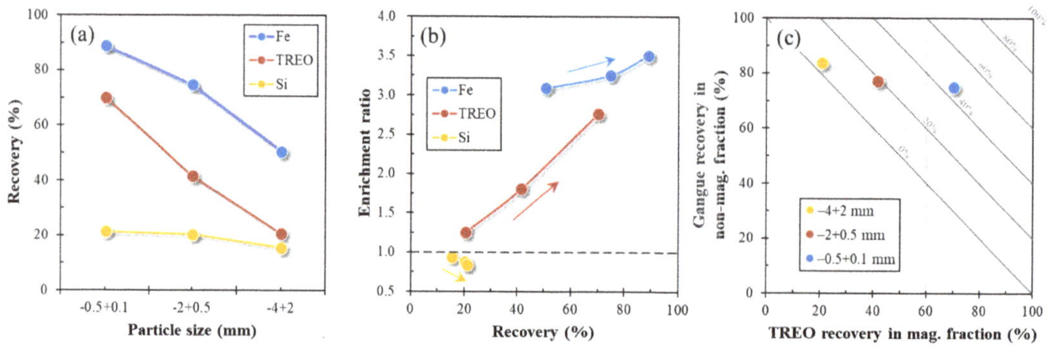

Figure 6. Effect of particle size on (**a**) the recovery of TREO, Fe, and Si from REE ore, (**b**) enrichment ratios of TREO, Fe, and Si as a function of recovery, and (**c**) relationship between TREO recovery in magnetic fraction and gangue recovery in non-magnetic fraction. Note: The arrows in Figure 6b indicate the particle size decreasing from −4 + 2 mm to −0.5 + 0.1 mm.

As shown in Figure 6a, TREO recovery increased from 20% to 70% with decreasing particle size. The Fe recovery also showed the same trend as that of TREO, but iron-bearing minerals were more strongly attracted by the rare earth roll magnet than REO; that is, Fe recovery increased from 50% to 89% as particle size decreased. These results are to be expected because of the moderate to strong attraction of iron-bearing minerals/materials to a magnetic field [30–32]. These results indicate that the smaller the particles, the greater the liberation of REE- and Fe-minerals locked within the quartz/albite matrix, and thus their recoveries could be improved. Although TREO recovery increased with decreasing particle size, Si recovery was almost kept constant (15%–20%) regardless of particle size, indicating that gangue minerals were not sufficiently separated from REE- and/or Fe-minerals even in the size fraction of −0.5 + 0.1 mm. As the recoveries of TREO and Fe were selectively increased while Si recovery was kept constant, the grades of TREO and Fe were enriched by 1.3–2.8 and 3.1–3.5 times, respectively (Figure 6b). Moreover, the separation efficiency of TREO from gangue minerals was remarkably increased from ~4.2% to ~45% with decreasing particle size (Figure 6c).

3.3. Wet High-Intensity Magnetic Separation

Although the dry high-intensity magnetic separator (DHIMS) concentrated REO by 2.8-fold, approximately 30% of REO was lost to the non-magnetic fraction due most likely to insufficient liberation. To improve the liberation of REO from the gangue minerals, finer grinding is required; however, particles finer than 100 μm are difficult to recover by

DHIMS [27,33,34]. Because of this, WHIMS—well-known for its ability to recover fine weakly magnetic minerals [33]—was employed. Figure 7 compares the performance of DHIMS (for −0.5 + 0.1 mm fraction) and WHIMS (for −106 + 75 μm fraction) on TREO recovery and the separation efficiency. Between the two, WHIMS showed better TREO recovery (~82%) and separation efficiency (~60%) than DHIMS (i.e., TREO recovery, ~70%; separation efficiency, ~45%). As the processing of finer particles via WHIMS showed promising results for improving TREO recovery as well as the separation efficiency of REO from gangue minerals compared with DHIMS, it was further investigated under various conditions (e.g., particle size, feed flow rate, and magnetic induction strength).

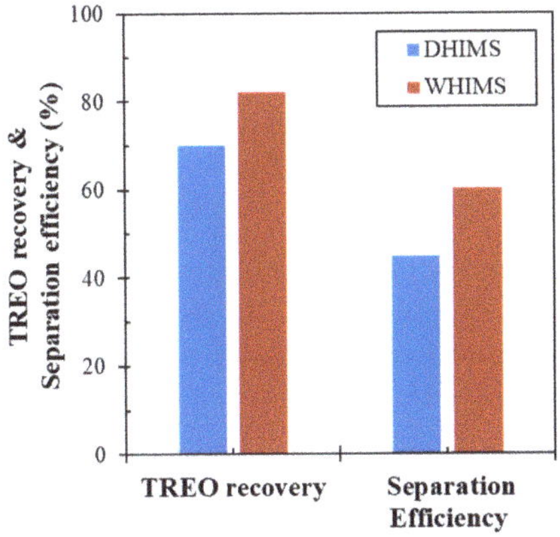

Figure 7. Comparison of TREO recovery and separation efficiency by DHIMS and WHIMS.

3.3.1. Effect of Particle Size

Figure 8a shows the effect of particle size on the recovery of TREO, Fe, and Si operated at a magnetic induction of 1 T and a feed flow rate of 3.2 L/min. The recoveries of TREO and Fe were 79–88% and 89–94%, respectively, both of which slightly decreased with decreasing particle size. The magnetic force acting on a particle is given by the following equation [33]:

$$\vec{F}_m = \frac{\kappa}{\mu_0} V|B|\mathrm{grad}\left|\vec{B}\right|, \tag{4}$$

where κ is the volume magnetic susceptibility, μ_0 is the permeability of free space (numerically equal to $4\pi \times 10^{-7}$ Hm^{-1}), V is the volume of the particle, and B and grad B are the magnetic induction and its gradient, respectively. As can be seen in Equation (4), F_m is proportional to the volume of the particle, which indicates that the recoveries of TREO and Fe were decreased because of the decreased F_m when particle size is reduced.

It is important to note that although the decrease in particle size slightly reduced the recovery of TREO, it was particularly beneficial in reducing the recovery of the major gangue minerals (e.g., quartz and other silicate minerals); that is, Si recovery was ~34% in the −212 + 106 μm size fraction but decreased to ~15% when particle size became <106 μm. Reduction of Si recovery from ~34% to ~15% translated into a better enrichment ratio of TREO from 2.4 to 3.8–3.9 (Figure 8b). Similarly, Fe enrichment ratio increased from 2.5 to 4.3–4.4 as particle size decreased, whereas Si enrichment ratio decreased from 0.9 to 0.7. As shown in Figure 8c, the separation efficiencies of TREO from gangue minerals at various particle sizes are in the following order: −106 + 75 μm (60%) ≈ −75 μm (59%) > −212 + 106 μm (51%). These results indicate that the −106 + 75 μm size fraction showed better REO upgrading and recovery compared with the other size fractions, so the particle size of the sample was fixed to −106 + 75 μm for the following experiments.

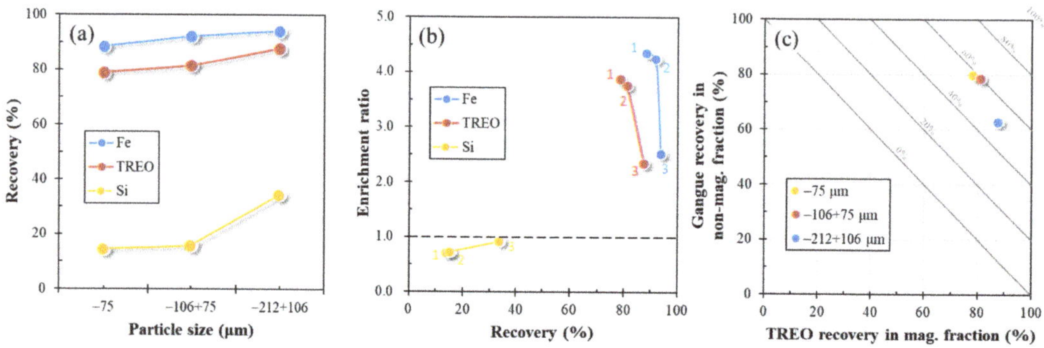

Figure 8. Effect of particle size on (**a**) the recovery of TREO, Fe, and Si from REE ore, (**b**) enrichment ratio as a function of recovery, and (**c**) relationship between TREO recovery in magnetic fraction and gangue recovery in non-magnetic fraction. Note: The numbers in Figure 8b denote that particle size (1: −75 μm, 2: −106 + 75 μm, 3: −212 + 106 μm).

3.3.2. Effect of Feed Flow Rate

The movement of mineral particles in the separation chamber of WHIMS is influenced by not only the magnetic force but also the fluid resistance (drag) force (F_d, Equation (5)) and the gravitational force (F_g, Equation (6)). The latter two physical forces acting on mineral particles are known as competing forces [33]:

$$\vec{F}_d = 3\pi\eta d\left(\vec{v}\left(\vec{r}\right) - \frac{d\vec{r}}{dt}\right), \tag{5}$$

$$\vec{F}_g = \frac{1}{6}\pi\left(\rho_p - \rho_f\right)d^3\vec{g}, \tag{6}$$

where η is the dynamic viscosity of the fluid, d is the particle diameter, v(r) is the velocity of the particle at position r, dr/dt is the velocity of the fluid, ρ_p and ρ_f are the densities of the particle and the fluid medium, and g is the gravitational acceleration. For a magnetic particle to be retained by the magnetized matrix, the magnetic force has to overcome all other physical forces (i.e., $F_m >> F_d + F_g$) [35]. As illustrated in Equation (5), the drag force is strongly influenced by the feed flow rate. At a low feed flow rate, retention of magnetic minerals by the magnetized matrix increases as the drag force is weakened, but the entrainment of gangue minerals is also increased, leading to lowering of the grade of products. At a high feed flow rate, on the other hand, the recovery of magnetic minerals, as well as the entrainment of gangue minerals, are both decreased. Thus, the operation at the appropriate feed flow rate is of importance for the recovery/grade of TREO.

Figure 9a shows the effect of feed flow rate on the recovery of TREO, Fe, and Si at a magnetic induction of 1 T. The recovery of magnetic minerals (i.e., REO and iron-minerals) slightly increased as the feed flow rate decreased; that is, the recoveries of TREO and Fe increased from 73% to 88% and from 87% to 98%, respectively. It is important to note that the recovery of Si was less than 23% at the range of 1.6–5.8 L/min but significantly increased to 49% at a feed flow rate of 0 L/min. This indicates that the lowering of the feed flow rate enhanced the recovery of not only REO but also gangue minerals. As displayed in Figure 9b, the enrichment ratios of magnetic minerals peaked at 3.2 L/min but decreased at feed flow rates of <3.2 L/min. Moreover, the separation efficiencies of TREO from gangue minerals at various feed flow rates (Figure 9c) are in the following order: 3.2 L/min (60%) > 1.6 L/min (57%) > 5.8 L/min (51%) > 0 L/min (41%). Thus, 3.2 L/min is the most effective feed flow rate for upgrading REO from the ore in this study.

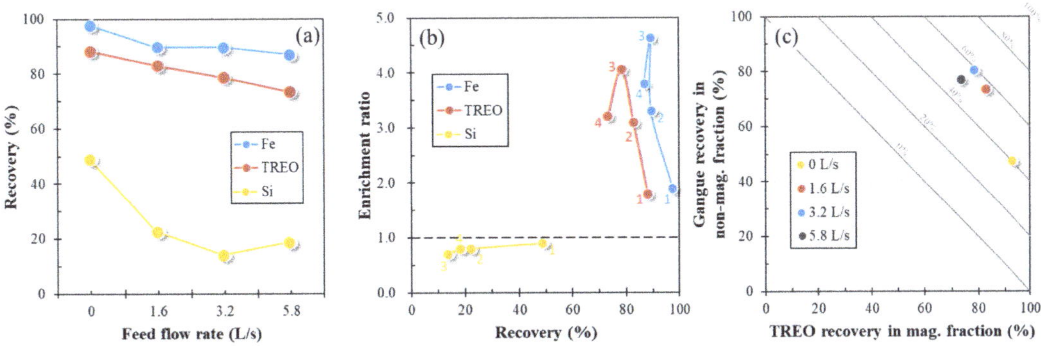

Figure 9. Effect of feed flow rate on (**a**) the recovery of TREO, Fe, and Si from REE ore, (**b**) enrichment ratio as a function of recovery, and (**c**) relationship between TREO recovery in magnetic fraction and gangue recovery in non-magnetic fraction. Note: The numbers in Figure 9b denote that the feed flow rate (1: 0 L/s, 2: 1.6 L/s, 3: 3.2 L/s, 4: 5.8 L/s).

3.3.3. Effect of Magnetic Induction

Magnetic induction is one of the most important factors controlling the recovery of target minerals during magnetic separation. As explained by Equation (4), the magnetic force (F_m) is proportional to the magnetic induction, which means that the recovery of magnetic minerals is promoted at higher magnetic induction. As illustrated in Figure 10a, magnetic induction has a significant effect on the recovery of TREO. At 0.4–0.6 T, the recovery of TREO was 66–67% that increased to 77–79% at higher magnetic induction (0.8–1.0 T). In contrast, the recovery of iron minerals was almost constant (84–90%) irrespective of magnetic induction. This difference could be explained by the higher magnetic susceptibilities of iron minerals (e.g., hematite, 0.4–7.0 $\times$ 10^{-3}; magnetite, 3.0 $\times$ 10^{0}) compared with REE minerals (e.g., bastnaesite, 0.2 $\times$ 10^{-3}) [36,37], which allow the former to be collected in the magnetized matrix even at a low magnetic induction. The recovery of Si was low (5–8%) irrespective of magnetic induction. As shown in Figure 10b, the enrichment ratio of TREO increased from 4.9 to 5.5 as magnetic induction increased to 0.8 T but slightly decreased to 5.3 at 1.0 T. Moreover, the separation efficiency was ~64% at 0.8–1.0 T and 53–54% at 0.4–0.6 T (Figure 10c). This indicates a magnetic induction of 0.8 T was the most effective in upgrading REO from the ore; that is, the grade of TREO could be enriched by approximately 5.5-fold (i.e., feed: 7240 ppm, magnetic concentrate: 39,800 ppm).

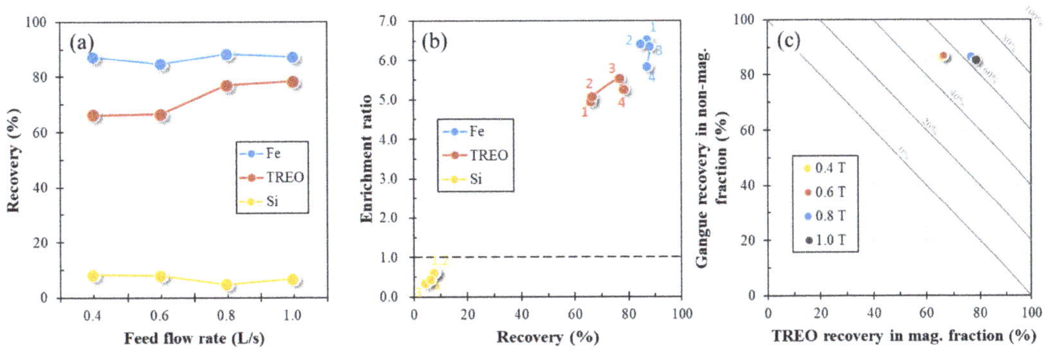

Figure 10. Effect of magnetic induction on (**a**) the recovery of TREO, Fe, and Si from REE ore, (**b**) enrichment ratio as a function of recovery, and (**c**) relationship between TREO recovery in magnetic fraction and gangue recovery in non-magnetic fraction. Note: The numbers in Figure 10b denote that the magnetic induction (1: 0.4 T, 2: 0.6 T, 3: 0.8 T, 4: 1.0 T).

4. Conclusions

This study investigated the dry/wet magnetic separation for upgrading REO from the Khalzan Buregtei ore deposit, Mongolia. The ore used in this study has a low total REE content (~6720 ppm) but contains a relatively high content of HREE (~3180 ppm). The ore is predominantly composed of quartz and aluminosilicate minerals (e.g., albite and microcline) with minor amounts of Fe-, Zr-, and Nb-bearing minerals (e.g., hematite, aegirine, arfvedsonite, zircon, and columbite-(Fe)) and trace amounts of REE-bearing minerals (e.g., bastnaesite, pyrochlore, and synchysite). The dry high-intensity magnetic separation of −0.5 + 0.1 mm size fraction concentrated REO by 2.8-fold with 70% recovery. Compared with DHIMS, the multistage WHIMS showed exceptional results (i.e., TREO recovery, ~80%; enrichment ratio, 5.5) under the following conditions: particle size, −106 + 75 μm; feed flow rate, 3.2 L/min; magnetic induction, 0.8 T. Based on the findings of this study, it has been confirmed that combining dry and wet magnetic separation is an effective pre-concentration stage where the grade of TREO could be enriched by more than 5-fold (~39,800 ppm) and ~90% of gangue minerals are removed from magnetic concentrates. To further upgrade REO from magnetic concentrates, finer grinding (<75 μm) followed by froth flotation is recommended.

Supplementary Materials: The following are available online at https://www.mdpi.com/article/10.3390/min11121432/s1, Figure S1: Back-scattered electron (BSE) images and EDS point analyses of REE ore thin-section.

Author Contributions: Conceptualization, I.P., M.I. and N.H.; methodology, I.P. and S.J.; investigation, Y.K., N.S. and P.G.; resources, P.G.; data curation, I.P. and Y.K.; writing—original draft preparation, I.P.; writing—review and editing, I.P., S.J., M.I., M.K.J., C.B.T. and N.H.; visualization, I.P.; funding acquisition, I.P. All authors have read and agreed to the published version of the manuscript.

Funding: This research was funded by Japan Society for the Promotion of Science (JSPS) Grant-in-Aid for Early-career Scientists (JP20K15214).

Data Availability Statement: The data presented in this study are available on request from the corresponding author.

Conflicts of Interest: The authors declare no conflict of interest.

References

1. Hodgkinson, J.H.; Smith, M.H. Climate Change and Sustainability as Drivers for the next Mining and Metals Boom: The Need for Climate-Smart Mining and Recycling. *Resour. Policy* **2018**, *74*, 101205. [CrossRef]
2. Miah, M.D.; Kabir, M.H.; Koike, M.; Akther, S. Major Climate-Change Issues Covered by the Daily Newspapers of Bangladesh. *Environmentalist* **2011**, *31*, 67–73. [CrossRef]
3. Tabelin, C.B.; Yoo, K.; Li, J. Editorial for Special Issue "Novel and Emerging Strategies for Sustainable Mine Tailings and Acid Mine Drainage Management. *Minerals* **2021**, *11*, 902. [CrossRef]
4. World Meteorological Organization (WMO). *State of the Global Climate 2020*; WMO: Geneva, Switzerland, 2020; pp. 1–38.
5. The Paris Agreement. Available online: https://unfccc.int/process-and-meetings/the-paris-agreement/the-paris-agreement (accessed on 1 November 2021).
6. United Nations Goal 13 | Department of Economic and Social Affairs. Available online: https://sdgs.un.org/goals/goal13 (accessed on 1 November 2021).
7. Tabelin, C.B.; Park, I.; Phengsaart, T.; Jeon, S.; Villacorte-Tabelin, M.; Alonzo, D.; Yoo, K.; Ito, M.; Hiroyoshi, N. Copper and Critical Metals Production from Porphyry Ores and E-Wastes: A Review of Resource Availability, Processing/Recycling Challenges, Socio-Environmental Aspects, and Sustainability Issues. *Resour. Conserv. Recycl.* **2021**, *170*, 105610. [CrossRef]
8. Hund, K.; La Porta, D.; Fabregas, T.P.; Laing, T.; Drexhage, J. *Minerals for Climate Action: The Mineral Intensity of the Clean Energy Transition*; The World Bank: Washington, DC, USA, 2020.
9. Tabelin, C.B.; Dallas, J.; Casanova, S.; Pelech, T.; Bournival, G.; Saydam, S.; Canbulat, I. Towards a Low-Carbon Society: A Review of Lithium Resource Availability, Challenges and Innovations in Mining, Extraction and Recycling, and Future Perspectives. *Miner. Eng.* **2021**, *163*, 106743. [CrossRef]
10. Alves Dias, P.; Bobba, S.; Carrara, S.; Plazzotta, B. *The Role of Rare Earth Elements in Wind Energy and Electric Mobility*; European Commission: Luxembourg, 2020.

11. Yoon, H.-S.; Kim, C.-J.; Chung, K.W.; Jeon, S.; Park, I.; Yoo, K.; Jha, M.K. The Effect of Grinding and Roasting Conditions on the Selective Leaching of Nd and Dy from NdFeB Magnet Scraps. *Metals* **2015**, *5*, 1306–1314. [CrossRef]
12. Choubey, P.K.; Singh, N.; Panda, R.; Jyothi, R.K.; Yoo, K.; Park, I.; Jha, M.K. Development of Hydrometallurgical Process for Recovery of Rare Earth Metals (Nd, Pr, and Dy) from Nd-Fe-B Magnets. *Metals* **2021**, *11*, 1987. [CrossRef]
13. Ballinger, B.; Schmeda-Lopez, D.; Kefford, B.; Parkinson, B.; Stringer, M.; Greig, C.; Smart, S. The Vulnerability of Electric-Vehicle and Wind-Turbine Supply Chains to the Supply of Rare-Earth Elements in a 2-Degree Scenario. *Sustain. Prod. Consum.* **2020**, *22*, 68–76. [CrossRef]
14. Alonso, E.; Wallington, T.; Sherman, A.; Everson, M.; Field, F.; Roth, R.; Kirchain, R. An Assessment of the Rare Earth Element Content of Conventional and Electric Vehicles. *SAE Int. J. Mater. Manuf.* **2012**, *5*, 473–477. [CrossRef]
15. Barteková, E. Chapter 10—The Role of Rare Earth Supply Risk in Low-Carbon Technology Innovation. In *Rare Earths Industry: Technological, Economic, and Environmental Implications*; de Lima, I.B., Filho, W.L., Eds.; Elsevier: Boston, MA, USA, 2016; pp. 153–169. ISBN 978-0-12-802328-0.
16. Jordens, A.; Cheng, Y.P.; Waters, K.E. A Review of the Beneficiation of Rare Earth Element Bearing Minerals. *Miner. Eng.* **2013**, *41*, 97–114. [CrossRef]
17. Garside, M. Rare Earth Mining—Global Distribution by Country 2020. Available online: https://www.statista.com/statistics/270277/mining-of-rare-earths-by-country/ (accessed on 27 August 2021).
18. Muff, R.; Tamiraa, A. *Rare Earths of Mongolia: Evaluation of Market Opportunities for the Principal Deposits of Mongolia*; Ulaanbaatar/Bundesanstalt für Geowissenschaften und Rohstoffe: Hannover, Germany, 2013; p. 31.
19. Gerel, O.; Majigsuren, Y.; Munkhtsengel, B. Rare Earth Mineral Deposits. In *Mineral Resources of Mongolia*; Gerel, O., Pirajno, F., Batkhishig, B., Dostal, J., Eds.; Modern Approaches in Solid Earth Sciences; Springer: Singapore, 2021; pp. 185–210. ISBN 9789811559433.
20. Gronen, L.H.; Sindern, S.; Katzmarzyk, J.L.; Bormann, U.; Hellmann, A.; Wotruba, H.; Meyer, F.M. Mineralogical and Chemical Characterization of Zr-REE-Nb Ores from Khalzan Buregtei (Mongolia)—Approaches to More Efficient Extraction of Rare Metals from Alkaline Granitoids. *Minerals* **2019**, *9*, 217. [CrossRef]
21. Kempe, U.; Möckel, R.; Graupner, T.; Kynicky, J.; Dombon, E. The Genesis of Zr–Nb–REE Mineralisation at Khalzan Buregte (Western Mongolia) Reconsidered. *Ore Geol. Rev.* **2015**, *64*, 602–625. [CrossRef]
22. Long, K.R. *The Future of Rare Earth Elements—Will These High-Tech Industry Elements Continue in Short Supply?* U.S. Geological Survey: Reston, VA, USA, 2011; pp. 1–43.
23. Rybak, A.; Rybak, A. Characteristics of Some Selected Methods of Rare Earth Elements Recovery from Coal Fly Ashes. *Metals* **2021**, *11*, 142. [CrossRef]
24. Goode, J.R. Rare Earth Elements. In *SME Mineral Processing & Extractive Metallurgy Handbook*; Society for Mining, Metallurgy & Exploration (SME): Englewood, CO, USA, 2019; Volume 2, pp. 1891–1916.
25. Abaka-Wood, G.B.; Zanin, M.; Addai-Mensah, J.; Skinner, W. Recovery of Rare Earth Elements Minerals from Iron Oxide–Silicate Rich Tailings—Part 1: Magnetic Separation. *Miner. Eng.* **2019**, *136*, 50–61. [CrossRef]
26. Kovalenko, V.I.; Tsaryeva, G.M.; Goreglyad, A.V.; Yarmolyuk, V.V.; Troitsky, V.A.; Hervig, R.L.; Farmer, G.L. The Peralkaline Granite-Related Khaldzan-Buregtey Rare Metal (Zr, Nb, REE) Deposit, Western Mongolia. *Econ. Geol.* **1995**, *90*, 530–547. [CrossRef]
27. Wills, B.A.; Finch, J.A. Magnetic and Electrical Separation. In *Wills' Mineral Processing Technology*; Butterworth Heinemann: Oxford, UK, 2016; pp. 381–408.
28. Hanchar, J.M.; van Westrenen, W. Rare Earth Element Behavior in Zircon-Melt Systems. *Elements* **2007**, *3*, 37–42. [CrossRef]
29. Burnham, A.D. Key Concepts in Interpreting the Concentrations of the Rare Earth Elements in Zircon. *Chem. Geol.* **2020**, *551*, 119765. [CrossRef]
30. Jeon, S.; Ito, M.; Tabelin, C.B.; Pongsumrankul, R.; Tanaka, S.; Kitajima, N.; Saito, A.; Park, I.; Hiroyoshi, N. A Physical Separation Scheme to Improve Ammonium Thiosulfate Leaching of Gold by Separation of Base Metals in Crushed Mobile Phones. *Miner. Eng.* **2019**, *138*, 168–177. [CrossRef]
31. Tabelin, C.B.; Resabal, V.J.T.; Park, I.; Villanueva, M.G.B.; Choi, S.; Ebio, R.; Cabural, P.J.; Villacorte-Tabelin, M.; Orbecido, A.; Alorro, R.D.; et al. Repurposing of Aluminum Scrap into Magnetic Al^0/ZVI Bimetallic Materials: Two-Stage Mechanical-Chemical Synthesis and Characterization of Products. *J. Clean. Prod.* **2021**, *317*, 128285. [CrossRef]
32. Choi, S.; Jeon, S.; Park, I.; Ito, M.; Hiroyoshi, N. Addition of Fe_3O_4 as Electron Mediator for Enhanced Cementation of Cd^{2+} and Zn^{2+} on Aluminum Powder from Sulfate Solutions and Magnetic Separation to Concentrate Cemented Metals from Cementation Products. *J. Environ. Chem. Eng.* **2021**, *9*, 106699. [CrossRef]
33. Svoboda, J. *Magnetic Methods for the Treatment of Minerals*; Elsevier: Amsterdam, The Netherlands, 1987; pp. 1–235.
34. Jordens, A.; Marion, C.; Langlois, R.; Grammatikopoulos, T.; Rowson, N.A.; Waters, K.E. Beneficiation of the Nechalacho Rare Earth Deposit. Part 1: Gravity and Magnetic Separation. *Miner. Eng.* **2016**, *99*, 111–122. [CrossRef]
35. Rasool, R.; Lieberwirth, H. A Continuum Based Numerical Modelling Approach for the Simulation of WHIMS. *Miner. Eng.* **2018**, *118*, 97–105. [CrossRef]

36. Jordens, A.; Sheridan, R.S.; Rowson, N.A.; Waters, K.E. Processing a Rare Earth Mineral Deposit Using Gravity and Magnetic Separation. *Miner. Eng.* **2014**, *62*, 9–18. [CrossRef]
37. Hrouda, F.; Chlupacova, M.; Chadima, M. *The Use of Magnetic Susceptibility of Rocks in Geological Exploration*; Terraplus: Brno, Czech Republic, 2009; pp. 1–27.

minerals MDPI

Case Report

Characterization and Removal Potential of Fluorine in Lignite from a Mine in Shaanxi Province, China: A Case Study

Chao Ni [1], Shaoqi Zhou [1], Xiangning Bu [1,*], Muhammad Bilal [2,*], Fawad Ul Hassan [2], Yuran Chen [3,*], Guangqian Xu [4] and Guangyuan Xie [1]

1 Key Laboratory of Coal Processing and Efficient Utilization (Ministry of Education), School of Chemical Engineering and Technology, China University of Mining and Technology, Xuzhou 221116, China; sunnichao@126.com (C.N.); ts18040100a31@cumt.edu.cn (S.Z.); xgywl@163.com (G.X.)
2 Department of Mining Engineering, Balochistan University of Information Technology Engineering and Management Sciences (BUITEMS), Quetta 87300, Pakistan; fawad-hassan@hotmail.com
3 School of Materials Science and Engineering, Zhengzhou University, Zhengzhou 450001, China
4 Shandong Energy Xinwen Mining Group Co., Ltd., Xinen 271000, China; 87ue@163.com
* Correspondence: xiangning.bu@foxmail.com or xiangning.bu@cumt.edu.cn (X.B.); bilalkhan4p@gmail.com (M.B.); cyr_hy@zzu.edu.cn (Y.C.)

Abstract: Fluorine appears in coal and is released into the atmosphere upon combustion, resulting in harmful impacts on the environment and life, which needs to be removed from coal before utilization. Coal can be processed by flotation and gravity separation to reduce its fluorine content. In this study, a lignite sample from a mine in Shaanxi Province, China, was characterized using the float–sink test, sieving test, X-ray diffraction (XRD), and polarized light microscopy. Mineralogical analysis indicated that the fluorine in coal is mainly contained in Muscovite and polylithionite, and partly in pyrite. The washability and floatability analyses were employed to evaluate the extent of fluorine removal from >0.5 and <0.5 mm size fractions of lignite, respectively. Compared to the raw sample that contained 347.74 μg/g fluorine content, the proposed combination of gravity-flotation separation process decreased the fluorine content to 90.14 μg/g, which meets the requisites of coal standards.

Keywords: fluorine; occurrence; removal; gravity-flotation separation; lignite

check for updates

Citation: Ni, C.; Zhou, S.; Bu, X.; Bilal, M.; Ul Hassan, F.; Chen, Y.; Xu, G.; Xie, G. Characterization and Removal Potential of Fluorine in Lignite from a Mine in Shaanxi Province, China: A Case Study. *Minerals* **2022**, *12*, 280. https://doi.org/10.3390/min12030280

Academic Editor: Luis Marcelo Tavares

Received: 31 January 2022
Accepted: 22 February 2022
Published: 23 February 2022

1. Introduction

Fluorine in coal is potentially hazardous to the environment and life [1–3]. During the coal combustion process, fluorine gets converted into gases (such as HF, SiF_4, and CF_4) and dust particles (such as SiF_6 and CaF_2), which enter into the atmospheric environment having detrimental environmental and ecological consequences [4–6]. The gaseous fluoride emissions such as CF_4, which are reported to have greenhouse potential (GWP) around 6500 times as compared to CO_2, can cause a significant greenhouse effect, while HF is multiple times more toxic to animals and plant life than SO_2 [7,8].

Interim Measures for Commercial Coal Quality Management (2014) stipulates that the maximum fluorine content of coal for commercial purposes must be less than 200 μg/g to ensure environmental protection. The fluorine content in Chinese coal ranges between 2 and 911 μg/g, and it occurs in various modes, as listed in Table 1. The majority of fluorine in coal comes from inorganic minerals. Notably, the fluorine content of different regions in China fluctuates greatly [9,10]. The fluorine content in Chinese coal is reported to follow a logarithmic normal distribution, having a range of 17–3088 mg/kg of fluorine fraction. In total, 73% of Chinese coal contains fluorine ranging from 50 to 300 mg/kg, averaging 202 mg/kg [11]. Power plants use nearly 1.76 billion tons of coal produced in China to generate electricity [12,13]. Considering the average fluorine concentration of coal at a low value of 100 μg/g, the fluorine emission from 1 t steam coal combustion into the atmosphere amounts to 8.3×10^{-5} t [14]. This can have a detrimental effect on humans, animals, and plants which is why it is important to minimize the harmful effects of fluorine generated from coal combustion.

Table 1. A summary of the fluorine content and occurrence modes in some Chinese coals.

Coal Sample	Fluorine Content (μg/g)	Fluorine Occurrence Mode	Ref.
A coal sample from Guizhou Province, China	357	Fluorine mainly occurred in residual-associated form and carbonate- and Fe/Mn-associated form.	[15]
	<200 μg/g	Most of these coals are low fluorine (<200 μg/g); however, the fluorine content of clay is as high as 1027.6 μg/g.	[16]
	715.20 (average)	The inorganic occurrence is the primary mode of occurrence of fluorine in coals from Guizhou province.	[17]
Haerwusu Surface Mine, Inner Mongolia, China	286	Boehmite and kaolinite are prime carriers of fluorine, but sometimes associated both with organic and inorganic matter.	[18,19]
Xiangning mining area, Shanxi Province, China, China	2–911	Fluorine in coal mainly exists in an inorganic bound state in forms such as fluorapatite and calcium fluoride.	[20]
Nantong coalfield, Chongqing Province, China	490 (average)	Fluorine in coals from Nantong coalfield existed in the organic and inorganic minerals simultaneously.	[21]
China coals	67.3–3145.4	Fluorine in coal mainly occurs in an inorganic form.	[22]

Various engineering measures, such as denitrification and desulfurization, can significantly reduce the amount of fluorine released during coal combustion, thereby reducing the associated environmental impacts [20]. The defluorination processes can be classified into defluorination before combustion, fluorine fixation during combustion, and flue gas defluorination after combustion [23,24]. The fluorine removal process depends on whether fluorine has an inorganic form, which is predominantly abundant, or organic form [9,20,25]. Both fluorine fixation and flue gas defluorination (the latter of which is often employed with coal-fired boilers in large power stations) can effectively remove both inorganic and organic fluorine from coal. These two methods are widely useful in particular for the removal of organic fluorine [4,14,20,26]. For coal with a higher content of inorganic fluorine, the defluorination of pulverized coal before combustion using various physical or physicochemical methods has a low operating cost compared to the other two processes [27]. As mentioned previously, the majority of fluorine in coal is in association with inorganic minerals, indicating that most of the fluorine has a strong affinity for ash materials (i.e., gangue). Thus, the commonly used coal preparation methods can be useful for the removal of fluorine from coal, such as gravity separation (physical method) and flotation (physicochemical method). Zhou et al. [15] reported that a total of 76.05% of the total fluorine was removed by froth flotation under an optimum condition. Lin et al. [20] demonstrated that froth flotation can reduce the amount of fluorine significantly. Mohanty et al. [28] found that a fine coal cleaning circuit using enhanced-gravity separation (EGS) and column flotation in a rougher–cleaner arrangement provides a more efficient reduction in ash and the trace element (i.e., Hg and Se) when compared to the results of single-stage cleaning operations. It can be implied that flotation has a high removal rate of fluorine, due to its strong interaction with ash, similar to Hg/Se.

In coal preparation plants, coal of size > 0.5 mm is commonly upgraded by gravity separation devices, such as teeter bed separators [29,30], spirals [31,32], dense medium cyclones [33,34], and jigs [35,36]. Flotation is an effective method to separate hydrophilic gangue from hydrophobic coal (<0.5 mm) [37,38]. As discussed, previous studies mainly focused on the removal of fluorine in fine coal (<0.5 mm) by using froth flotation. However, there is a lack of literature reporting the removal of fluorine in coarse coal (>0.5 mm) by using gravity separation. To bridge this gap, gravity separation as well as flotation have been utilized to remove fluorine from >0.5 and <0.5 mm coal size fractions, respectively. The first step was to study the modes of occurrence of fluorine in coal. Then, fluorine removal efficiency by gravity-separation and flotation was investigated. Finally, a flowsheet to remove fluorine from <50 mm coal was proposed.

2. Materials and Methods

2.1. Materials

Coal samples smaller than 50 mm in size were collected from a mine in Shaanxi Province, China. Sec-octyl alcohol (analytical reagent) and kerosene (commercially pure) were used as the frother and collector, respectively. Zinc chloride, benzene (carbon tetrachloride), and cyclohexane were obtained from China National Pharmaceutical Group Co., Ltd. (Beijing, China) and were used for float–sink tests.

2.2. Sieve Analysis

The sieve analysis was carried out according to Chinese Standard GB/T 477-2008 (Method for size analysis of coal) using sieves with mesh sizes of 50, 25, 13, 6, 3, 1, and 0.5 mm. The coal sample used for the sieving was air-dried under prescribed conditions. After the sieving test, the retained size fractions of 25–50, 13–25, 6–13, 3–6, 1–3, 0.5–1, and <0.5 mm were weighed. Afterwards, the sub-samples were prepared according to Chinese Standard GB/T 474-2008 (Method for preparation of coal sample). The obtained sub-samples were further used for the determination of fluorine, ash, and total sulfur content.

2.3. Float–Sink Analysis

Representative coal samples were prepared for each size fraction. Float–sink analyses were conducted according to Chinese Standard GB/T 478-2008 (Method for float and sink analysis of coal). To perform a float–sink test on >0.5 mm coal, water was mixed with pure zinc chloride (as a liquid sorting medium), to obtain a solution of a specific density. Organic heavy media with different densities were prepared to perform the float–sink test of <0.5 mm coal by mixing benzene (carbon tetrachloride) and cyclohexane.

The float–sink products were dried and weighed. After that, the sub-samples were prepared according to Chinese Standard GB/T 474-2008 (Method for preparation of coal sample) for further determination of fluorine, ash, and total sulfur content.

2.4. Batch Flotation Test

The procedure of the batch flotation test was carried out according to Chinese Standard GB/T 4757-2013 (Methods for the batch flotation testing of fine coal) using a standard laboratory RK/FD-II sub-aeration flotation cell (volume = 1.5 dm^3). The pulp with a 60 g/L solid concentration was conditioned for 3 min. The required amount of kerosene was added and conditioned for two more minutes. After that, the required amount of sec-octyl alcohol was added, and the slurry was conditioned for another 0.5 min. Then, the air valve was opened at 4.17 dm^3/ min air flow rate, and the froth was collected for 3 min. In each test, tap water was added to maintain a constant pulp level and a froth layer of 1 cm. The agitation speed was kept constant at 1900 rpm. A detailed description of the working process of the mechanical flotation cell is reported in the literature [39]. The release analysis test was performed to determine the floatability of the coking coal fines according to Chinese Standard GB/T 36167-2018 (Methods for coal preparation laboratory timed-release flotation analysis). Flotation release tests were performed based on published literature [40–42]. The flotation time was constant at 3 min for different flotation stages of release flotation. The release analysis procedure used in this study is similar to "reverse release analysis" introduced by Randolph [43], where the tailings are collected in the cleaning stage of flotation.

2.5. XRD Test

The ground sample (<0.074 mm) obtained using an XPM-φ 120 × 3 three-headed grinding machine (Nanchang Source of Mining and Metallurgy Equipment Co., Ltd.,

Nanchang, China) was subjected to XRD analysis using a D8 Advance X-ray diffractometer (Bruker, Germany).

2.6. Determination of Ash, Fluorine, Sulfur Contents, and Calorific Value

The proximate analysis for the determination of ash content was carried out according to Chinese Standards (Table S1) GB/T 212-2008 (Proximate analysis of coal), using a CTM100 muffle furnace (Tairui Company, Xuzhou, China). The ultimate analysis was carried out according to GB/T 31391-2015 (Ultimate analysis of coal). The total sulfur content was determined according to Chinese Standard GB/T 214-2007 (Determination of total sulfur in coal) using the CTS7000 automatic sulfur detector (Tairui Company, Xuzhou, China). The fluorine content in coal was analyzed by pyro hydrolysis/fluorine ion-selective electrode method [26] according to Chinese Standard 4633-2014 (Determination of fluorine in coal). The calorific value of coal was determined according to Chinese Standard GB/T 213-2008 (Determination of the calorific value of coal) using an auto-calorimetry instrument (CT5000, Weike Tech. Company, Xuzhou, China).

3. Results and Discussion

3.1. Mineralogical Characterization

Interim Measures for Commercial Coal Quality Management (2014) states that the maximum ash, total sulfur, and fluorine contents of coal should not exceed 30%, 1.5%, or 200 mg/g, respectively. Results of the proximate and ultimate analyses are given in Table 2. According to Chinese Standard GB/T5751-2009 (Chinese classification of coals), the coal under consideration is classified as lignite. The ash content, total sulfur content, fluorine content, and net calorific values are 38.88%, 2.95%, 241 μg/g, and 16.07 MJ/kg, respectively. Thus, the quality of this coal needs to be upgraded to meet those requirements.

Table 2. Proximate and ultimate analysis data.

Proximate Analysis (%)				Ultimate Analysis (%)					F_{ad} (μg/g)	$Q_{net.ar}$ (MJ/kg)
M_{ad}	A_d	V_{daf}	FC_d	C_{daf}	H_{daf}	O_{daf}	N_{daf}	$S_{t.d}$		
4.68	38.88	44.46	33.95	74.39	4.67	13.80	1.08	2.95	241	16.07

Note: in Table 2, M_{ad} and F_{ad} are the moisture content (M) and the fluorine content (F) in the coal on an air-dried (ad) basis; A_d and FC_d are the ash content (A) and the fixed carbon content (FC) on a dry (d) basis; V_{daf}, C_{daf}, H_{daf}, O_{daf}, N_{daf}, and $S_{t.d}$ are the carbon, hydrogen, oxygen, nitrogen and total sulfur, on dry ash-free (daf) basis, respectively; $Q_{net.ar}$ is the net calorific value (at constant pressure) as received.

The ash content, total sulfur content, and fluorine content of different size fractions are shown in Figure 1. The amount of product in each size group and their ash, sulfur, and fluorine contents and distribution are given in the Table S2. It is evident that there is a direct relation between ash and fluorine contents. As the most active non-metallic element known, fluorine has a strong oxidation affinity and can combine with almost any other to form fluoride [25]. According to previous studies [3,10,19,25,26], fluorine in coal mainly exists in the form of inorganic substances. Fluorapatite minerals may be the main occurrence mode of fluorine in coal. In addition, other inorganic substances include fluorspar, tourmaline, mica, and clay minerals.

3.2. Fluorine and Sulfur Distributions in Different Density Fractions of >0.5 mm Coal

As shown in Figure 2, the fluorine content increases significantly when the density fraction increases. This phenomenon indicates that the occurrence of fluorine in this coal sample is mainly inorganic. Figure 3 shows the relationships between ash content, total sulfur content, and fluorine content for different size fractions. For a given size fraction when the density fraction is increased the ash content increased significantly. As observed from Figure 3, the higher the ash content of the coal, the higher the content of fluorine and sulfur. According to the literature [20], fluorine in coal might be adsorbed in fluorine-containing minerals (fluorapatite) and clay minerals such as muscovite and kaolinite in ionic state. The XRD result of the high-density fraction (>2.0 g/cm^3) is shown in Figure 4. As shown in Figure 4, the main minerals are muscovite-3T, polylithionite, nimite-1MIIb, clinochlore-1MIIb, pyrite, and quartz. Fluorine mainly exists as muscovite-3T and polylithionite, while pyrite is the main form of inorganic sulfur.

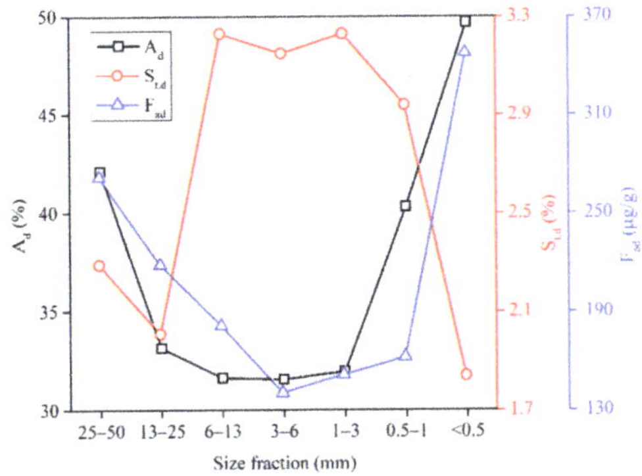

Figure 1. The ash content, total sulfur content, and fluorine content of different size fractions.

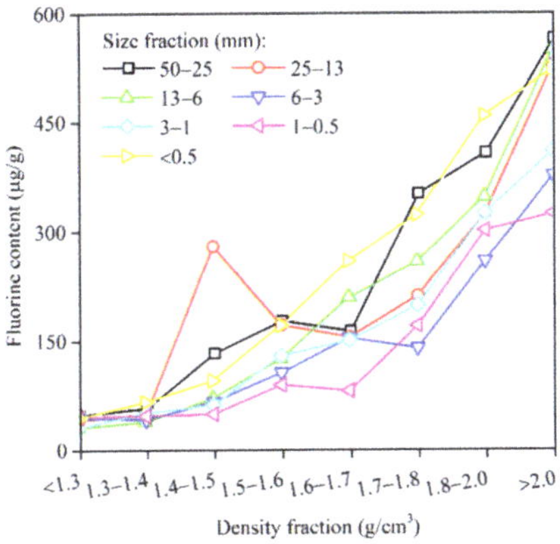

Figure 2. The fluorine content of different density fractions.

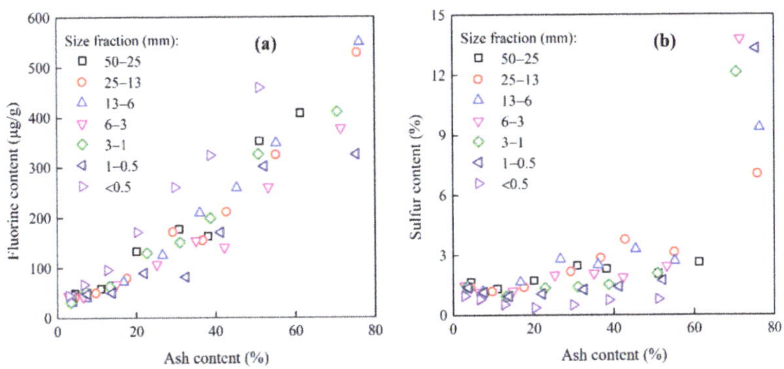

Figure 3. The relationships between ash content, total sulfur content, and fluorine content for different size fractions. (**a**) ash content vs. fluorine content; (**b**) ash content vs. total sulfur content. The high ash content corresponds to the fraction with a high density.

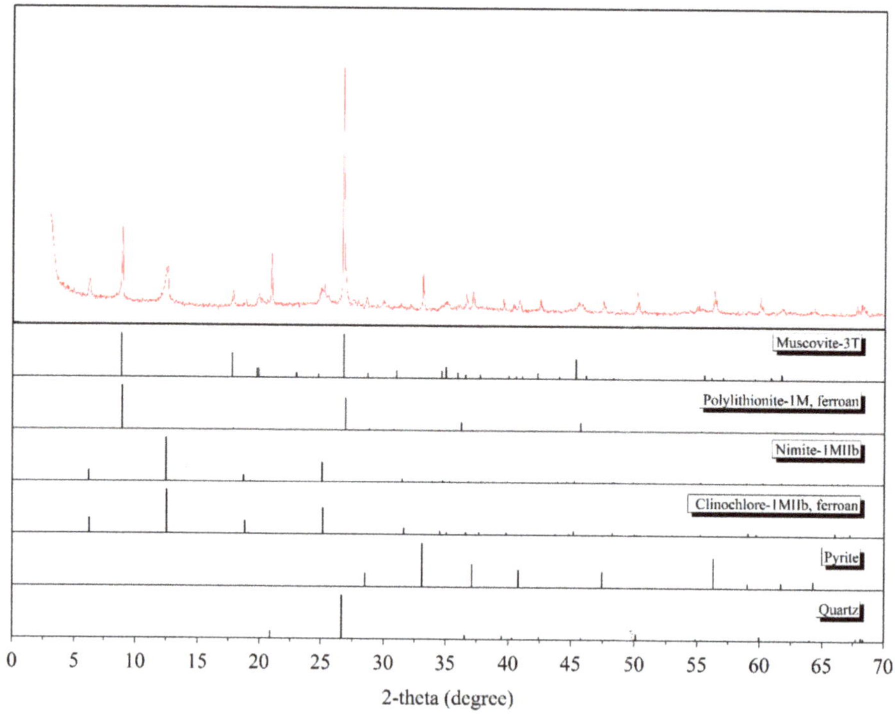

Figure 4. XRD result of the high-density fraction (>1.8 g/cm^3).

The washability curves for >0.5 mm coal are given in Figure 5. The degrees of washability of clean coal as a function of ash content and the corresponding values of fluorine, sulfur, and net calorific value in clean coal are presented in Table 3. The cumulative yield of clean coal is 72.97% when the ash content is 15%.

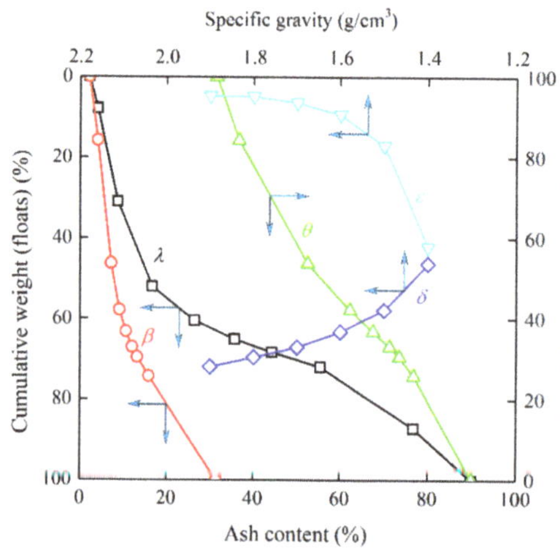

Figure 5. Classical washability curves for >0.5 mm coal. λ—primary curve; β—cumulative float curve; θ—cumulative sink curve; δ—density distribution curve; ε—near-gravity material with ±0.1 g/cm^3 specific gravity (s. g.).

Table 3. The degrees of washability of clean coal at the different given ash contents [1].

A_d (%)	Yield (%)	F_{ad} (μg/g)	$S_{t.d}$ (%)	$Q_{net.ar}$ (MJ/kg)	δ_s (g/cm^3)	$\varepsilon_{\pm 0.1}$ (%)	Degree of Washability
10.00	61.44	64.65	1.33	23.05	1.52	18.55	Moderate separation
11.00	64.70	69.00	1.37	22.73	1.59	11.04	
12.00	67.27	72.58	1.40	22.42	1.66	7.66	Simple separation
13.00	69.56	77.68	1.44	22.10	1.75	5.69	
14.00	71.18	83.39	1.46	21.80	1.80	5.31	
15.00	72.97	89.23	1.49	21.53	1.86	4.94	

[1] δ_s is the separation gravity, and $\varepsilon_{\pm 0.1}$ is the near-gravity material with ± 0.1 g/cm^3 that corresponds to δ_s. The degree of washability was determined according to Chinese Standard GB/T16417-2011 (Method for evaluating the washability of coal).

3.3. Fluorine and Sulfur Distributions in Release Flotation of <0.5 mm Coal

The ash, fluorine, and sulfur contents, after release flotation tests, are shown in Figure 6. For release analysis, the kerosene dosage and the sec-Octyl alcohol dosage were 9000 and 4500 g/t, respectively. The solid concentration was 100 g/L, the airflow rate was 0.25 m^3/h, and the impeller speed was 1900 rpm. As shown in Figure 6, the fluorine content has a direct correlation with the ash content. This indicates that fluorine exists in <0.5 mm coal in the form of inorganic matter. However, results of tailings indicate that the occurrence form of sulfur is mainly organic. Classical floatability curves for <0.5 mm coal are given in Figure 7. The degrees of floatability of flotation concentrates at different given ash contents are summarized in Table 4. It can be seen that the degree of floatability is difficult-to-float for all given percentages of the ash.

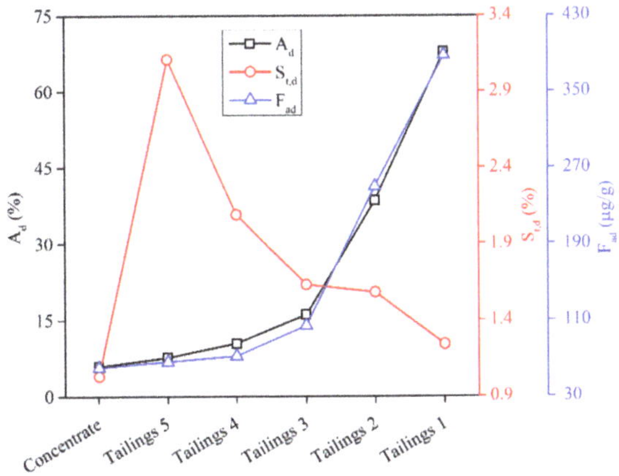

Figure 6. Ash contents, fluorine contents, sulfur contents, and net calorific values for different products of release flotation tests.

Table 4. The degrees of floatability of flotation concentrates at different given ash contents [1].

A_d (%)	Yield (%)	F_{ad} (μg/g)	$S_{t.d}$ (%)	$Q_{net.ar}$ (MJ/kg)	Combustible Recovery (%)	Degree of Floatability
10.00	27.16	79.02	1.76	23.02	48.34	Difficult-to-float
11.00	28.37	84.54	1.76	22.72	49.93	
12.00	29.72	91.41	1.75	22.44	51.72	
13.00	30.93	97.06	1.74	22.15	53.21	
14.00	32.14	103.34	1.73	21.85	54.66	
15.00	33.36	108.82	1.73	21.56	56.07	

[1] combustible recovery = $yield.(100 - A_{d,con})/100 - A_{d,<0.5\text{mm}}$; $A_{d,con}$ is the given ash.

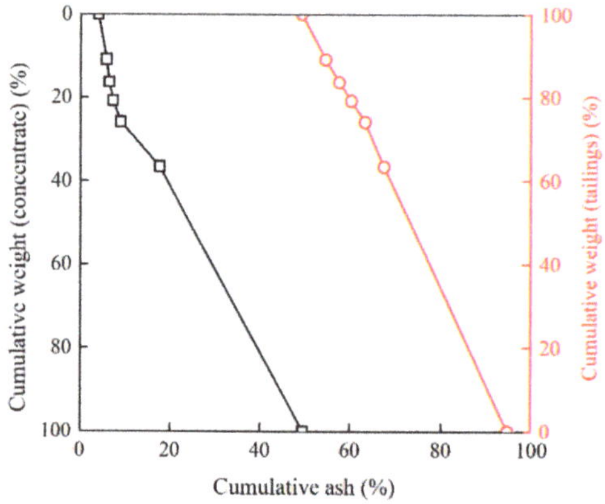

Figure 7. Classical floatability curves of <0.5 mm coal.

The flotation performance of lignite is known to be poor since several hydrophilic functional groups are present, such as hydroxyl, carboxyl, methoxyl, and carbonyl [44–46]. For the comparison, the float–sink test of <0.5 mm coal was performed. The upgrading curves can be used for the analysis of the separation process [47–50]. The comparison of float–sink and release flotation tests is given in Figure 8. The content of fluorine in clean coal increases as yields increase. It is observed that the yield of clean coal using gravity separation is higher than that of flotation separation. However, the fluorine content at a given yield (around 35%) is significantly lower than that of gravity separation, which makes the flotation technique more suited to remove fluorine from <0.5 mm size fractions.

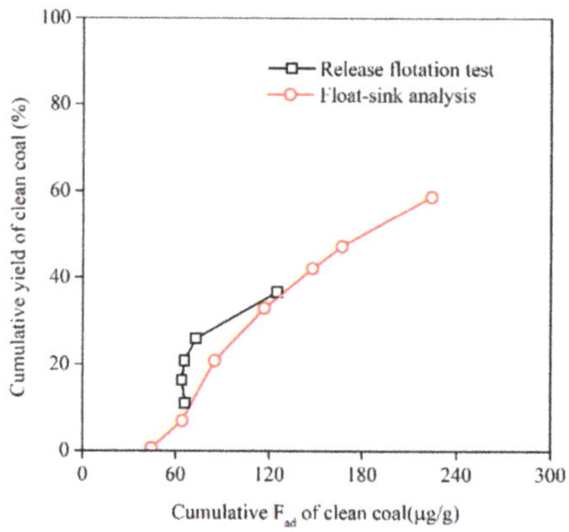

Figure 8. Comparison of upgrading curves (F_{ad} vs. yield) of <0.5 mm coal between float–sink and release flotation tests.

3.4. *A Proposed Beneficiation Circuit for <50 mm Coal*

It was determined that >0.5 mm coal can be straightforwardly separated using dense medium cyclones and jigs due to the degree of washability, however, for <0.5 mm the efficiency of using dense medium separation diminishes [34,51]. On the other hand, the efficiency of flotation for <0.5 mm is far greater than dense medium separation, making it more suitable for fluorine removal. It can be concluded from a simple approximation calculation (see Figure 9A) that fine coal (347.74 μg/g) can be directly mixed with coarse clean coal to obtain the final product with a qualified quality (122.14 μg/g). To further decrease the fluorine content of fine coal (see Figure 9B) flotation separation can be used before mixing it with coarse clean coal. As a result, the fluorine content in the final product is reduced down to 90.14 μg/g.

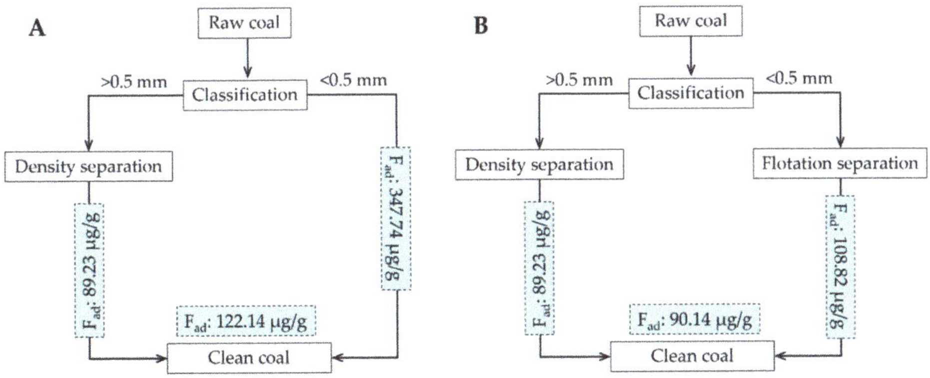

Figure 9. The proposed beneficiation circuits for <50 mm coal. (**A**) No treatment of <0.5 mm coal; (**B**) <0.5 mm coal treated by flotation separation.

4. Conclusions

In this study, gravity separation/sink–float and flotation methods were investigated for the removal of fluorine from coal (0.001–0.5 mm in size). Based on the results of proximate and ultimate analyses, the coal was classified as lignite. The results of XRD analysis showed that the majority of fluorine exists as Muscovite-3T and Polylithionite and inorganic sulfur occurs as pyrite primarily. Ash content and fluorine were found to be directly related. A qualified coal product can be obtained by gravity separation treatment of coarse coal (>0.5 mm). Considering the extent of fluorine removal, for coal <0.5 mm in size, flotation is more efficient than dense medium separation. Furthermore, the mixture of clean coarse coal (float–sink) and clean fine coal (flotation) contains significantly less fluorine than that of the mixture of clean coarse coal (float–sink) and untreated fine coal. The proposed circuit combining the float-sink–flotation process can significantly reduce the fluorine content of coal, eventually reducing fluorine emissions to the atmosphere by utilizing a fine coal fraction cleaned with flotation, covering the inefficient fluorine removal from fine coal fraction by dense medium separation.

Supplementary Materials: The following supporting information can be downloaded at: https://www.mdpi.com/article/10.3390/min12030280/s1, Table S1. List of Chinese Standards, Table S2. The amount of product in each size group and their ash, sulfur and fluorine contents and distribution. All information are mentioned in the reference [52] here in the text and in the Supplementary Material file.

Author Contributions: Conceptualization, X.B. and Y.C.; methodology, C.N. and S.Z.; software, validation, and formal analysis, G.X. (Guangqian Xu); investigation, S.Z.; resources, G.X. (Guangyuan Xie); data curation, X.B.; writing—original draft preparation, X.B., F.U.H. and M.B.; writing—review and editing and visualization, X.B., Y.C., M.B. and F.U.H.; supervision, C.N.; project administration, X.B.; funding acquisition, C.N. All authors have read and agreed to the published version of the manuscript.

Funding: This work was supported by the Fundamental Research Funds for the Central Universities (2018QNA16).

Institutional Review Board Statement: Not applicable.

Informed Consent Statement: Not applicable.

Acknowledgments: The authors acknowledge the contribution to the manuscript from Li Pan.

Conflicts of Interest: The authors declare no conflict of interest.

References

1. Gao, G.; Yan, B.; Yang, L. Determination of total fluorine in coal by the combustion-hydrolysis/fluoride-ion selective electrode method. *Fuel* **1984**, *63*, 1552–1555. [CrossRef]
2. Greta, E.; Dai, S.; Li, X. Fluorine in Bulgarian coals. *Int. J. Coal Geol.* **2013**, *105*, 16–23. [CrossRef]
3. Liu, G.; Zhang, H.; Gao, L.; Zheng, L.; Peng, Z. Petrological and mineralogical characterizations and chemical composition of coal ashes from power plants in Yanzhou mining district, China. *Fuel Process. Technol.* **2004**, *85*, 1635–1646. [CrossRef]
4. Liu, J.; Qi, Q.; Zhou, J.; Cao, X.; Cen, K. Distribution of fluoride in the combustion products of coal. *Huan Jing Ke Xue = Huanjing Kexue* **2003**, *24*, 127–130.
5. Sredović, I.; Rajaković, L. Pyrohydrolytic determination of fluorine in coal: A chemometric approach. *J. Hazard. Mater.* **2010**, *177*, 445–451. [CrossRef]
6. Guo, S.; Yang, J.; Liu, Z. The fate of fluorine and chlorine during thermal treatment of coals. *Environ. Sci. Technol.* **2006**, *40*, 7886–7889. [CrossRef]
7. Pisupati, S.V.; Wasco, R.S.; Scaroni, A.W. An investigation on polycyclic aromatic hydrocarbon emissions from pulverized coal combustion systems. *J. Hazard. Mater.* **2000**, *74*, 91–107. [CrossRef]
8. Thomas, G.; Fariborz, G. Effect of geological processes on coal quality and utilization potential: Review with examples from western Canada. *J. Hazard. Mater.* **2000**, *74*, 109–124. [CrossRef]
9. Luo, K.; Ren, D.; Xu, L.; Dai, S.; Cao, D.; Feng, F.; Tan, J.A. Fluorine content and distribution pattern in Chinese coals. *Int. J. Coal Geol.* **2004**, *57*, 143–149. [CrossRef]
10. Dai, S.; Ren, D. Fluorine concentration of coals in China—An estimation considering coal reserves. *Fuel* **2006**, *85*, 929–935. [CrossRef]
11. Qi, Q.J.; Yu, G.S.; Li, F.W.; LIU, J.Z.; CEN, K.F. Inorganic/organic affinity and washing floatation characteristic of fluorine in coal. *J. Liaoning Tech. Univ. Nat. Sci. Ed.* **2006**, *25*, 481–484.
12. Bu, X.; Chen, Y.; Ma, G.; Sun, Y.; Ni, C.; Xie, G. Differences in dry and wet grinding with a high solid concentration of coking coal using a laboratory conical ball mill: Breakage rate, morphological characterization, and induction time. *Adv. Powder Technol.* **2019**, *30*, 2703–2711. [CrossRef]
13. Bu, X.; Chen, Y.; Ma, G.; Sun, Y.; Ni, C.; Xie, G. Wet and dry grinding of coal in a laboratory-scale ball mill: Particle-size distributions. *Powder Technol.* **2020**, *359*, 305–313. [CrossRef]
14. Luo, K.; Xu, L.; Li, R.; Xiang, L. Fluorine emission from combustion of steam coal of North China Plate and Northwest China. *Chin. Sci. Bull.* **2002**, *47*, 1346–1350. [CrossRef]
15. Zhou, C.-C.; Liu, C.; Zhang, N.-N.; Cong, L.-F.; Pan, J.-H.; Peng, C.-B. Fluorine in Coal: The Modes of Occurrence and its Removability by Froth Flotation. *Int. J. Coal Prep. Util.* **2018**, *38*, 149–161. [CrossRef]
16. Dai, S.; Ren, D.; Ma, S. The cause of endemic fluorosis in western Guizhou Province, Southwest China. *Fuel* **2004**, *83*, 2095–2098. [CrossRef]
17. Liu, X. Evolution Of Migration Behavior of Fluofine in Coal and Sulfur-Containing Coal Molecular Structure. Ph.D. Thesis, Huazhong University of Science and Technology, Wuhan, China, 2009.
18. Wang, X.; Dai, S.; Sun, Y.; Li, D.; Zhang, W.; Zhang, Y.; Luo, Y. Modes of occurrence of fluorine in the Late Paleozoic No. 6 coal from the Haerwusu Surface Mine, Inner Mongolia, China. *Fuel* **2011**, *90*, 248–254. [CrossRef]
19. Dai, S.; Li, D.; Chou, C.-L.; Zhao, L.; Zhang, Y.; Ren, D.; Ma, Y.; Sun, Y. Mineralogy and geochemistry of boehmite-rich coals: New insights from the Haerwusu Surface Mine, Jungar Coalfield, Inner Mongolia, China. *Int. J. Coal Geol.* **2008**, *74*, 185–202. [CrossRef]
20. Guohua, L.; Qiyan, F.; Xiaoli, D.; Yahong, C.; Wenbo, L.; Hui, W.; Bo, G.; Lai, Z.; Xin, W. Geochemical characteristics of fluorine in coal within Xiangning mining area, China, and associated mitigation countermeasures. *Energy Explor. Exploit.* **2019**, *37*, 1737–1751. [CrossRef]
21. Liu, Z. The Research of Content of Distribution and Accurrence Form of Fluorine Element in Coals from Nantong Coalfield in Chongqing, China. Master's Thesis, Anhui University of Science and Technology, Huainan, China, 2016.
22. Wu, X. Fluorin Eemission Behavior and Conversion Characteristic in Industrial and Power Plant Boiler during Coal Combustion. Master's Thesis, Liaoning Technical University, Fuxin, China, 2005.
23. Prenger, F.C.; Stewart, W.F.; Hill, D.D.; Avens, L.R.; Worl, L.A.; Schake, A.; Aguero, K.J.d.; Padilla, D.D.; Tolt, T.L. *High Gradient Magnetic Separation Applied to Environmental Remediation*; Plenum Press: New York, NY, USA, 1994.
24. Liu, S.; Ye, X.; He, K.; Chen, Y.; Hu, Y. Simultaneous removal of Ni(II) and fluoride from a real flue gas desulfurization wastewater by electrocoagulation using Fe/C/Al electrode. *J. Water Reuse Desalination* **2016**, *7*, 288–297. [CrossRef]
25. Liu, X.F.; Zheng, C.G.; Song, D.Y.; Liu, J.; Zhang, J.Y. Migration behavior and washability of fluorine in Guizhou coal during cleaning process. In Proceedings of the 6th International Symposium on Coal Combustion, Wuhan, China, 1–5 December 2007; pp. 593–599.
26. Liu, D.; Sakai, Y.; Yamamoto, M.; Sadakata, M. Behavior of Fluorine in the Combustion of Chinese Coal in Small Furnaces. *Energy Fuels* **2006**, *20*, 1406–1410. [CrossRef]
27. Zhang, Z.; Yan, G.; Zhu, G.; Zhao, P.; Ma, Z.; Zhang, B. Using microwave pretreatment to improve the high-gradient magnetic-separation desulfurization of pulverized coal before combustion. *Fuel* **2020**, *274*, 117826. [CrossRef]
28. Mohanty, M.K.; Honaker, R.Q.; Mondal, K.; Paul, B.C.; Ho, K. Trace Element Reductions in Fine Coal Using Advanced Physical Cleaning. *Coal Prep.* **1998**, *19*, 195–211. [CrossRef]

29. Bu, X.; Ni, C.; Xie, G.; Peng, Y.; Ge, L.; Sha, J. Preliminary study on foreign slime for the gravity separation of coarse coal particles in a teeter bed separator. *Int. J. Miner. Processing* **2017**, *160*, 76–80. [CrossRef]
30. Galvin, K.P.; Pratten, S.J.; Nicol, S.K. Dense medium separation using a teetered bed separator. *Miner. Eng.* **1999**, *12*, 1059–1081. [CrossRef]
31. Jain, P.K. An analytical approach to explain complex flow in spiral concentrator and development of flow equations. *Miner. Eng.* **2021**, *174*, 107027. [CrossRef]
32. Ma, G.; Bu, X.; Xie, G.; Peng, Y.; Sha, J.; Xia, W.; Wu, E. Comparative Study of Separation Performance of a Spiral and Dense-medium Cyclone on Cleaning Coal. *Int. J. Coal Prep. Util.* **2021**, *41*, 108–116. [CrossRef]
33. Chen, J.; Chu, K.W.; Zou, R.P.; Yu, A.B.; Vince, A.; Barnett, G.D.; Barnett, P.J. How to optimize design and operation of dense medium cyclones in coal preparation. *Miner. Eng.* **2014**, *62*, 55–65. [CrossRef]
34. Napier-Munn, T. The dense medium cyclone—Past, present and future. *Miner. Eng.* **2018**, *116*, 107–113. [CrossRef]
35. Galvin, K.P.; Iveson, S.M. 10-Cleaning of coarse and small coal. In *The Coal Handbook: Towards Cleaner Production*; Osborne, D., Ed.; Woodhead Publishing: Cambridge, UK, 2013; Volume 1, pp. 263–300.
36. Tripathy, A.; Panda, L.; Sahoo, A.K.; Biswal, S.K.; Dwari, R.K.; Sahu, A.K. Statistical optimization study of jigging process on beneficiation of fine size high ash Indian non-coking coal. *Adv. Powder Technol.* **2016**, *27*, 1219–1224. [CrossRef]
37. Bu, X.; Xie, G.; Chen, Y.; Ni, C. The Order of Kinetic Models in Coal Fines Flotation. *Int. J. Coal Prep. Util.* **2017**, *37*, 113–123. [CrossRef]
38. Bu, X.; Zhou, S.; Danstan, J.K.; Bilal, M.; Ul Hassan, F.; Chao, N. Prediction of coal flotation performance using a modified deep neural network model including three input parameters from feed. *Energy Sources Part A Recover. Util. Environ. Eff.* **2022**, 1–13. [CrossRef]
39. Zhou, S.; Wang, X.; Bu, X.; Wang, M.; An, B.; Shao, H.; Ni, C.; Peng, Y.; Xie, G. A novel flotation technique combining carrier flotation and cavitation bubbles to enhance separation efficiency of ultra-fine particles. *Ultrason. Sonochem.* **2020**, *64*, 105005. [CrossRef]
40. Wang, X.; Bu, X.; Ni, C.; Zhou, S.; Yang, X.; Zhang, J.; Alheshibri, M.; Peng, Y.; Xie, G. Effect of scrubbing medium's particle size on scrubbing flotation performance and mineralogical characteristics of microcrystalline graphite. *Miner. Eng.* **2021**, *163*, 106766. [CrossRef]
41. Bu, X.; Zhang, T.; Peng, Y.; Xie, G.; Wu, E. Multi-Stage Flotation for the Removal of Ash from Fine Graphite Using Mechanical and Centrifugal Forces. *Minerals* **2018**, *8*, 15. [CrossRef]
42. Zhang, Q.; Niu, C.; Bu, X.; Bilal, M.; Ni, C.; Peng, Y. Enhancement of Flotation Performance of Oxidized Coal by the Mixture of Laurylamine Dipropylene Diamine and Kerosene. *Minerals* **2021**, *11*, 1271. [CrossRef]
43. Randolph, J.M. Characterizing Flotation Response: A Theoretical and Experimental Comparison of Techniques. Master's Thesis, Virginia Polytechnic Institute and State University, Blacksburg, VA, USA, 1997.
44. Xia, W.; Yang, J.; Liang, C. A short review of improvement in flotation of low rank/oxidized coals by pretreatments. *Powder Technol.* **2013**, *237*, 1–8. [CrossRef]
45. Ni, C.; Bu, X.; Xia, W.; Liu, B.; Peng, Y.; Xie, G. Improving lignite flotation performance by enhancing the froth properties using polyoxyethylene sorbitan monostearate. *Int. J. Miner. Process.* **2016**, *155*, 99–105. [CrossRef]
46. Mao, Y.; Bu, X.; Peng, Y.; Tian, F.; Xie, G. Effects of simultaneous ultrasonic treatment on the separation selectivity and flotation kinetics of high-ash lignite. *Fuel* **2020**, *259*, 116270. [CrossRef]
47. Drzymala, J.; Ahmed, H.A.M. Mathematical equations for approximation of separation results using the Fuerstenau upgrading curves. *Int. J. Miner. Process.* **2005**, *76*, 55–65. [CrossRef]
48. Drzymala, J.; Kowalczuk, P.B.; Oteng-Peprah, M.; Foszcz, D.; Muszer, A.; Henc, T.; Luszczkiewicz, A. Application of the grade-recovery curve in the batch flotation of Polish copper ore. *Miner. Eng.* **2013**, *49*, 17–23. [CrossRef]
49. Bu, X.; Zhang, T.; Chen, Y.; Xie, G.; Peng, Y. Comparative study of conventional cell and cyclonic microbubble flotation column for upgrading a difficult-to-float Chinese coking coal using statistical evaluation. *Int. J. Coal Prep. Util.* **2020**, *40*, 359–375. [CrossRef]
50. Bu, X.; Zhang, T.; Chen, Y.; Peng, Y.; Xie, G.; Wu, E. Comparison of mechanical flotation cell and cyclonic microbubble flotation column in terms of separation performance for fine graphite. *Physicochem. Probl. Miner. Process.* **2018**, *54*, 732–740. [CrossRef]
51. Paul, S.R.; Bhattacharya, S. Size by Size Separation Characteristics of a Coal Cleaning Jig. *Trans. Indian Inst. Met.* **2018**, *71*, 1439–1444. [CrossRef]
52. Chinese Standard GB/T, GBT, GB. Available online: www.ChineseStandard.net (accessed on 14 January 2022).

minerals MDPI

Article

A Kinetic Study on Enhanced Cementation of Gold Ions by Galvanic Interactions between Aluminum (Al) as an Electron Donor and Activated Carbon (AC) as an Electron Mediator in Ammonium Thiosulfate System

Sanghee Jeon [1,*], Sharrydon Bright [2], Ilhwan Park [1], Akuru Kuze [3], Mayumi Ito [1] and Naoki Hiroyoshi [1]

[1] Division of Sustainable Resources Engineering, Faculty of Engineering, Hokkaido University, Sapporo 060-0808, Japan; i-park@eng.hokudai.ac.jp (I.P.); itomayu@eng.hokudai.ac.jp (M.I.); hiroyosi@eng.hokudai.ac.jp (N.H.)

[2] Department of Mining Chemical and Metallurgical Engineering, Faculty of Engineering, University of Zimbabwe, Harare 11334, Zimbabwe; sharrydonbright@frontier.hokudai.ac.jp

[3] Division of Sustainable Resources Engineering, School of Engineering, Hokkaido University, Sapporo 060-0808, Japan; accle.4842@gmail.com

* Correspondence: shjun1121@eng.hokudai.ac.jp; Tel.: +81-11-706-6918

Abstract: The enhanced cementation technique by galvanic interaction of aluminum (Al; electron donor) and activated carbon (AC; electron mediator) to recover gold (Au) ions from the ammonium thiosulfate solution is a promising technique to eliminate the challenges of poor recovery in the system. This study presents the kinetics of Au ion cementation in an ammonium thiosulfate lixiviant as functions of initial Au concentration, size/amount of Al and AC, temperature, and shaking speed. The recovery results basically followed first order kinetics and showed that the cementation rate increased with a higher initial concentration of Au, smaller electron donor size, greater both electron donor and mediator quantity, decrease in temperature, and higher shaking speed in the system, while size of electron mediator did not significantly affect Au recovery.

Keywords: gold; cementation; galvanic interaction; aluminum; activated carbon; kinetics; ammonium thiosulfate solution; mass transfer

check for updates

Citation: Jeon, S.; Bright, S.; Park, I.; Kuze, A.; Ito, M.; Hiroyoshi, N. A Kinetic Study on Enhanced Cementation of Gold Ions by Galvanic Interactions between Aluminum (Al) as an Electron Donor and Activated Carbon (AC) as an Electron Mediator in Ammonium Thiosulfate System. *Minerals* **2022**, *12*, 91. https://doi.org/10.3390/min12010091

Academic Editor: Kenneth N. Han

Received: 1 December 2021
Accepted: 12 January 2022
Published: 13 January 2022

1. Introduction

Climate change is an urgent global issue affecting industries and communities alike. To reduce greenhouse gas emissions for achieving a climate-neutral world, countries worldwide aim to fulfill a Paris Agreement aligned to a target of transitioning to low/zero-carbon energy sources by promoting next-generation vehicles and developing mobility business by the mid-century [1,2]. For the initiatives towards decarbonization, mineral resources/metallurgy fields try to strengthen the exposure to precious metals for supporting electric/hydrogen vehicles, especially gold (Au).

In Au-hydrometallurgy, copper (Cu)-catalyzed ammonium thiosulfate leaching has gained increasing attention as an alternative to the conventional cyanide solvent due to its non-toxicity, low corrosiveness, and high selectivity for Au [3–5]. Although the leaching in this eco-friendly solvent, ammonium thiosulfate, is well established [6–9], the system is, however, commercially underdeveloped so far, because there are few acceptable methods to recover Au ions from the pregnant solutions. The adsorption of Au ions onto activated carbon (AC) has been the mainstay of Au-hydrometallurgy for several decades in cyanide-based lixiviants due to its high efficiency, relatively low cost, and high purity of the products [10]; however, its application is not preferred in the ammonium thiosulfate system. Gallagher with co-authors reported that the effectiveness of AC in adsorbing Au ions from aqueous solutions decreases in the following ligand order:

$SCN^- > SC(NH_2)_2 > CN^- >> S_2O_3{}^{2-}$, indicating that the Au thiosulfate complex does not effectively adsorb onto AC [11], hence the carbon-in-pulp (CIP) or carbon-in-leach (CIL) is not properly applicable to the thiosulfate system. Cementation (i.e., reductive precipitation), a conventional recovery process whereby Au ions are reduced to metallic Au via electron transfer, can be employed by using zero-valent base metals (i.e., cementing agents). Zero-valent copper, zinc (ZVZn), aluminum (Al), and iron (ZVI) are reasonable choices in cyanide-based lixiviants [10,12–14] but their application to thiosulfate solution is difficult because of the dissolution of the cementing agents (i.e., Cu and Zn) or the formation of the oxide/sulfide layers on the cementing agents (i.e., Al and Fe). The abundant sulfur and Cu ions in the solution also restrict the application of solvent extraction and electrowinning as well due to the contamination of the products and the increased energy requirements [10].

A previous study of the authors developed a new recovery technique, which uses the galvanic interactions between Al as an electron donor and AC as an electron mediator to the Au thiosulfate complex [15]. The results showed that when only Al or AC was employed to recover Au ions from the ammonium thiosulfate solution, the recovery of Au in that system was negligible. On the other hand, when both Al and AC were employed in the recovery process, over 99% of Au ions could be recovered under the following conditions [15]: 0.15 g Al and 0.15 g of AC with a solution containing 1 M of $Na_2S_2O_3$, 0.5 M of NH_4OH, 0.25 M of $(NH_4)_2SO_4$ and 10 mM of $CuSO_4$ (pH between 9.5–10) with 100 mg/L of Au ions at 25 °C for 24 h with oxygen condition. Although the possibility of Au ion recovery from the ammonium thiosulfate medium by this simple and highly efficient technique is well established, the kinetics considering the various parameters that affect Au ion recovery (i.e., initial Au concentration, size and amount ratio of Al and AC, temperature, and shaking speed) have not been studied to date. Furthermore, there are many kinetic studies on cementation of Au ions by a single cementation agent (e.g., Al, Zn, Cu or Fe) in an ammonium thiosulfate system [16,17], whereas there remains some uncertainty regarding the cementation of Au ions enhanced by galvanic interaction between two materials. A better understanding of the Al and AC galvanic interaction on recovery of Au ions is essential before its application to industrial mining, and this will lead to better design on an industrial scale for Au mining employing ammonium thiosulfate solvent as an eco-friendly solution.

To this end, the present study investigated the kinetics of enhanced Au ions cementation by galvanic interaction between Al and AC in ammonium thiosulfate lixiviant with batch-type experiments as functions of initial Au concentration, size of Al and AC particles, their mixing ratio, and temperature as well as shaking speed, and a morphology study on the cemented Au was also carried out.

2. Materials and Methods

The ammonium thiosulfate solutions containing Au ions (i.e., Au-ammonium thiosulfate solution) were prepared by dissolving Au powder (99.999%, Wako Pure Chemical Industries, Ltd., Osaka, Japan) in ammonium thiosulfate solution containing 1 M of $Na_2S_2O_3$, 0.5 M of NH_3, 0.25 M of $(NH_4)_2SO_4$ and 10 mM of $CuSO_4$ (pH between 9.5 and 10) by a thermostat water bath shaker at 25 °C for 24 h with constant shaking amplitude and frequency of 40 mm and 120 min^{-1}, respectively.

The recovery of Au ions was carried out using a mixture of Au-ammonium thiosulfate solution and Al (99.99%, Wako Pure Chemical Industries, Ltd., Osaka, Japan) with AC (99.99%, Wako Pure Chemical Industries, Ltd., Osaka, Japan) in 50-mL Erlenmeyer flasks at 25 °C for 1 h under the nitrogen purging condition at thermostat water bath shaker (shaking amplitude of 40 mm and frequency of 120 min^{-1}). Subsequently, the filtrate and the residue were separated, the latter was washed with deionized (DI) water (18 MΩ·cm, Mill-Q® Integral Water Purification System, Merck Millipore, Burlington, MA, USA), dried in a vacuum oven at 40 °C for 24 h, and analyzed by scanning electron microscopy with energy dispersive X-ray spectroscopy (SEM-EDX, Superscan SSX-550, Shimadzu Corporation,

Kyoto, Japan). The filtrate was analyzed by inductively coupled plasma atomic emission spectroscopy (ICP-AES, ICPE-9820, Shimadzu Corporation, Japan) (margin of error = ±2%).

3. Results and Discussion

3.1. Recovery of Au Ions with Varying Initial Gold Concentrations

The cementation of Au ions from the ammonium thiosulfate solution with varying initial concentrations of Au (i.e., 1, 10, 50, 100 mg/L) was examined using the Al/AC mixture. The standard conditions were as follows: 10 mL solution containing 1 M of $Na_2S_2O_3$, 0.5 M of NH_3, 0.25 M of $(NH_4)_2SO_4$ and 10 mM of $CuSO_4$ (pH between 9.5 and 10) with 100, 50, 10, or 1 mg/L of Au, and 0.3 g mixture (0.15 g Al (−45 μm) + 0.15 g AC (−45 μm)) at 25 °C. In this recovery system, Al likely acted as the primary electron donor (i.e., anode) and the attached AC served as an electron mediator from Al to Au-thiosulfate complex ($Au(S_2O_3)_2^{3-}$), a configuration that promoted both galvanic interactions and Au recovery, cementation [15]. Generally, the cementation of Au ions has been found to obey first-order kinetics with the rate controlled by mass transfer of metal ions in the solution phase [14,16,18–20]. As reported in the previous study relating to the present system, Au was generally cemented on AC attached to Al and it exists with Cu in the same area (deposition order: Au-Cu-AC-Al), indicating that the cementation reaction rate is expected to be a function of the surface area of Al according to the following first-order expression:

$$\ln([Au]_t/[Au]_0) = -kAt/V, \quad (1)$$

where $[Au]_t$ is the Au concentration at time t (mg/L), $[Au]_0$ is the initial Au concentration (mg/L), k is the rate constant (cm/s), A is initial surface area of substrate (electron donor: Al)) (cm^2), t is time (s), and V is a solution volume (cm^3).

Figure 1a shows the variations for different initial Au concentrations in the solution with time, and the slope of $\ln([Au]_t/[Au]_0)$ vs. time indicates the reaction constant. The results showed that, as many researchers have mentioned, the cementation of Au ions using single materials (e.g., Cu or Zn) followed first-order kinetics [16,20]. The Au cementation in the present system using galvanic interaction of Al and AC also obeyed first-order kinetics and showed 83.7% of the recovery at 30 min for 1 mg/L of initial Au concentration, while the recovery reached up to about 99.8% of the recovery at 30 min for 100 mg/L of initial Au concentration, indicating that the rate increases with increasing the initial concentration of Au ions in the solution. The reaction rates of Au cementation were in the following order: 100 mg/L > 50 mg/L > 10 mg/L > 1 mg/L of initial Au concentration.

To calculate the rate constant, the surface area of the substrate (A) is required, which was calculated based on the following assumption: the shape of the Al is a sphere, and all the precipitates are present on Al. The calculated rate constants of Au ions cementation onto Al−AC are (cm/s): $k_{100\ mg/L} = 2.08 \times 10^{-4}$, $k_{50\ mg/L} = 1.61 \times 10^{-4}$, $k_{10\ mg/L} = 1.19 \times 10^{-4}$, and $k_{1\ mg/L} = 6.05 \times 10^{-5}$ (Table 1), and the reaction rate also showed a positive correlation with initial Au concentration as shown in Figure 1a,b.

Table 1. Summary of rate constant for the cementation of Au onto the Al as a function of initial Au concentration (base conditions: 10 mL solution containing 1 M of $Na_2S_2O_3$, 0.5 M of NH_3, 0.25 M of $(NH_4)_2SO_4$ and 10 mM of $CuSO_4$, and 0.3 g mixture (0.15 g Al (−45 μm) and 0.15 g AC (−45 μm)) at 25 °C).

Initial Au Conc.	Al Area (cm^2)	Rate Constant (cm/s)	Reaction Rate (ppm/s)	R^2
1 ppm	166.7	6.05×10^{-5}	-1.08×10^{-1}	0.9996
10 ppm	166.7	1.19×10^{-4}	-2.11×10^{-1}	0.9997
50 ppm	166.7	1.61×10^{-4}	-2.87×10^{-1}	0.9997
100 ppm	166.7	2.08×10^{-4}	-3.70×10^{-1}	0.9998

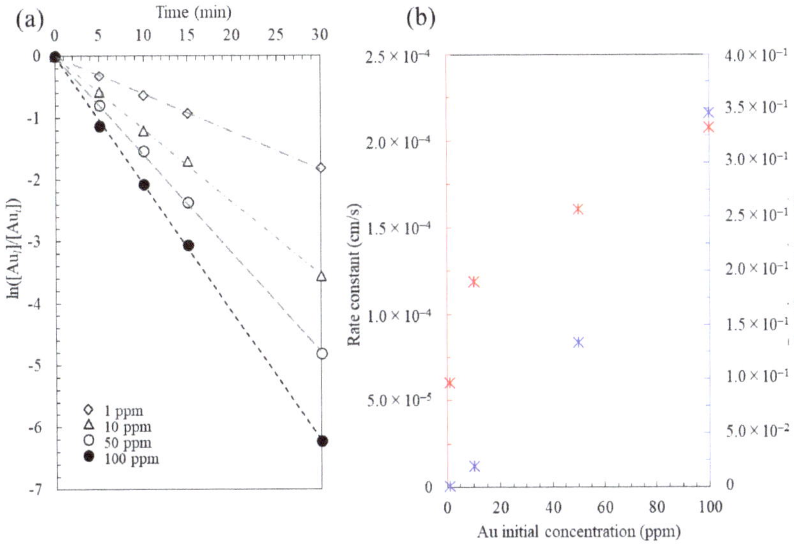

Figure 1. (**a**) The effects of initial Au concentration on the recovery of Au ions, and (**b**) the plot of rate constant/reaction rate vs. initial Au concentration for the cementation reaction by galvanic interaction of Al and AC in the ammonium thiosulfate system (Note that reaction rate has a minus value).

3.2. Recovery of Au Ions with Varying Size of Al

Recovery of Au ions from ammonium thiosulfate solution using different sizes of the electron donor Al, and a constant size of AC was examined under the following conditions: 1 M of $Na_2S_2O_3$, 0.5 M of NH_3, 0.25 M of $(NH_4)_2SO_4$ and 10 mM of $CuSO_4$ with 100 mg/L of Au, and 0.15 g of Al (size: −45, +45−75, +75−106, +106−150 μm) and 0.15 g of AC (size: −45 μm) at 25 °C. The results are expected to be a function of the surface area of the Al according to the first order, as illustrated in Equation (1). The results evidently showed that the smaller the size of Al particles, the faster the reaction rate, indicating that the reaction rate was related to the surface area of electron donor, Al, for the electron transfer (Figure 2a,b). For the −45 μm size of Al powder, the reduction of Au ions onto the Al−AC obeys first-order kinetics, and ~99.8% Au recovery was achieved for 30 min. As the range of Al particle size increased to +106−150 μm, the cementation reaction also obeyed first-order kinetics and had a recovery up to about 98.4%, which is slightly decreased compared to the result using the −45 μm size of Al. The rate constants (k) with different sizes of Al, i.e., −45, +45−75, +75−106, +106−150 μm, were calculated to be 2.08×10^{-4}, 2.45×10^{-4}, 6.82×10^{-4}, and 8.86×10^{-4} cms−1, respectively (Table 2). As the Al particle size increased, the rate constant indicating the rate at which cementation occurs per unit area, also increased, while the total cementation rate decreased (Figure 2b). The results suggested that as the surface area of the electron donor increased, this could more readily transfer the electrons from the Al−AC to the Au-thiosulfate complex ($Au(S_2O_3)_2^{3-}$) in an ammonium thiosulfate medium, while the cementation reaction rate per unit area decreased. Similarly, Hiskey and Lee [16] performed Au cementation from a thiosulfate solution using a different size of cementing agent—copper powder (150 × 200 and −200 mesh)—and reported that as the particle size of the cementing agent increased, the cementation rate of Au ions decreased. The SEM-EDX results showed that particles located in an area brighter than the background (i.e., the surface of the Al) present Au and Cu [15], and the larger the surface area, the more Au was noticeably recovered onto the Al surface (inset, Figure 2a).

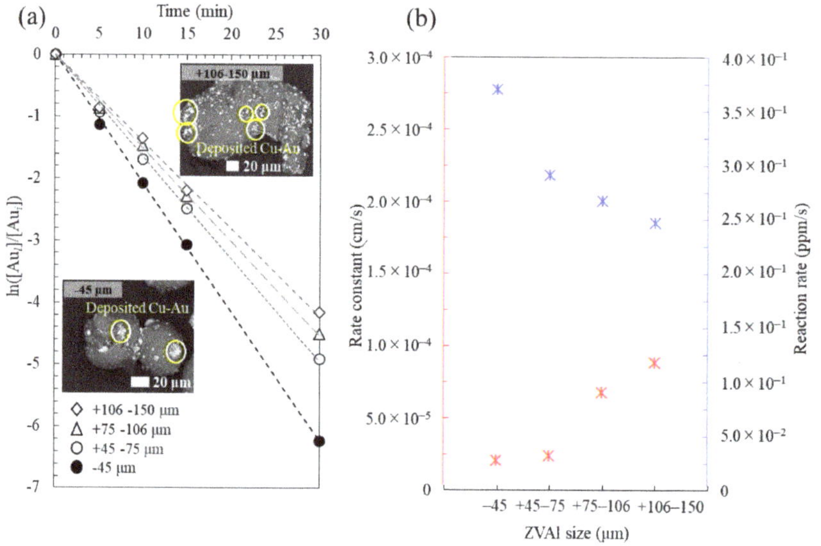

Figure 2. (**a**) The effects of Al particle size (i.e., −45, +45−75, +75−106, +106−150 µm) on recovery of Au ions with the SEM photomicrographs, and (**b**) the plot of rate constant/rate vs Al size for the cementation reaction by galvanic interaction of Al and AC in an ammonium thiosulfate system (Note that reaction rate has a minus value).

Table 2. Summary of rate constant for the cementation of Au onto the Cu-AC-Al as a function of electron donor (Al) size (base conditions: 10 mL solution containing 1 M of $Na_2S_2O_3$, 0.5 M of NH_3, 0.25 M of $(NH_4)_2SO_4$ and 10 mM of $CuSO_4$, and 0.3 g mixture (0.15 g Al and 0.15 g AC (−45 µm)) at 25 °C).

Al Size	Al Area (cm^2)	Rate Constant (k, cm/s)	Reaction Rate (ppm/s)	R^2
−45 µm	166.7	2.08×10^{-4}	-3.70×10^{-1}	0.9998
+45−75 µm	111.1	2.45×10^{-4}	-2.92×10^{-1}	0.9996
+75−106 µm	36.63	6.82×10^{-4}	-2.68×10^{-1}	0.9994
+106−150 µm	26.04	8.86×10^{-4}	-2.47×10^{-1}	0.9985

3.3. Recovery of Au Ions with Varying Size of AC

Recovery of Au ions from an ammonium thiosulfate solution using different sizes of the electron mediator, AC, and a constant size of Al particles was examined under the following conditions: 1 M of $Na_2S_2O_3$, 0.5 M of NH_3, 0.25 M of $(NH_4)_2SO_4$ and 10 mM of $CuSO_4$ with 100 mg/L of Au, and 0.15 g of Al (size: −45 µm) and 0.15 g of AC (size: −45 µm, +0.2−0.5 mm, +1.0−2.0 mm, +4.0−5.0 mm) at 25 °C. The results continued to exhibit first-order kinetics and showed similar results to those of the previous section; that is, the cementation rate using the smallest particle size of AC (−45 µm) showed the fastest reaction rate (Figure 3). Although, as the particle size of AC was increased, the reaction rate seemed to be less responsive, indicating that the effects of electron mediator size are not so significant in the system. For the −45 µm size of AC, the recovery showed up to about 99.8% at 30 min, while for the upper size ranges of AC, the recovery irregularly showed about 99.4−99.5%. This can be explained by the fact that once the activated carbon was added in the system, small AC particles (about 0.2−2 µm) detached from the original AC, became attached to the surface of Al [15], and acted as an electron mediator, making the original particle size of the AC not so important to the reaction rate. The residue analysis results also showed that there are no significant differences between −45 μm and +4.0−5.0 mm size of AC. The calculated rate constants of Au ions cementation onto Al−AC are (cms−1): $k_{-45\ \mu m} = 2.08 \times 10^{-4}$, $k_{+0.2-0.5\ mm} = 1.79 \times 10^{-4}$, $k_{+1.0-2.0\ mm} = 1.71 \times 10^{-4}$, and $k_{+4.0-5.0\ mm} = 1.76 \times 10^{-4}$ (Table 3).

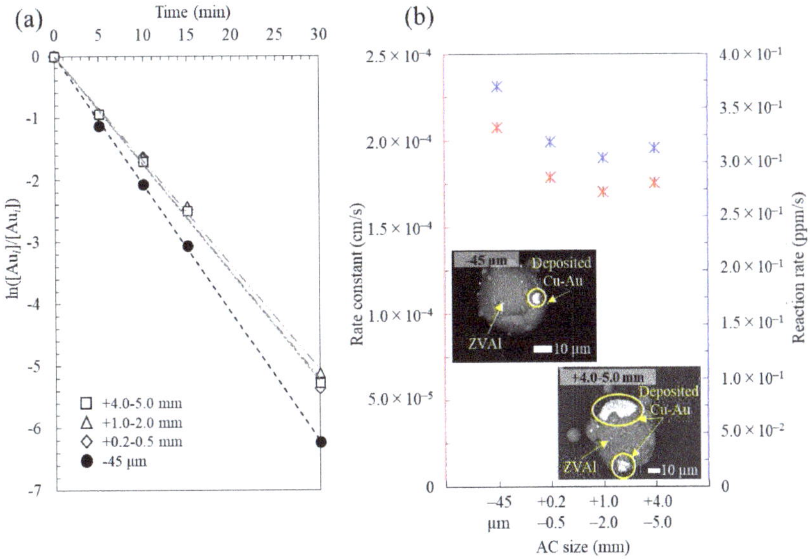

Figure 3. (**a**) The effects of AC size on the recovery of Au ions, and (**b**) the plot of rate constant/rate vs AC size for the cementation reaction by galvanic interaction of Al and AC in ammonium thiosulfate system (Note that reaction rate has a minus value).

Table 3. Summary of rate constant for the cementation of Au onto the Cu-AC-Al as a function of AC size (base conditions: 10 mL solution containing 1 M of $Na_2S_2O_3$, 0.5 M of NH_3, 0.25 M of $(NH_4)_2SO_4$ and 10 mM of $CuSO_4$, and 0.3 g mixture (0.15 g Al and 0.15 g AC) at 25 °C).

AC Size	Al Area (cm^2)	Rate Constant (k, cm/s)	Reaction Rate (ppm/s)	R^2
−45 μm	166.7	2.08×10^{-4}	-3.67×10^{-1}	0.9998
+0.2−0.5 mm	166.7	1.79×10^{-4}	-3.19×10^{-1}	0.9986
+1.0−2.0 mm	166.7	1.71×10^{-4}	-3.04×10^{-1}	0.9993
+4.0−5.0 mm	166.7	1.76×10^{-4}	-3.13×10^{-1}	0.9995

3.4. *Recovery of Au Ions with Varying Quantity of Al and AC*

Subsequently, the effects of Al quantity (0.01, 0.05, 0.1, 0.15 g) and AC quantity (0.01, 0.05, 0.1, 0.15 g) on the kinetics of Au ion cementation onto AC−Al are presented in Figure 4a−d under the following conditions: 1 M of $Na_2S_2O_3$, 0.5 M of NH_3, 0.25 M of $(NH_4)_2SO_4$ and 10 mM of $CuSO_4$ with 100 mg/L of Au, and both −45 μm particle size of Al and AC at 25 °C. As shown in Figure 4a,c, the Au cementation by galvanic interaction obeys first-order kinetics, and the quantity of Al and AC affects cementation of Au ions; especially, the cementation of Au ions rapidly increased even when only a small amount of Al was present together with AC in the system. For the mixture of 0.01 g AC and 0.15 g of Al (Figure 4c), the reduction of Au ions was about 97.8% recovery at 30 min, while for the mixture of 0.15 g AC and 0.01 g of Al (Figure 4a), the recovery showed about 99.4% at 30 min. Meanwhile, the maximum reduction of Au was achieved with 0.15 g Al and

0.15 g AC where 99.8% of Au was recovered. It can be explained by the fact that electron transfer is readily enabled when the electron donor quantity increases, leading to high Au recovery, and this suggests that the electron donor quantity is a more pronounced mediator in cementation of Au ions in this galvanic system

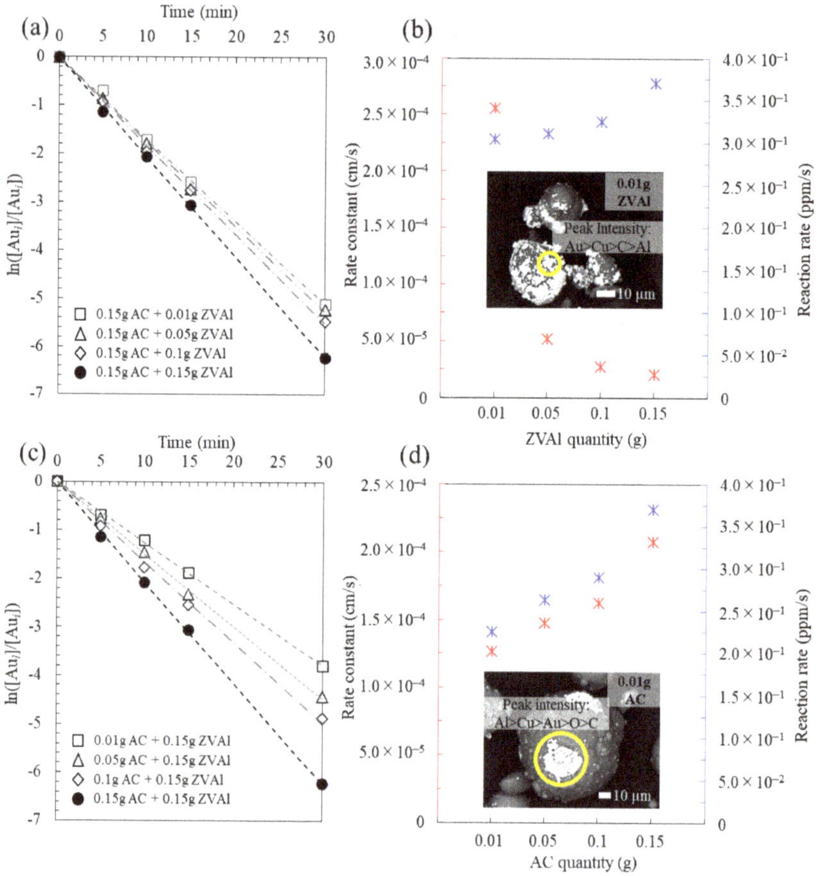

Figure 4. The effects of (**a**) Al and (**b**) AC quantity on recovery of Au ions, and the plot of rate constant/rate vs (**c**) Al and (**d**) AC quantity for the cementation reaction by galvanic interaction of Al and AC in an ammonium thiosulfate system (Note that reaction rate has a minus value).

The rate constant in the system for the different quantity of Al (i.e., 0.15 g AC + 0.01, 0.05, 0.1, or 0.15 g of Al) were calculated to be 2.56×10^{-4}, 5.22×10^{-4}, 2.74×10^{-4}, and 2.08×10^{-4} cms−1, while for the different quantity of AC (i.e., 0.01, 0.05, 0.1, or 0.15 g of AC + 0.15 g of Al), the calculated rate constants were 1.27×10^{-4}, 1.48×10^{-4}, 1.63×10^{-4}, and 2.08×10^{-4} cms−1, respectively (Table 4). The insets in Figure 4b,d are the back-scattered electron (BSE) photomicrographs of the residues obtained with 0.15 g AC + 0.01 g Al (inset, Figure 4b) and 0.01 g AC + 0.15 g Al (inset, Figure 4d) after 30 min cementation, respectively. The EDX point analysis of the residue with 0.15 g AC + 0.01 g Al (inset, Figure 4b) had strong signals of Au, Cu, AC and Al, while the residue with 0.01 g AC + 0.15 g Al (inset, Figure 4d) had Al and Cu as major signals and Au and C as minor signals, both of which were relatively lower than those observed at point in Figure 4b. This supports the aforementioned explanation of the momentous role of the electron donor in creating a high and fast cementation rate for Au ions by the galvanic interactions between Al and AC in an ammonium thiosulfate system.

Table 4. Summary of rate constant for the cementation of Au onto the Cu-AC-Al as a function of Al and AC quantity (base conditions: 10 mL solution containing 1 M of $Na_2S_2O_3$, 0.5 M of NH_3, 0.25 M of $(NH_4)_2SO_4$ and 10 mM of $CuSO_4$, and 0.3 g mixture (Al and AC) at 25 °C).

Al Quantity	Al Area (cm^2)	Rate Constant (k, cm/s)	Reaction Rate (ppm/s)	R^2
0.01 g	11.11	2.56×10^{-4}	-3.04×10^{-1}	0.9993
0.05 g	55.56	5.22×10^{-4}	-3.11×10^{-1}	0.9997
0.1 g	111.1	2.74×10^{-4}	-3.25×10^{-1}	0.9996
0.15 g	166.7	2.08×10^{-4}	-3.70×10^{-1}	0.9998
AC quantity	**Al area (cm^2)**	**Rate constant (k, cm/s)**	**Reaction rate (ppm/s)**	**R^2**
0.01 g	166.7	1.27×10^{-4}	-2.26×10^{-1}	0.9996
0.05 g	166.7	1.48×10^{-4}	-2.64×10^{-1}	0.9995
0.1 g	166.7	1.63×10^{-4}	-2.90×10^{-1}	0.9991
0.15 g	166.7	2.08×10^{-4}	-3.70×10^{-1}	0.9998

3.5. *Recovery of Au Ions with Varying Temperature*

The conditions of the recovery experiments under varying temperature were the same as the standard conditions as follows: 1 M of $Na_2S_2O_3$, 0.5 M of NH_3, 0.25 M of $(NH_4)_2SO_4$ and 10 mM of $CuSO_4$ with 100 mg/L of Au, and both −45 μm particle size of Al and AC. Figure 5 shows the effects of temperature on the kinetics at values ranging from 25 °C to 50 °C. The results obey first-order kinetics, and the rate were observed to decrease with increase in temperature under the current condition: For the reaction temperature at 25 °C, the reduction of Au ions was about 99.8%, while it showed 95.8% and 93.7% of the recovery at 35 °C and 50 °C, respectively. This is understandable for the following reasons: In general, cementation results show positive proportional increase with temperature [21]. The present cementation results by the galvanic interaction between Al and AC, however, showed that reaction rate decreased as temperature increased. This can be explained by the fact that there was an improved formation of the oxide layer on the aluminum surface [22]: Wahab et al. (2013) mentioned that the formation of the insulating film, i.e., oxyhydroxide layer on the surface of Al (electron donor) can be enhanced as the temperature increases, which contributes greatly to electrochemical reactions, i.e., cementation [22]. The rate constant in the system for different temperatures (i.e., 25, 35, and 50 °C) were calculated to be 2.06×10^{-4}, 1.06×10^{-4}, and 9.24×10^{-5}, respectively (Table 5).

Table 5. Summary of rate constant for the cementation of Au as a function of temperature (base conditions: 10 mL solution containing 1 M of $Na_2S_2O_3$, 0.5 M of NH_3, 0.25 M of $(NH_4)_2SO_4$ and 10 mM of $CuSO_4$, and 0.3 g mixture (0.15 g Al and 0.15 g AC) at 25, 35, and 50 °C).

Temperature	Al Area (cm^2)	Rate Constant (k, cm/s)	Reaction Rate (ppm/s)	R^2
25 °C	166.7	2.08×10^{-4}	-3.67×10^{-1}	0.9998
35 °C	166.7	1.06×10^{-4}	-1.88×10^{-1}	0.9912
50 °C	166.7	9.24×10^{-5}	-1.65×10^{-1}	0.9987

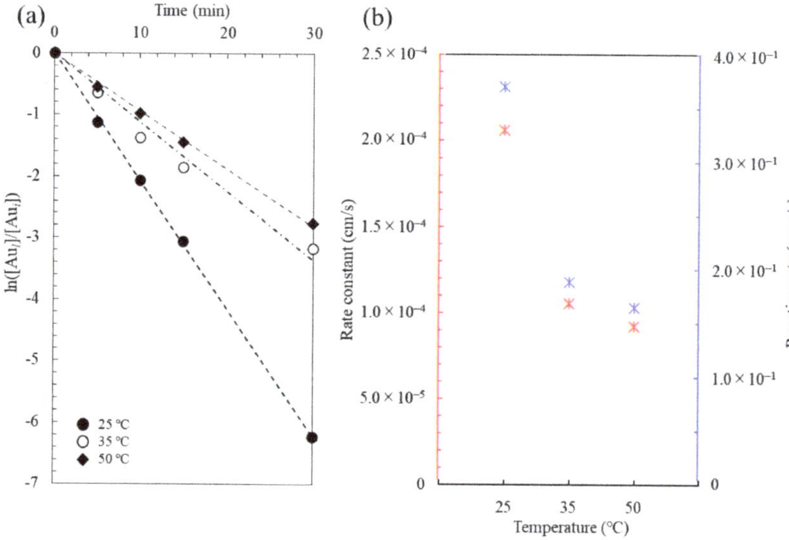

Figure 5. (**a**) The effects of temperature on recovery of Au ions, and (**b**) the plot of rate constant/rate vs temperature for the cementation reaction by galvanic interaction of Al and AC in ammonium thiosulfate system (Note that reaction rate has a minus value.

3.6. Recovery of Au Ions with Varying Shaking Speed

Kinetic results are presented in Figure 6 for ln($[Au]_t/[Au]_0$) as a function of time for shaking speeds from 80 to 200 rpm. The basic experimental conditions were as follows: 1 M of $Na_2S_2O_3$, 0.5 M of NH_3, 0.25 M of $(NH_4)_2SO_4$ and 10 mM of $CuSO_4$ with 100 mg/L of Au, and 0.15 g of Al and AC (both sizes: −45 μm) at 25 °C. The results also obeyed first-order kinetics, and the reaction rate increased as the shaking speed increased: the recovery results showed that about 94.9% of Au was recovered at 80 rpm for 30 min and it increased to 99.9% at 160 rpm. The rate constant (k) at 80, 120, and 160 rpm were calculated to be 0.99×10^{-4}, 2.08×10^{-4}, and 2.19×10^{-4}, respectively (Table 6).

Table 6. Summary of rate constant for the cementation of Au as a function of shaking speed (base conditions: 10 mL solution containing 1 M of $Na_2S_2O_3$, 0.5 M of NH_3, 0.25 M of $(NH_4)_2SO_4$ and 10 mM of $CuSO_4$, and 0.3 g mixture (0.15 g Al and 0.15 g AC) at 25°C).

Shaking Speed	Al Area (cm^2)	Rate Constant (k, cm/s)	Reaction Rate (ppm/s)	R^2
80 rpm	166.7	0.99×10^{-4}	-1.13×10^{-1}	0.9035
120 rpm	166.7	2.08×10^{-4}	-3.67×10^{-1}	0.9998
160 rpm	166.7	2.19×10^{-4}	-5.14×10^{-1}	0.9919

Finally, the rate constant and mass transfer coefficient values based on the recovery results in the present study were plotted in Figure 7, and clearly showed a positive correlation, especially as a function of electron size, quantity, and shaking speed compared to the function of electron mediator size, quantity and temperature, which indicates that the reactions were controlled by mass transfer. Guerra and Dreisinger (1999) and Hiskey and Lee (2003) who researched the kinetic study of Au(I) thiosulfate complex were also concluded that the cementation reactions were controlled by mass transfer [6,16].

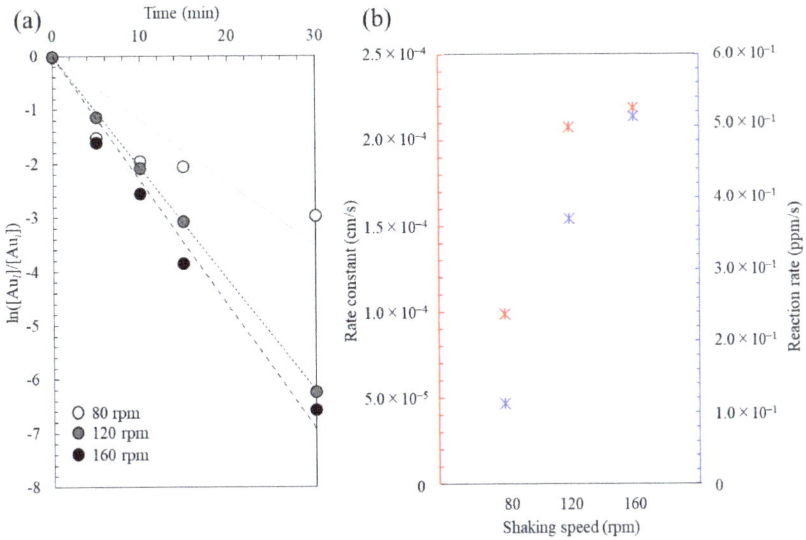

Figure 6. (**a**) The effects of shaking speed on recovery of Au ions, and (**b**) the plot of rate constant/rate vs shaking speed for the cementation reaction by galvanic interaction of Al and AC in ammonium thiosulfate system (Note that reaction rate has a minus value).

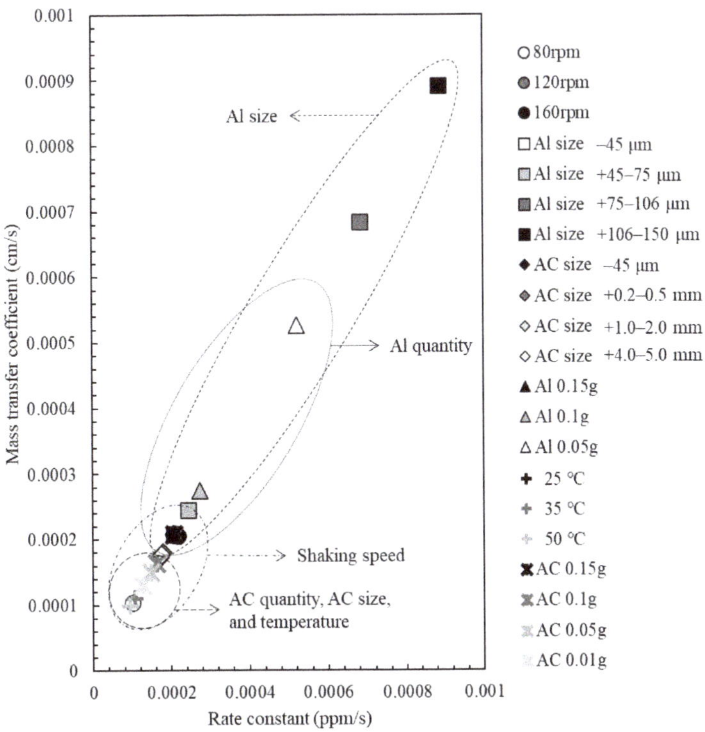

Figure 7. The plot of rate constant and mass transfer coefficient based on all recovery results in the present study.

4. Conclusions

This study described the kinetics of enhanced cementation of Au ions by galvanic interaction between Al (i.e., electron donor) and AC (i.e., electron mediator) as functions of various parameters (i.e., initial Au concentration, Al and AC size, Al and AC quantity, temperature, and shaking speed) in an ammonium thiosulfate system. The results showed that the Au cementation by Al and AC basically followed first order kinetics, and the rate increased with an increase in initial Au concentration and Al/AC quantity, as well as a decrease in Al size (i.e., surface area ↑), temperature, and shaking speed while AC size did not have significant effects on the reaction rate. These results could give a better understanding of the Au recovery technique using Al and AC before it to is applied to industrial mining, and this will help to establish a better design on an industrial scale for Au mining employing an eco-friendly solution.

Author Contributions: Conceptualization, S.J., I.P., M.I. and N.H.; methodology, S.J. and N.H.; investigation, S.J., S.B. and A.K.; writing—original draft preparation, S.J.; writing—review and editing, S.J., I.P., M.I. and N.H.; project administration, S.J., I.P., M.I. and N.H.; funding acquisition, S.J. All authors have read and agreed to the published version of the manuscript.

Funding: This study was financially supported by Japan Oil, Gas and Metals National Corporation (JOGMEC, grant number: PC2520A080).

Data Availability Statement: Data available on request due to restrictions, as the research is ongoing.

Conflicts of Interest: The authors declare no conflict of interest.

References

1. The Paris Agreement. Available online: https://unfccc.int/process-and-meetings/the-paris-agreement/the-paris-agreement (accessed on 22 November 2021).
2. Park, I.; Kanazawa, Y.; Sato, N.; Galtchandmani, P.; Jha, M.K.; Tabelin, C.B.; Jeon, S.; Ito, M.; Hiroyoshi, N. Beneficiation of Low-Grade Rare Earth Ore from Khalzan Buregtei Deposit (Mongolia) by Magnetic Separation. *Minerals* **2021**, *11*, 1432. [CrossRef]
3. Arima, H.; Fujita, T.; Uen, W.T. Using Nickel as a Catalyst in Ammonium Thiosulfate Leaching for Gold Extraction. *Mater. Trans.* **2004**, *45*, 516–526. [CrossRef]
4. Tabelin, C.B.; Park, I.; Phengsaart, T.; Jeon, S.; Tabelin, M.V.; Alonozo, D.; Yoo, K.; Ito, M.; Hiroyoshi, N. Copper and critical metals production from porphyry ores and E-wastes: A review of resource availability, processing/recycling challenges, socio-environmental aspects, and sustainability issues. *Resour. Conserv. Recycl.* **2021**, *170*, 105610. [CrossRef]
5. Jeon, S.; Ito, M.; Tabelin, C.B.; Pongsumrankul, R.; Kitajima, N.; Park, I.; Hiroyoshi, N. Gold recovery from shredder light fraction of E-waste recycling plant by flotation-ammonium thiosulfate leaching. *Waste Manag.* **2018**, *77*, 195–202. [CrossRef]
6. Guerra, E.; Dreisinger, D.B. A study of the factors affecting copper cementation of gold from ammoniacal thiosulphate solution. *Hydrometallurgy* **1999**, *51*, 155–172. [CrossRef]
7. Jeffrey, M.I.; Brunt, S.D. The quantification of thiosulfate and polythionates in gold leach solutions and on anion exchange. *Hydrometallurgy* **2007**, *89*, 52–60. [CrossRef]
8. Jeon, S.; Tabelin, C.B.; Park, I.; Nagata, Y.; Ito, M.; Hiroyoshi, N. Ammonium thiosulfate extraction of gold from printed circuit boards (PCBs) of end-of-life mobile phones and its recovery from pregnant leach solution by cementation. *Hydrometallurgy* **2020**, *191*, 105214. [CrossRef]
9. Lampinen, M.; Laari, A.; Turunen, I. Ammoniacal thiosulfate leaching of pressure oxidized sulfide gold concentrate with low reagent consumption. *Hydrometallurgy* **2015**, *151*, 1–9. [CrossRef]
10. Grosse, A.C.; Dicinoski, G.W.; Shaw, M.J.; Haddad, P.R. Leaching and recovery of gold using ammoniacal thiosulfate leach liquors (a review). *Hydrometallurgy* **2003**, *69*, 1–21. [CrossRef]
11. Gallagher, N.P.; Hendrix, J.L.; Milosavljevic, E.B.; Nelson, J.H.; Solujic, L. Affinity of activated carbon towards some gold(I) complexes. *Hydrometallurgy* **1990**, *25*, 305–316. [CrossRef]
12. Kenna, C.C.; Ritchie, I.M.; Singh, P. The cementation of gold by iron from cyanide solutions. *Hydrometallurgy* **1990**, *23*, 263–279. [CrossRef]
13. Miller, J.D.; Wan, R.Y.; Parga, J.R. Characterization and electrochemical analysis of gold cementation from alkaline cyanide solution by suspended zinc particles. *Hydrometallurgy* **1990**, *24*, 373–392. [CrossRef]
14. Nguyen, H.H.; Tran, T.; Wong, P.L.M. A kinetic study of the cementation of gold from cyanide solutions onto copper. *Hydrometallurgy* **1997**, *46*, 55–69. [CrossRef]
15. Jeon, S.; Tabelin, C.B.; Takahashi, H.; Park, I.; Ito, M.; Hiroyoshi, N. Enhanced cementation of gold via galvanic interactions using activated carbon and zero-valent aluminum: A novel approach to recover gold ions from ammonium thiosulfate medium. *Hydrometallurgy* **2020**, *191*, 105165. [CrossRef]

16. Hiskey, J.; Lee, J. Kinetics of gold cementation on copper in ammoniacal thiosulfate solutions. *Hydrometallurgy* **2003**, *69*, 45–56. [CrossRef]
17. Karavasteva, M. Kinetics and deposit morphology of gold cemented on magnesium, aluminum, zinc, iron, and copper from ammonium thiosulfate-ammonia solutions. *Hydrometallurgy* **2010**, *104*, 119–122. [CrossRef]
18. Dönmez, B.; Sevim, F.; Saraç, H. A kinetic study of the cementation of copper from sulphate solutions onto a rotating aluminum disc. *Hydrometallurgy* **1999**, *53*, 145–154. [CrossRef]
19. Oo, M.T.; Tran, T. The effect of lead on the cementation of gold by zinc. *Hydrometallurgy* **1991**, *26*, 61–74. [CrossRef]
20. Zhang, H.G.; Doyle, J.A.; Kenna, C.C.; La Brooy, S.R.; Hefter, G.T.; Ritchie, I.M. A kinetic and electrochemical study of the cementation of gold onto mild steel from acidic thiourea solutions. *Electrochim. Acta* **1996**, *41*, 389–395. [CrossRef]
21. Demirkiran, N.; Ekmekyapar, A.; Kunkul, A.; Baysar, A. A kinetic study of copper cementation with zinc in aqueous solutions. *Int. J. Miner. Process.* **2007**, *82*, 80–85. [CrossRef]
22. Wahab, J.A.; Derman, M.N.; Daud, Z.C. The effects of electrolyte temperature on formation of porous aluminum oxide films in anodizing process. *Adv. Environ. Biol.* **2013**, *7*, 3708–3713.

X-ray powder diffraction (XRD) in combination with the Rietveld method was used to determine the phases in the FNS. The BGMN program was applied for Rietveld refinement [33]. Figure 1 shows the mineralogical compositions of the sample determined by Rietveld refinement. The R_{wp} value was 7.88%. Forsterite is the main component with a small amount of calcite and quartz with values of 1.41% and 0.08%. In addition, the low background of the XRD pattern indicates that the ferronickel slag mineral has a large crystallinity, which makes the process of ore dissolution more difficult.

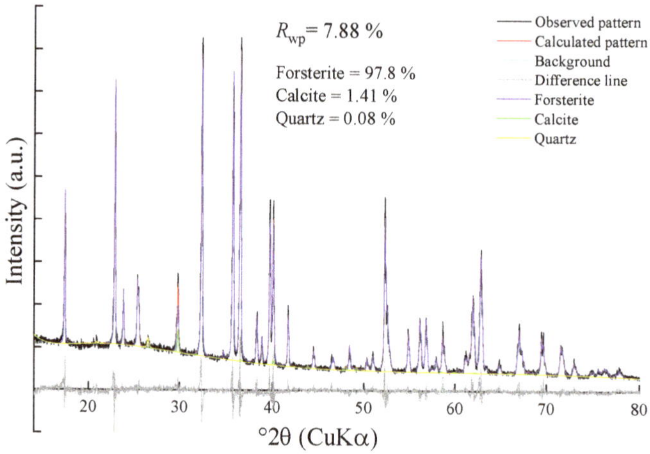

Figure 1. XRD pattern of ferronickel slag.

2.2. Methods

2.2.1. Sulfuric Acid Leaching

Ferronickel slag was first broken into fine particles with a size less than 2 mm by double teeth roll crusher then the uniformly mixed fine particles were ground through a sampling machine, and the collected powder was used as the raw sample used for the primary leaching experiment. In the leaching experiments, a specific concentration of sulfuric acid solution was poured into a conical flask, which was placed in a constant temperature water bath and heated to a specified temperature. Then, 20 g of FNS samples was soaked in sulfuric acid with a specific liquid-solid ratio (L/S) and magnetically stirred at a specified speed for a time while keeping the flask sealed. After the reaction was completed, the obtained mixture was carefully filtered with a vacuum filter, and the leaching residue was washed twice with hot sulfuric acid solution and thoroughly washed with deionized water until no SO_4^{2-} was detected in the filtrate (tested by $BaCl_2$). The filtrate was collected for the subsequent experiment, while the residue was dried at 100 °C for 2 h, and the mass of the dried residue was measured. According to the China industry standard-named determination of 22 metal elements in solid waste (HJ 781-2016), the dried residue was completely digested on the electric heating plate in the mixed acid solution, and the Mg content was measured by atomic absorption spectrophotometry to determine the leaching percentage, which was calculated using the following equations:

$$\eta_1 = \frac{m_0 \times \alpha_0 - m_1 \times \alpha_1}{m_0 \times \alpha_0} \times 100\% \quad (1)$$

$$\eta_2 = \frac{m_1 \times \alpha_1 - m_2 \times \alpha_2}{m_1 \times \alpha_1} \times 100\% \quad (2)$$

$$\eta^* = \frac{m_0 \times \alpha_0 - m_2 \times \alpha_2}{m_0 \times \alpha_0} \times 100\% \quad (3)$$

where η_1, η_2, and η^* are the Mg of primary, secondary, and comprehensive leaching percentage (%), respectively, m_1, m_2, and m_0 are the mass (g) of the first-stage leaching residue, second-stage leaching residue, and raw FNS, respectively, and α_1, α_2, and α_0 are the mass percent (%) of Mg in them, respectively.

The leaching residues with the highest Mg leaching ratio in primary and secondary leaching were analyzed by XRF and XRD, and the results were compared with the initial slag sample.

2.2.2. Purification of Magnesium

During the separation of Mg from impurities, 100 mL of acid leaching solution was first mixed with 1.5 mL of 15% H_2O_2 in a 250 mL beaker, then the breaker was heated at 30 °C and simultaneously stirred at a speed of 200 rpm for 10 min to convert Fe^{2+} into Fe^{3+}. A sensitivity test was performed to assess whether Fe^{2+} was completely oxidized by using 0.1 mol/L potassium ferricyanide. Then, the pH value of the solution was adjusted to pH 7 using about 3 mL aqueous ammonia, so that the impurity ions such as Fe^{3+}, Al^{3+}, and Cr^{3+} in the leachate could be converted into hydroxide precipitation. Next, the pH of the solution was adjusted sequentially with aqueous ammonia in order to further remove the hybrids and obtain a purer magnesium sulfate solution and precipitate.

At the end of the purification process, the magnesium sulfate solution and precipitate were thoroughly separated by vacuum filtration. To evaluate the purifying efficiency, the magnesium filtrate was precipitated as hydroxide magnesium ($Mg(OH)_2$) by sodium hydroxide; the higher the $Mg(OH)_2$ content, the higher the impurity removal rate and the better the impurity removal effect. At the same time, the Mg content of filter residue was analyzed by XRF to calculate the loss ratio of magnesium according to Equation (4):

$$\eta_L = \frac{m' \times \beta}{V \times c} \times 100\% \quad (4)$$

where η_L is the loss ratio of magnesium, m' is the mass (g) of filter residue after impurity removal, β is the Mg content (%) in it, and V and c are the solution volume (L) and its Mg content (g/L) before impurity removal, respectively.

2.2.3. Preparation of Magnesium Oxide

In this study, the alkaline magnesium carbonate calcination process was used to produce highly pure magnesium oxide. Firstly, the magnesium was precipitated into basic magnesium carbonate: 100 mL of purified magnesium sulfate solution was transferred into a 250 mL beaker, and a specified amount of ammonia water and ammonium bicarbonate were added. After stirring at a constant temperature for a certain period of time, a basic magnesium carbonate precipitation was obtained (the main reactions during the process are shown in Equation (5) [34]). After vacuum filtration, cleaning, and drying, basic magnesium carbonate powder was obtained. The magnesium precipitation effect is evaluated by the purity of $4MgCO_3 \bullet Mg(OH)_2 \bullet 4H_2O$ (Equation (6)) and the metal conversion rate (Equation (7)):

$$5MgSO_4 + 6NH_3 \cdot H_2O + 4NH_4HCO_3 \rightarrow 4MgCO_3 \bullet Mg(OH)_2 \bullet 4H_2O + 5(NH_4)_2SO_4 \quad (5)$$

$$\eta_p = \frac{\gamma}{\gamma_0} \times 100\% \quad (6)$$

$$\eta_c = \left(\frac{m - m_x}{m}\right) \times 100\% \quad (7)$$

where η_p is the purity of $4MgCO_3 \bullet Mg(OH)_2 \bullet 4H_2O$, γ is the content of Mg found in the obtained $4MgCO_3 \bullet Mg(OH)_2 \bullet 4H_2O$, γ_0 is the theoretical amount of Mg in $4MgCO_3 \bullet Mg(OH)_2 \bullet 4H_2O$, η_c is the conversion percentage of metals, and m_x and m are the total amount of metals after precipitation and in the original solution, respectively.

Secondly, the obtained basic magnesium carbonate was placed in a muffle furnace at a rate of 10 °C/min and heated to 900 °C and then calcined at a constant temperature for 90 min (the main chemical reaction of calcination is shown in Equation (8)). Then, the furnace was cooled to room temperature, and the obtained powder was light magnesium oxide. Figure 2 shows the process flow diagram of recovering Mg from FNS to prepare MgO.

$$4MgCO_3{\bullet}Mg(OH)_2{\bullet}4H_2O \rightarrow 5MgO + 4CO_2\uparrow + 6H_2O \quad (8)$$

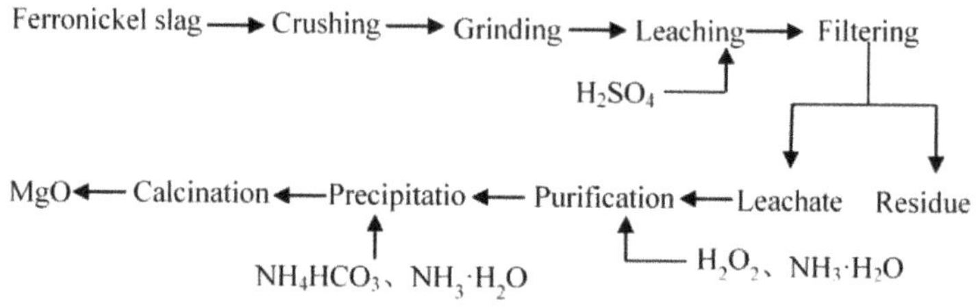

Figure 2. Process flow of recovering Mg from FNS to prepare MgO.

2.3. Characterization Methods

The Mg content in all leaching residues was determined by atomic absorption spectrophotometry (Z-2000, Tokyo, Japan) under the following instrument operating conditions: acetylene flow of 3.0 L/min, lamp current of 4 mA, slit width of 1.0 nm. The chemical compositions of ferronickel slag and leaching residues were examined using an X-ray fluorescence spectrometer (Kyoto, Japan, XRF-1800), which operated at 60 kV voltage and 140 mA current with a scanning speed of 300°/min. The phase composition of the samples was determined by the Rietveld method using X-ray diffraction (XRD) data. The measurements were performed on a Rigaku Ultima IV instrument (Kyoto, Japan, at 40 mA and 40 kV) using CuKα radiation with a wavelength of 1.5406 Å in 10° to 80° 2θ range with an angular speed of 20°/min. The BGMN program was applied for Rietveld refinement. The morphology of the samples was characterized by a scanning electron microscope (SEM, MV-RR-CNJ-0010, Shanghai, China) equipped with an energy dispersive spectroscopy (EDS) detector. All chemicals used in the study were from the laboratory of the University of Science and Technology, Beijing, and were analytical grade. Deionized water was used throughout the experiments and analytical measurements.

3. Results and Discussion

3.1. The Acid Leaching Process of Ferronickel Slag

The effects of FNS grinding fineness, acid concentration, liquid-to-solid ratio, leaching temperature, leaching time, and stirring speed on the magnesium leaching percentage were studied, and the results are shown in Figure 3.

Figure 3a indicates that the leaching efficiency of magnesium increased gradually with decreasing particle size. This is attributed to the increase in the specific surface areas of the samples, which provides a greater reaction interface between the sulfuric acid and FNS particles. However, a smaller particle size does not necessarily lead to a higher extraction rate and may even have a negative influence [35]. Therefore, the optimum grinding fineness was determined to be −0.074, mm accounting for 94.5%.

Figure 3b shows that a modest increase in acid concentration was beneficial to increase the Mg leaching rate. Around 63.90% of magnesium can be extracted when the concentration of acid reaches 2.8 mol/L, and the reasons for the decrease at a higher concentration were because (1) the viscosity of the slurry increased, (2) the mass transfer efficiency was reduced, and (3) the soluble sulfuric magnesium was captured in the residue. A previous study also concluded that an appropriate sulfuric acid concentration was identified to ensure a reasonable final pH level to inhibit the generation of colloidal silica and produce acceptable solid-liquid separation characteristics of the leaching slurry [18]. Hence, a lower sulfuric acid concentration is practically acceptable.

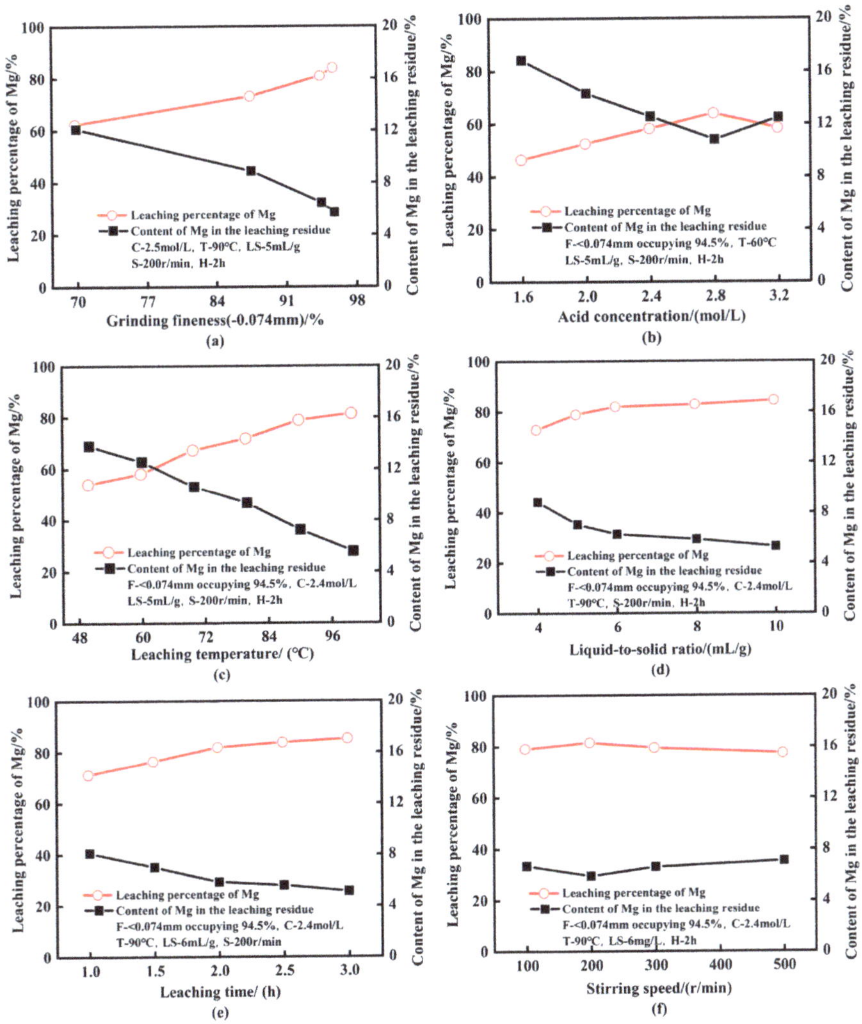

Figure 3. Effect of variables on the efficiency of leaching magnesium from ferronickel slag (**a**) grinding fineness; (**b**) acid concentration; (**c**) leaching temperature; (**d**) liquid-to-solid ration; (**e**) leaching time; (**f**) stirring speed (F—grinding fineness; C—acid concentration; T—leaching temperature; LS—liquid-to-solid ratio; H—leaching time; S—stirring speed).

The results of Figure 3c show that the extraction rate of magnesium increased by 27.39% when the temperature increased from 50 °C to 100 °C. Correspondingly, the content of Mg in the leaching residue decreased from 12.55% to 5.56%, which indicates that the leaching rate of magnesium is highly dependent on temperature. An increase in temperature will increase the available energy in the reaction system and accelerate the thermal movement of H^+ and crystal lattice molecules of forsterite. With a faster diffusion rate, the effective number of collisions between H^+ and FNS particles increases, which makes it easier for H^+ ions to abolish the crystal lattice of forsterite, and thus, enhances the Mg^{2+} leaching rate. A higher leaching temperature is advantageous for the increase in Mg leaching velocity and leaching ratio. However, H_2SiO_3 may be formed when leaching at a temperature above 99 °C, which is harmful to the leaching of Mg, because H_2SiO_3 can increase the solution viscosity by forming a colloidal substance. SiO_2 may be the stable form of Si when the leaching occurs at a temperature lower than 99 °C, which is less harmful [31]. Therefore, the sulfuric acid leaching temperature of FNS should be controlled lower than 99 °C.

Figure 3d illustrates that increasing the liquid-to-solid ratio is beneficial to the extraction of magnesium. As the liquid-to-solid ratio increases, the slurry density gradually decreases, accelerating mass transfer, and therefore, providing benefits to mineral dissolution. Moreover, as the solution viscosity decreases, FNS particles can more efficiently contact sulfuric acid. However, it was noteworthy that a relatively low degree of extraction improvement occurred when the liquid-to-solid ratio exceeded 6 mL/g.

Figure 3e shows the tendency of magnesium extraction when the leaching time gradually increases from 60 min to 180 min. Moreover, it was observed that magnesium extractions increased significantly with the increase in leaching time until 120 min, and then gradually reached a stable state. This can be attributed to the fact that only a small amount of unreacted FNS particles remained after leaching for 120 min, and the probability of intermolecular collisions decreased with the consumption of sulfuric acid. In addition, amorphous silica was produced simultaneously, which was easy to accumulate on the surface of the reactor and hinder the reaction.

As shown in Figure 3f, the stirring speed has an indistinctive effect on the leaching rate. The stirring should ensure that the particles are effectively suspended, but cannot be too fast; this is because excessive mixing speed causes the material in the leaching system to rotate, which makes it easy to agglomerate on the stirring bar, resulting in a poor mass transfer effect and slow leaching rate.

Considering the leaching efficiency, filterability of leaching slurry, and energy consumption, the proper technical conditions for primary leaching were determined as follows: raw material grinding fineness of −0.074 mm occupied 94.5%, acid concentration of 2.4 mol/L, leaching temperature of 90 °C, liquid-to-solid ratio of 6 mL/g, leaching time of 120 min, stirring speed of 200 r/min, and leaching percentage of Mg was 82.05%.

The second leaching of the primary residue was operated for more Mg recovery. The influence of the main factors on the corresponding Mg extractions and Mg contents are presented in Figure 4.

It is not difficult to find that the influence of the main factors on the magnesium leaching rate was similar in the two leaching steps. The optimum sulfuric acid concentration for the second leaching step was also 2.4 mol/L. The influence of temperature on the secondary leaching process was stronger, and the leaching rate increased by 23.03% when the temperature was raised from 90 °C to 95 °C. The optimum liquid-solid ratio of the secondary process was 3 mL/g, which was lower than that of the primary acid leaching process, which occurred because there are fewer substances that can react with sulfuric acid in the primary acid leaching residue than the original ferronickel slag. The improvement in the second-stage acid leaching process by prolonging the leaching time was more obvious than in the first-stage acid leaching process. The leaching rate of the second-stage acid leaching process increased by 3.37% as the leaching time increased from 120 min to 150 min. Overall, the optimum factors for second leaching were: acid concentration of 2.4 mol/L, leaching temperature of 95 °C, liquid-to-solid ratio of 3 mL/g, leaching time of 150 min, stirring speed of 200 r/min, and secondary Mg leaching percentage of 68.21%, then the comprehensive leaching percentage of FNS was 94.29%.

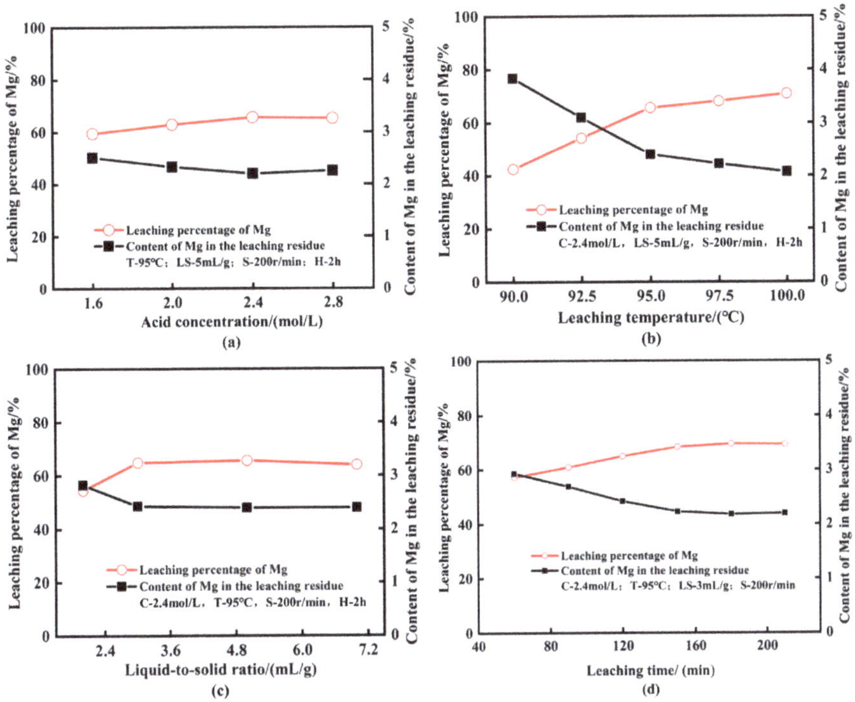

Figure 4. Influence of the main factors on the leaching rate of Mg in secondary acid leaching (**a**) acid concentration; (**b**) leaching temperature; (**c**) liquid-to-solid ration; (**d**) leaching time.

3.2. Analysis of the Leaching Residues

The XRF analysis results of leaching residues obtained under the optimized conditions are presented in Table 2. The main component in the residues changed to SiO_2 after acid leaching, and the contents of other impurities were substantially low. The X-ray diffraction pattern of the residues (Figure 5) demonstrates that after the primary acid leaching process, the intensity of the diffraction peak of magnesium olivine evidently decreases, which indicates that most crystalline and glassy phases in the ferronickel slag have been leached out by acid. After the secondary acid leaching process, there was only a trace of quartz remaining, and the peak of magnesia olivine disappeared, which indicated that the minerals containing magnesium and iron in FNS basically reacted completely. The main component of the secondary acid leaching residue is purer amorphous silica, which seems to be more environmentally friendly and suitable for further disposal.

Table 2. XRF comparative analysis of the two stages acid leaching residue.

Component	SiO_2	CaO	MgO	Fe_2O_3	Al_2O_3	Cr_2O_3	MnO	TiO_2	K_2O	Na_2O	NiO
The primary	80.73	5.71	4.17	4.17	3.56	0.69	0.32	0.15	0.12	0.11	0.02
The secondary	84.20	5.54	2.28	3.32	3.14	0.59	0.27	0.13	0.11	0.09	0.01

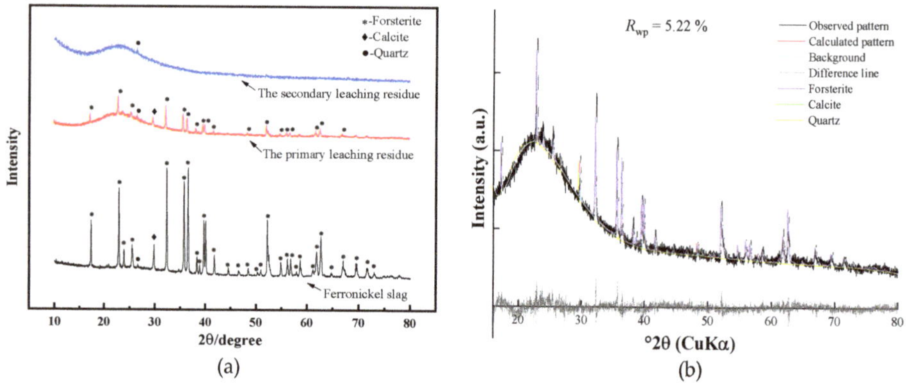

Figure 5. (**a**) XRD comparison of the FNS and acid leaching residues; (**b**) Rietveld XRD diagram of the primary leaching residue.

3.3. Response Surface Optimization

In order to optimize the process conditions and explore the main process factors and their interactions, response surface optimization was performed. In this study, the optimization of significant factors was performed via the response surface methodology (RSM). Table 3 summarizes the range of independent variables and their levels. Experiments were carried out according to the Box-Behnken design (BBD), as shown in Table 4.

Table 3. Experimental variables and levels used in the response surface design.

Code	Variables	Unit	Level			Factor		
A	acid concentration	mol/L	2.0	2.4	2.8	−1	0	+1
B	reaction temperature	°C	80	90	100	−1	0	+1
C	reaction time	min	90	120	150	−1	0	+1

Table 4. Experimental scheme and results from the response surface designed experiment.

Number	Variables			Leaching Percentage of Mg/%
	A	B	C	
1	2.8	90	150	83.49
2	2.4	100	90	84.78
3	2	100	120	83.67
4	2.4	90	120	82.05
5	2.4	100	150	86.77
6	2.8	80	120	76.04
7	2	80	120	75.26
8	2.4	80	90	72.78
9	2.8	100	120	84.35
10	2.4	90	120	81.85
11	2.8	90	90	80.33
12	2	90	150	80.56
13	2.4	80	150	76.22
14	2	90	90	77.56
15	2.4	90	120	83.01

According to the experimental results in Table 4, the Design Expert software (Stat-Ease, Inc., 1300 Godward Street Northeast; Version 12.0.3) was used to execute the regression analysis, and the response surface equation of Mg leaching rate R was fitted as follows:

$$R = -151.07583 + 34.28750A + 3.18963B + 0.350486C - (6.250 \times 10^{-3})AB - (3.333 \times 10^{-3}) AC - (1.208 \times 10^{-3})\,BC - 6.64323A^2 - 0.014104B^2 - (8.39 \times 10^{-4})\,C^2$$

(formula: A—acid concentration, mol/L; B—reaction temperature, C—reaction time, min).

The coefficient of determination R^2 of the above model is 0.9689, which indicates that the fitted model is highly correlated [36].

Analysis of variance (ANOVA) was used to confirm the significance and adequacy of the quadratic response surface model. All observations with ANOVA are presented in Table 5. The F value of the regression model is 31.13 and p-value is 0.0029. The results showed that the model was significant (highly significant: $p < 0.001$; significant: $p < 0.05$; not significant; $p > 0.10$). Therefore, under a given level of the considered factors, the degree of influence of the three factors on the leaching rate of Mg is as follows: reaction temperature > reaction time > acid concentration. The change in sulfuric acid concentration (from 2 mol/L to 2.8 mol/L) has a weak effect on the Mg leaching rate and increasing the reaction temperature (from 80 °C to 100 °C) has a highly significant effect on the improvement in the Mg leaching rate. It can be considered that increasing the reaction temperature and appropriately extending the reaction time are the technical keys to improving the leaching effect of Mg, while controlling the sulfuric acid concentration above 2 mol/L.

Table 5. ANOVA for the response surface quadratic model.

Source	Sum of Square	Degree of Freedom	Mean Square	F-Value	p-Value	Significance Level
Model	228.48	9	25.39	17.31	0.0029	significant
A-Acid concentration	6.41	1	6.41	4.37	0.0908	
B-Reactiontemperature	192.77	1	192.77	131.46	<0.0001	
C-Reaction time	16.79	1	16.79	11.45	0.0196	
AB	0.0025	1	0.0025	0.0017	0.9687	
AC	0.0064	1	0.0064	0.0044	0.9499	
BC	0.5256	1	0.5256	0.3585	0.5754	
A^2	4.17	1	4.17	2.84	0.1525	
B^2	7.35	1	7.35	5.01	0.0754	
C^2	2.11	1	2.11	1.44	0.2843	
Residual	7.33	5	1.47			
Lack of Fit	6.56	3	2.19	5.69	0.1531	not significant
Pure Error	0.7691	2	0.3845			
Cor Total	235.81	14				

The relationships between the three variables were resolved with the help of response surface plots, and the contour plots and 3D plots are displayed in Figure 6a–c. The significance of the interaction between the factors can be deduced by the shape of the contour lines (significant: elliptical; not significant: rounded), and the optimum range of the factors is reflected by the curvature of the response surface. According to Figure 6a, there is an optimal range for Mg leaching efficiency during an acid concentration of 2.4~2.6 mol/L and a temperature of 95~100 °C; moreover, there is a certain interaction between the two factors. The contour plots of Figure 6b are elliptical, indicating that the interaction between the reaction temperature and time is significant, but the reaction time has a much milder effect. Figure 6c shows that the extraction rate of magnesium has difficulty reaching over 84% when the reaction temperature is fixed at 90 °C. The interaction between acid concentration and reaction time is not significant because of the rounded contour plots.

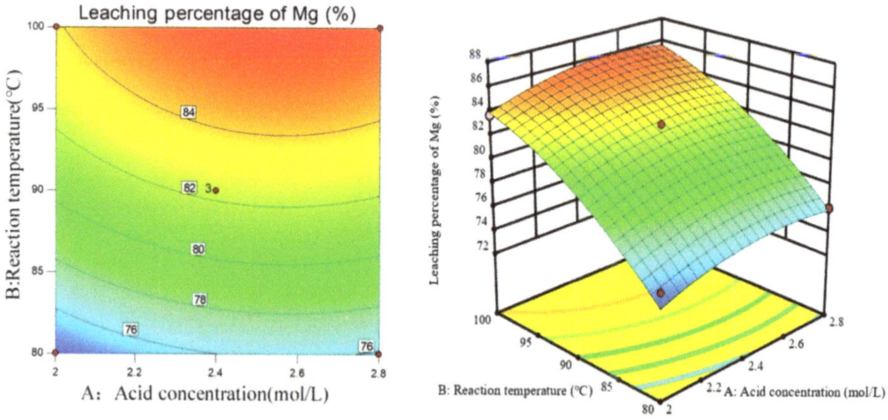

(a) Fixed reaction time 120 min

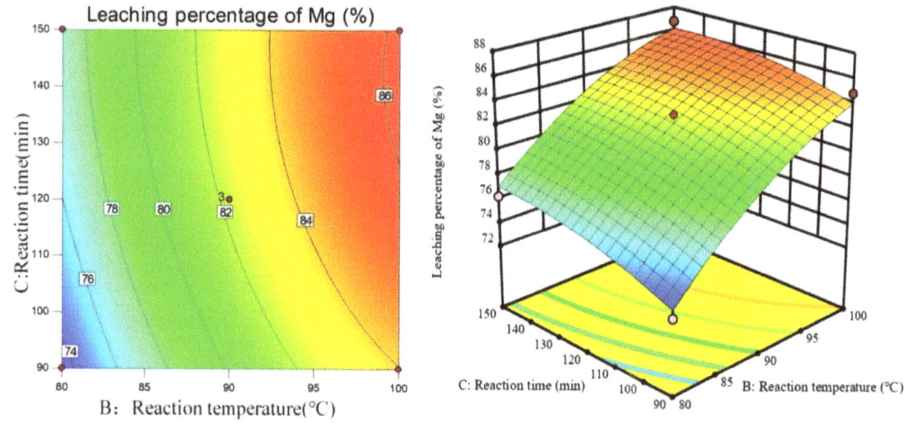

(b) Fixed acid concentration 2.4 mol/L

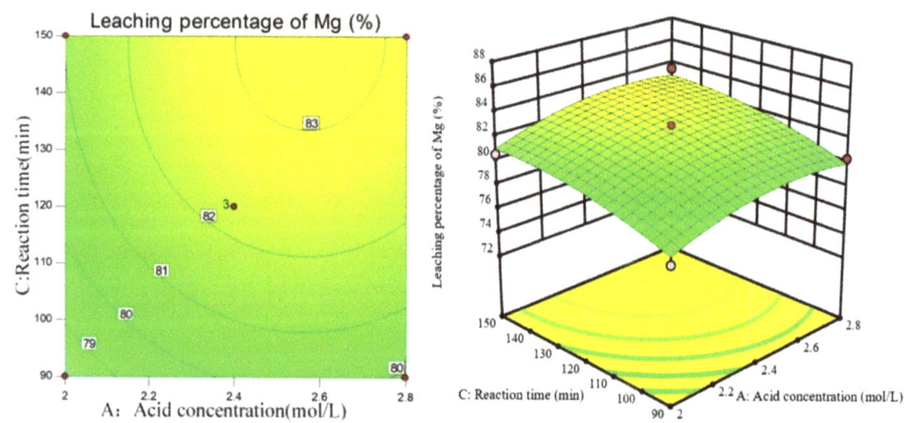

(c) Fixed reaction temperature 90 °C

Figure 6. The 3D plots and contour plots showing the effect of (**a**) acid concentration and reaction temperature, (**b**) reaction temperature and reaction time, and (**c**) reaction time and acid concentration.

The most suitable conditions for primary acid leaching recommended by the response surface method are: acid concentration of 2.44 mol/L, reaction temperature of 96.11 °C, reaction time of 116.15 min, and the predictive leaching rate of magnesium is 84.68%. Considering the convenience of practical operation, the optimal process conditions were adjusted as follows: acid concentration of 2.4 mol/L, reaction temperature of 96 °C, and reaction time of 116 min. The average leaching percentage of magnesium was 84.97% after three replicate validation experiments. Secondary leaching of the primary residue was also carried out, and the comprehensive Mg leaching percentage of FNS reached 95.82%. The optimization was appreciable.

3.4. Analysis of Leaching Kinetics

Visual observations from the experiments showed that the leaching process followed the shrinkage-nuclear-reaction model. According to the leaching rate with reaction time at different reaction temperatures experimental data, the kinetics of magnesium leaching is analyzed, as shown in Figure 7a. Therefore, under the assumption that the sample is a homogeneous spherical solid phase, the chemical-reaction-controlled and diffusion-controlled models can be represented by the following equations [37,38]:

$$1 - (1 - \alpha)^{1/3} = k_1 t \tag{9}$$

$$1 - 2\alpha/3 - (1 - \alpha)^{2/3} = k_2 t \tag{10}$$

where α is the leaching efficiency of magnesium, %, t is the leaching time, min, k_1 is the chemical reaction rate constant, min^{-1}, and k_2 is the apparent rate constant of the internal diffusion-controlled model, min^{-1}.

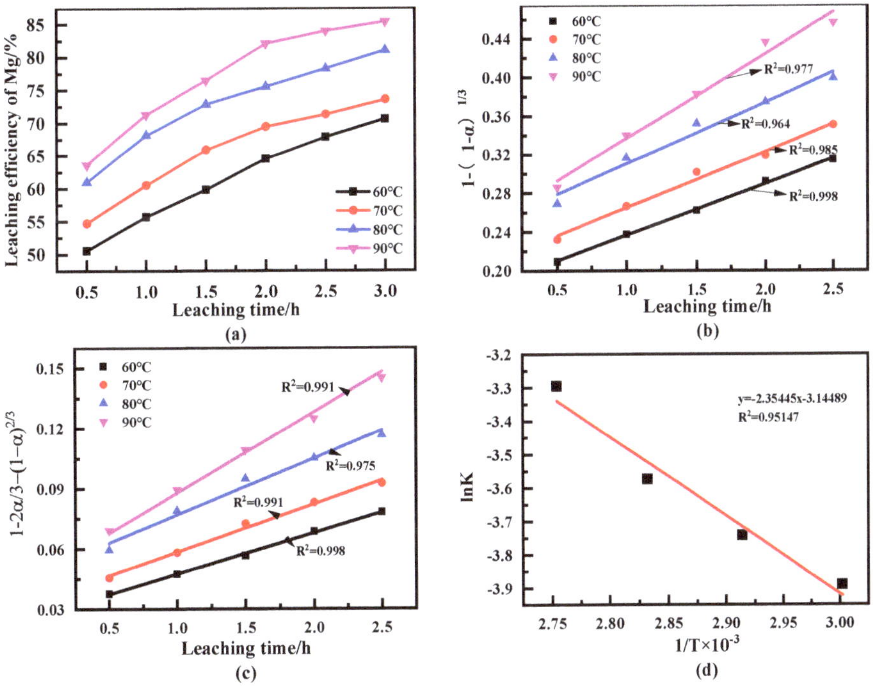

Figure 7. (**a**) Relationship between Mg leaching efficiency and leaching time; (**b**) plots of $1 - (1 - \alpha)^{1/3}$ vs. time for magnesium leaching; (**c**) plots of $1 - 2\alpha/3 - (1 - \alpha)^{2/3}$ vs. time for magnesium leaching; (**d**) Arrhenius plot of the solid membrane diffusion-controlled model.

The curves in Figure 7b exhibit a higher degree of deviation, indicating that the magnesium leaching process does not agree with the chemical reaction-controlled model. As shown in Figure 7c, during the initial 2.5 h leaching process, four straight lines that approach the zero point are observed, which are closer to the origin at low temperature (60 °C) and deviate from the origin at high temperature, demonstrating that the extraction is more likely controlled by diffusion through the solid membrane.

To calculate the activation energy of magnesium leaching, the Arrhenius equation was applied [39]:

$$\ln K = \ln A - E/RT \quad (11)$$

where K is the reaction rate constant, min^{-1}, E is the activation energy, $J \cdot mol^{-1}$, T is the reaction temperature, K, R is the gas constant with a value of 8.314 $J \cdot K^{-1} \cdot mol^{-1}$, and A is the frequency factor.

A plot of lnK vs. $1/T$ for the diffusion-controlled model is linear, as presented in Figure 7d. The apparent activation energy (E) of 19.57 $kJ \cdot mol^{-1}$ was calculated from the slope of the line. It is inferred that the magnesium leaching process obeys composite control [39]. At low temperatures, the acid leaching reaction rate is slower and mainly controlled by chemical reactions. When the temperature rises to a certain value, the chemical reaction rate increases, resulting in an increase in the thickness of the product layer; the control step changes from chemical control to solid film diffusion control (mainly manifested as internal diffusion control). Therefore, thinning the solid film or accelerating the diffusion, such as by increasing the lixiviant concentration or decreasing the particle size of the solid, can be used to enhance the extraction of magnesium.

3.5. Purification of Magnesium

The two-stage acid leachates were mixed, and their metal ion composition is shown in Table 6. Besides Mg^{2+}, the leaching solution contains Fe^{2+}/Fe^{3+}, Al^{3+}, and Cr^{2+}, which are impurities in the preparation of magnesium products. The quality of magnesium product vitally depends on the purity of the magnesium in the leachate; therefore, impurities should be removed as much as possible. Precipitation is a traditional impurity removal method; the theoretical basis is that different ions will be converted to metal hydroxide precipitation at different pH values of the solution, as shown in Table 7 [32]. Irons such as Fe^{3+}, Al^{3+}, and Cr^{3+} are relatively easy to be separated with Mg^{2+} due to their large and varying precipitation pH values, while Fe^{2+} and Mn^{2+} prevent the purification and recovery of Mg, since the precipitation pH values of Fe^{2+} and Mn^{2+} overlap with that of Mg^{2+}. Considering the low content of impurities, except for Fe, it is difficult to remove all of them sequentially and the cost is high. Therefore, centralized removal is adopted in this study. Fe^{2+} and Mn^{2+} were first oxidized to Fe^{3+} and MnO_2 with H_2O_2, then the leachate pH value was gradually adjusted to 7 by aqueous ammonia. Accordingly, most of the foreign ions could be separated as sediments from Mg^{2+}, which remained in the liquid. Following this, the leachate underwent deep impurity removing by optimizing the pH value, react temperature, and time. The results are shown in Figure 8.

Table 6. Metal ion concentrations in leachates (g/L).

Metal Ion	Mg^{2+}	Fe^{2+}/Fe^{3+}	Al^{3+}	Cr^{3+}	Mn^{2+}
Concentration	9.76	1.72	0.8–0.6	0.2–0.1	<0.1

Table 7. pH of different metal ions upon hydroxide precipitation (25 °C).

Ions	Fe^{3+}	Al^{3+}	Cr^{3+}	Fe^{2+}	Mn^{2+}	Mg^{2+}
pH	1.5–3.5	3.1–5.1	4–7	6.3–9.3	7.4–10.4	8.4–11.4

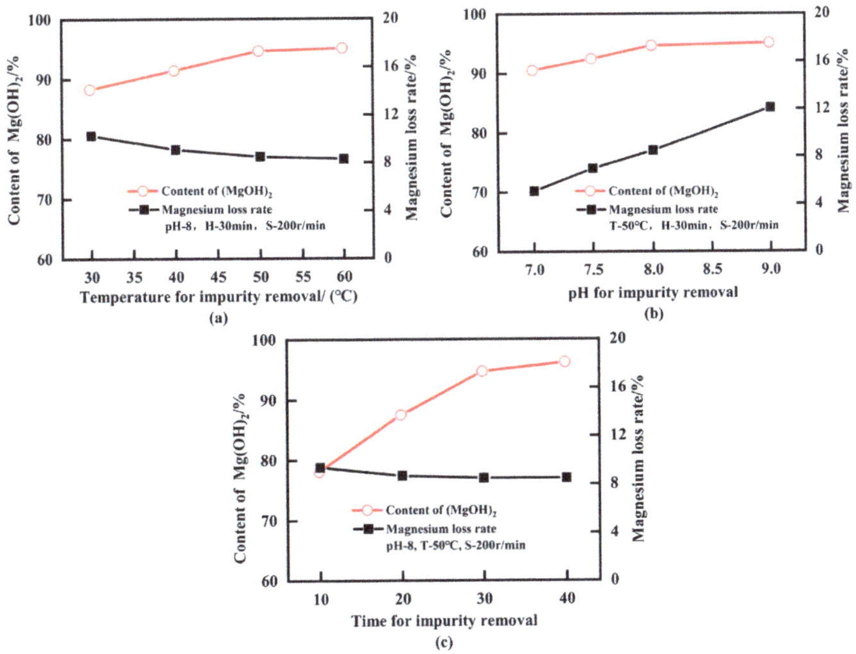

Figure 8. Impurity removal results of acid leachate (**a**) temperature for impurity removal; (**b**) pH for impurity removal; (**c**) time for impurity removal.

As shown in Figure 8a, increasing the reaction temperature can accelerate the precipitation and promote the generation of large-scale precipitates, thereby reducing the adsorption of Mg^{2+} by the precipitates and improving the subsequent solid-liquid separation effect. Thus, with the reaction temperature increase, Mg loss decreases, and the content of the intermediate product of $Mg(OH)_2$ increases. However, this effect became considerably weak after 50 °C. Therefore, the appropriate temperature for impurity removal was 50 °C.

Figure 8b illustrates that the pH value of the solution should be modest—the impurity ion precipitated incompletely at a lower pH, resulting in a lower $Mg(OH)_2$ content. Furthermore, when the pH was too high, a large amount of Mg^{2+} precipitated and the Mg loss in the solution increased. The optimum pH value for impurity removal was 8.0.

From Figure 8c, it can be concluded that the impurity removal time affected the sedimentation of impurity ions while having minimal effect on the Mg loss rate. The $Mg(OH)_2$ content decreased when the removal time was no longer enough—owing to incomplete precipitation of impurity ions and large amounts of impurity ions remaining in the magnesium sulfate solution. After 30 min of impurity removal, the $Mg(OH)_2$ content tended to be stable; therefore, the impurity removal time was fixed at 30 min.

The chemical composition of the impurity removal residue is summarized in Table 8 after deep impurity removal under the above optimum conditions. The metal ion concentration of leachate after metal purification is shown in Table S1. It is shown that Fe, Al, Cr, and other elements are enriched in the debris, which creates conditions for their progressive recovery and utilization. After deep purification, the content of the intermediate product $Mg(OH)_2$ of magnesium sulfate was 95.10%, and the Mg loss rate was 8.5%

Table 8. Elemental composition of impurity removal residue (%).

Component	Fe_2O_3	Al_2O_3	SiO_2	MgO	CaO	MnO	Cr_2O_3	SO_3	Others
Content	44.48	11.06	2.92	10.06	0.64	1.03	4.80	24.77	0.24

3.6. Preparation of Magnesium Oxide

The alkaline magnesium carbonate calcination process was used to overcome the problems of high S content and bulk density of MgO products prepared by the conventional magnesium hydroxide calcination process [34]. Figure 9 illustrates the effect of the main technological conditions on the preparation of basic magnesium carbonate from purified magnesium sulfate solution.

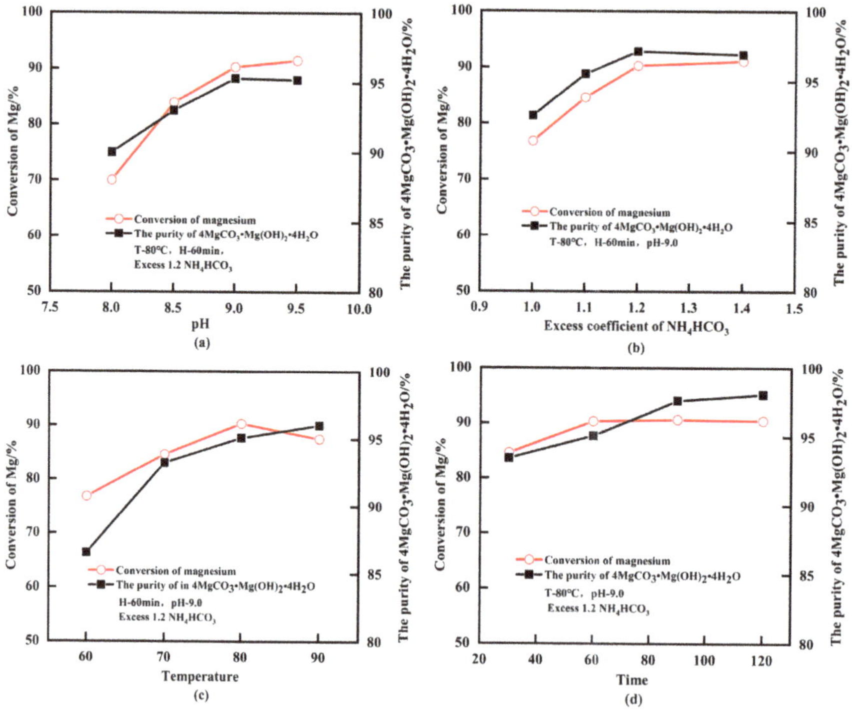

Figure 9. Effect of the main process conditions on magnesium precipitation efficacy (**a**) pH; (**b**) excess coefficient of NH_4HCO_3; (**c**) temperature; (**d**) time. The purity of the precipitated products was calculated by the content of Mg found in the obtained precipitate vs the theoretical amount of Mg in $4MgCO_3{\bullet}Mg(OH)_2{\bullet}4H_2O$.

The Mg conversion rate and purity of the precipitated products gradually increased with the increase in pH value. When the pH reached 9.0, the Mg conversion rate was 90.3%, and the purity of the product was 95.08%. Continuously increasing pH resulted in a slight increase in Mg conversion but a decrease in product purity (Figure 9a), indicating that low pH and an insufficient amount of ammonia led to incomplete precipitation of Mg^{2+}, low product purity, and Mg conversion. However, if the pH is large enough, increasing the ammonia dosage will again promote the production of $Mg(OH)_2$ precipitate (milk-white precipitate has been observed at pH = 9.5), and the adsorption of ions in the solution by $Mg(OH)_2$ exist in colloidal form, which will affect the production and quality of basic magnesium carbonate, which is also unfavorable to the subsequent product filtration and washing [34]. The optimal pH was determined to be 9.0, corresponding to an ammonia addition of 2 mL.

The effect of the NH_4HCO_3 content on the production of basic magnesium carbonate is shown in Figure 9b. When the excess coefficient of NH_4HCO_3 is 1.0–1.2, purity and magnesium conversion of the precipitates increase with the excess coefficient. Moreover, when the excess coefficient reached 1.3, the purity in the precipitates decreased and the magnesium conversion was basically unchanged. It can be observed that sufficient NH_4HCO_3 is a necessity for the complete conversion of Mg^{2+} into basic magnesium carbonate; however, when NH_4HCO_3 is in excess, it causes incomplete dissolution, which is unfavorable to the reaction. Therefore, an excess coefficient of 1.2 was chosen for NH_4HCO_3.

Figure 9c shows that the magnesium precipitation reaction needs to be carried out at a certain temperature. Low temperature led to a slow reaction speed, and ammonia volatilized retarding the reaction when the temperature was too high (Equation (5)). In addition, the formation of basic magnesium carbonate was hindered due to a portion of Mg^{2+} converting to $Mg(HCO_3)_2$ or $MgCO_3{\bullet}3H_2O$ [34]. These all led to a decrease in Mg conversion and product purity; therefore, 80 °C is the comparative suitable reaction temperature.

The results shown in Figure 9d suggest that increasing the reaction time had a beneficial effect on the purity of the product and the conversion of magnesium; however, the effect on the product's purity was more obvious. After 60 min reaction, the Mg^{2+} in the solution completely participated in the reaction, and the Mg conversion no longer changed. When the reaction time was continued to 90 min, and the reaction was dominated by the conversion of intermediate products such as $Mg(HCO_3)_2$ or $MgCO_3{\bullet}3H_2O$ rather than basic magnesium carbonate, the purity of the products was further improved. An optimal reaction time of 90 min was experimentally determined.

In summary, the purity of the magnesium precipitation product obtained under optimal conditions was 97.62%, and the Mg conversion rate was 90.6%. The XRD and SEM analysis of magnesium precipitation products are shown in Figure 10. The XRD patterns were in agreement with PDF card 25-0513 ($4MgCO_3{\bullet}Mg(OH)_2{\bullet}4H_2O$), which proved that the product was basic magnesium carbonate. In addition, the intensity of XRD peaks was high and the shape of peaks was sharp, indicating high purity. From the SEM image, it can be clearly seen that this basic magnesium carbonate is a spherical particle with a diameter of about 510 μm; the particle size was uniform, the product was purer, and there were fewer residual impurities on the particle surface. The produced basic magnesium carbonate can be used as a raw material for the preparation of MgO, and can also be used as a final product with a flexible process.

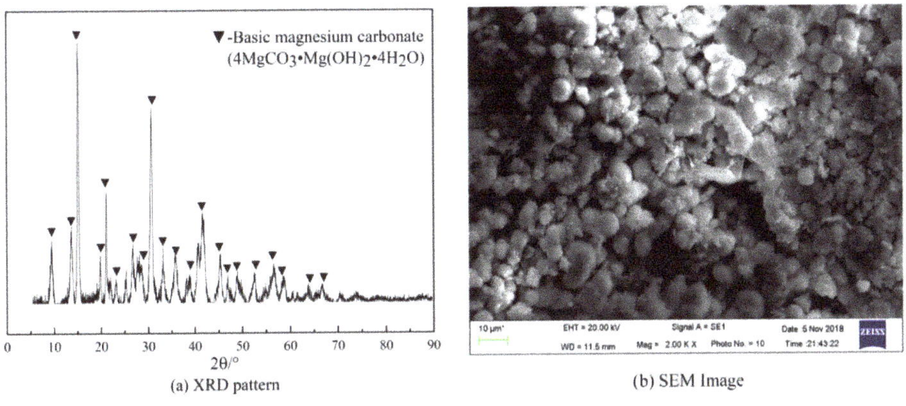

Figure 10. Analysis of magnesium precipitation product (**a**) XRD pattern; (**b**) SEM image.

A powdery material was obtained by calcining the basic magnesium carbonate at 900 °C for 90 min. The color of the powder was white, which is consistent with the appearance characteristics of magnesium oxide. The XRD and SEM of the powder are shown in Figure 11. The XRD pattern was highly consistent with the PDF card -#45-0946 (MgO), and no other impurity peaks were found, indicating that the main chemical composition of the powder was MgO. The SEM images of the powder show that the size of the powder particles is about 2–5 μm with a rod shape, and the surface of the particles is loose, which implies the bulk density of the powder is relatively low. The result of EDS analysis further confirmed that the powder was light magnesium oxide with high purity. Other quality indicators of MgO powder are shown in Table 9. The quality of the powder meets the national industrial magnesium oxide standard, and the profit of this preparation process is considerable (Table S2).

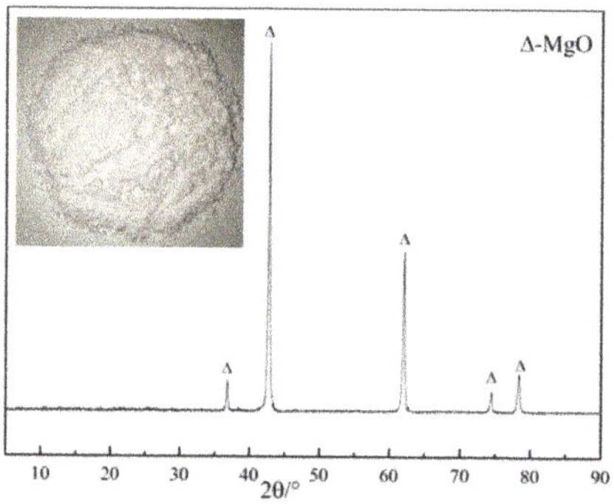

(a) MgO XRD pattern

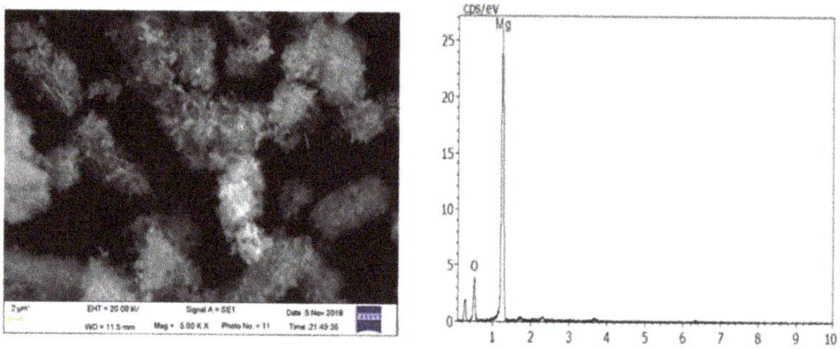

(b) SEM-EDS Images

Figure 11. Analysis of calcined product (**a**) MgO XRD pattern; (**b**) SEM-EDS images.

Table 9. Chinese industry standard for first-class magnesia index and product indexes.

Project	MgO /%	CaO /%	HCl Insoluble/%	Sulfate /%	150 μm Residue on Sieve/%	Fe /%	Mn /%	Loss on Ignition/%	Bulk Density /(g/mL)
First-class	≥93.0	≤1.5	≤0.20	≤0.6	≤0.03	≤0.06	≤0.01	≤5.0	≤0.20
Product	94.85	0.8	trace	0.48	trace	0.014	0.008	3.10	0.16

4. Conclusions

The present study proposed a mild and efficient route for the recovery of magnesium from ferronickel slag to prepare magnesium oxide.

(1) The results of the mineralogical analysis showed that the main chemical components of the FNS were SiO_2, MgO, and Fe_2O_3, mainly in forsterite form, with strong crystallinity and leaching out difficulty. The single factor acid leaching experiments showed that grinding fineness, acid concentration, liquid-to-solid ratio, leaching temperature, leaching time, and stirring speed all had an impact on the leaching of Mg. The influence of acid leaching temperature, acid leaching time, and acid concentration were more prominent. The maximum leaching rate of Mg in the primary acid leaching was 82.05% and increased by 12.24% after the secondary acid leaching. Therefore, a two-step leaching process is recommended.

(2) The response surface analysis showed that when the sulfuric acid concentration was controlled above 2 mol/L, the degree of influence of the following three factors on the magnesium leaching rate were: reaction temperature > reaction time > acid concentration. In addition, there is a certain interaction between the leaching temperature and the leaching time. Combining the single factor experiment with response surface optimization, the optimum process conditions for acid leaching are as follows: in the two-step leaching process, the FNS grinding fineness of −0.074 mm accounted for 94.5%, the acid concentration was 2.4 mol/L, the leaching temperature was 96 °C, and the stirring speed was 200 r/min; the liquid-to-solid ratios in the primary and secondary processes were 6 mL/g and 3 mL/g; the leaching times in the first and second stage were 120 min and 150 min, respectively. The comprehensive magnesium leaching rate from ferronickel slag reached 95.82% (84.97% for primary leaching and 72.19% for secondary leaching).

(3) The acid leaching kinetics study showed that the atmospheric sulfuric acid leaching of magnesium from FNS satisfied the mixed control model. The surface chemical is dominant at low temperatures and turns to film diffusion control at high temperatures. Therefore, reducing the particle size of the FNS and increasing the reaction temperature are important for improving the Mg leaching effect.

(4) Due to the low content of other impurities in the acid leaching solution, (except for Mg), this study prefers to use a centralized exclusion process to remove impurities. After pre-oxidation, pre-purification, and deep impurity removal (50 °C, pH = 8, 30 min) processes, most of the impurities were removed; the content of intermediate product $Mg(OH)_2$ of magnesium sulfate solution was 95.10%, and the Mg loss was 8.5%. Basic magnesium carbonate was prepared by the mixed ammonia precipitation method, and the optimum process conditions were as follows: pH = 9.0, NH_4HCO_3 excess coefficient of 1.2, 80 °C, 90 min, and the Mg conversion was 90.6%. The results of XRD and SEM analyses showed that the alkali magnesium carbonate obtained in the experiment consisted of spherical particles with a diameter of 5–10 μm, and with high purity (MgO of 42.3%). The white powders obtained by 900 °C roasting 90 min of basic magnesium carbonate are MgO, which is characterized by a diameter of about 2–5 μm, with rod shape and loose surface. The quality of the powder reached the first-grade standard of MgO for industrial use in China.

Supplementary Materials: The following are available online at https://www.mdpi.com/article/10.3390/min11121375/s1, Table S1: metal ion concentrations for the leachate after metals purification (g/L); Table S2: table of economic summary for the treatment of 1 t ferronickel slag.

Author Contributions: Writing, formal analysis, and original draft preparation, J.Y.; conceptualization, review, and revision, X.D.; design and experiments, L.L.; supervision, H.Y.; investigation, X.J. This was a joint work of the five authors; each author was in charge of their expertise and capability. All authors have read and agreed to the published version of the manuscript.

Funding: This research received no external funding.

Acknowledgments: We would like to express our sincere appreciation to the anonymous reviewers for their insightful comments, which helped us improve the quality of this paper. We thanked Tangshan Boquan Industrial Co., Ltd. for its experimental raw materials support in project research.

Conflicts of Interest: The authors declare no conflict of interest.

References

1. Luo, J.; Li, G.; Rao, M.; Zhang, Y.; Peng, Z.; Zhi, Q.; Jiang, T. Evaluation of Sintering Behaviors of Saprolitic Nickeliferous Laterite Based on Quaternary Basicity. *JOM* **2015**, *67*, 1966–1974. [CrossRef]
2. Saha, A.K.; Sarker, P.K. Expansion due to alkali-silica reaction of ferronickel slag fine aggregate in OPC and blended cement mortars. *Constr. Build. Mater.* **2016**, *123*, 135–142. [CrossRef]
3. Peng, Z.; Wang, L.; Gu, F.; Tang, H.; Rao, M.; Zhang, Y.; Li, G.; Jiang, T. Recovery of chromium from ferronickel slag: A comparison of microwave roasting and conventional roasting strategies. *Powder Technol.* **2020**, *372*, 578–584. [CrossRef]
4. Zhang, Z.; Zhu, Y.; Yang, T.; Li, L.; Zhu, H.; Wang, H. Conversion of local industrial wastes into greener cement through geopolymer technology: A case study of high-magnesium nickel slag. *J. Clean. Prod.* **2017**, *141*, 463–471. [CrossRef]

5. Saha, A.K.; Sarker, P.K. Sustainable use of ferronickel slag fine aggregate and fly ash in structural concrete: Mechanical properties and leaching study. *J. Clean. Prod.* **2017**, *162*, 438–448. [CrossRef]
6. Choi, Y.C.; Choi, S. Alkali-silica reactivity of cementitious materials using ferro-nickel slag fine aggregates produced in different cooling conditions. *Constr. Build. Mater.* **2015**, *99*, 279–287. [CrossRef]
7. Saha, A.K.; Sarker, P.K.; Golovanevskiy, V. Thermal properties and residual strength after high temperature exposure of cement mortar using ferronickel slag aggregate. *Constr. Build. Mater.* **2019**, *199*, 601–612. [CrossRef]
8. Pan, J.; Zheng, G.L.; Zhu, D.Q.; Zhou, X.L. Utilization of nickel slag using selective reduction followed by magnetic separation. *Trans. Nonferrous Met. Soc. China* **2013**, *23*, 3421–3427. [CrossRef]
9. Komnitsas, K.; Zaharaki, D.; Bartzas, G. Effect of sulphate and nitrate anions on heavy metal immobilisation in ferronickel slag geopolymers. *Appl. Clay Sci.* **2013**, *73*, 103–109. [CrossRef]
10. Ljatifi, E.; Kamusheva, A.; Grozdanov, A.; Paunović, P.; Karamanov, A. Optimal thermal cycle for production of glass–ceramic based on wastes from ferronickel manufacture. *Ceram. Int.* **2015**, *41*, 11379–11386. [CrossRef]
11. Karamanov, A.; Paunović, P.; Ranguelov, B.; Ljatifi, E.; Kamusheva, A.; Načevski, G.; Karamanova, E.; Grozdanov, A. Vitrification of hazardous Fe-Ni wastes into glass-ceramic with fine crystalline structure and elevated exploitation characteristics. *J. Environ. Chem. Eng.* **2017**, *5*, 432–441. [CrossRef]
12. Wang, W.; Chen, J.; Yu, J.; Zhou, L.; Dai, S.; Tian, W. Adjusting the melting and crystallization behaviors of ferronickel slag via partially replacing of SiO2 by B2O3 for mineral wool production. *Waste Manag.* **2020**, *111*, 34–40. [CrossRef]
13. Gu, F.; Peng, Z.; Zhang, Y.; Tang, H.; Ye, L.; Tian, W.; Liang, G.; Rao, M.; Li, G.; Jiang, T. Facile Route for Preparing Refractory Materials from Ferronickel Slag with Addition of Magnesia. *ACS Sustain. Chem. Eng.* **2018**, *6*, 4880–4889. [CrossRef]
14. Peng, Z.; Tang, H.; Augustine, R.; Lee, J.; Tian, W.; Chen, Y.; Gu, F.; Zhang, Y.; Li, G.; Jiang, T. From ferronickel slag to value-added refractory materials: A microwave sintering strategy. *Resour. Conserv. Recycl.* **2019**, *149*, 521–531. [CrossRef]
15. Gu, F.; Peng, Z.; Zhang, Y.; Tang, H.; Tian, W.; Lee, J.; Rao, M.; Li, G.; Jiang, T. Promoting spinel formation and growth for preparation of refractory materials from ferronickel slag. *Int. J. Appl. Ceram. Technol.* **2020**, *17*, 1701–1712. [CrossRef]
16. Liu, K. *The Study on the Preparation of Forsterite Refractories from High Magnesium Nickel Slag*; University of Science and Technology: Beijing, China, 2018; pp. 1–76.
17. Wang, Y.; Zhu, R.; Chen, Q.; Wei, G.; Hu, S.; Guo, Y. Recovery of Fe, Ni, Co, and Cu from nickel converter slag through oxidation and reduction. *ISIJ Int.* **2018**, *58*, 2191–2199. [CrossRef]
18. Huang, F.; Liao, Y.; Zhou, J.; Wang, Y.; Li, H. Selective recovery of valuable metals from nickel converter slag at elevated temperature with sulfuric acid solution. *Sep. Purif. Technol.* **2015**, *156*, 572–581. [CrossRef]
19. Tian, D.; Shen, X.; Zhai, Y.; Xiao, P.; Webley, P. Extraction of iron and aluminum from high-iron bauxite by ammonium sulfate roasting and water leaching. *J. Iron Steel Res. Int.* **2019**, *26*, 578–584. [CrossRef]
20. Zhai, X.J.; Li, N.J.; Zhang, X.; Fu, Y.; Jiang, L. Recovery of cobalt from converter slag of Chambishi Copper Smelter using reduction smelting process. *Trans. Nonferrous Met. Soc. China* **2011**, *21*, 2117–2121. [CrossRef]
21. Li, Y.; Perederiy, I.; Papangelakis, V.G. Cleaning of waste smelter slags and recovery of valuable metals by pressure oxidative leaching. *J. Hazard. Mater.* **2008**, *152*, 607–615. [CrossRef]
22. Gbor, P.K.; Ahmed, I.B.; Jia, C.Q. Evaluation of contributions of acid and ligand to Ni, Co, and Fe dissolution from nonferrous smelter slags in aqueous sulfur dioxide. *Ind. Eng. Chem. Res.* **2002**, *41*, 1861–1867. [CrossRef]
23. Perederiy, I.; Papangelakis, V.G.; Buarzaiga, M.; Mihaylov, I. Co-treatment of converter slag and pyrrhotite tailings via high pressure oxidative leaching. *J. Hazard. Mater.* **2011**, *194*, 399–406. [CrossRef] [PubMed]
24. Curlook, W.; Baghalha, M.; Papangelakis, V.; Curlook, W.; Baghalha, M.; Papangelakis, V. Process for the Recovery of Residual Metal Values from Smelter Waste Slags, and from Converter Slags. CA2363969 C, 13 January 2001.
25. Altinkaya, P.; Mäkinen, J.; Kinnunen, P.; Kolehmainen, E.; Haapalainen, M.; Lundström, M. Effect of biological pretreatment on metal extraction from flotation tailings for chloride leaching. *Miner. Eng.* **2018**, *129*, 47–53. [CrossRef]
26. Bulaev, A.G.; Muravyov, M.I.; Pivovarova, T.A.; Fomchenko, N.V.; Kondrat'eva, T.F. Bioprocessing of mining and metallurgical wastes containing nonferrous and precious metals. *Adv. Mater. Res.* **2013**, *825*, 301–304. [CrossRef]
27. Zhang, X.; Gu, F.; Peng, Z.; Wang, L.; Tang, H.; Rao, M.; Zhang, Y.; Li, G.; Jiang, T.; Wang, Y. Recovering Magnesium from Ferronickel Slag by Vacuum Reduction: Thermodynamic Analysis and Experimental Verification. *ACS Omega* **2019**, *4*, 16062–16067. [CrossRef]
28. Zhao, C.; Cai, Y.; Ning, Z.; Wang, G.; Kang, S.; Zhang, C.; Zhai, Y. Recovery of MgO from laterite nickel slag through roasting by ammonium sulfate. *Zhongnan Daxue Xuebao (Ziran Kexue Ban)/J. Cent. South Univ. (Sci. Technol.)* **2017**, *48*, 1972–1978. [CrossRef]
29. Prasetyo, A.B.; Rahadian, D.; Mayangsari, W.; Febriana, E.; Permana, S.; Maksum, A.; Soesaptri, O.; Firdiyono, F.; Soedarsono, J.W. Reverse leaching of magnesium from ferronickel slag using alkali solvent naoh. *Eastern-Eur. J. Enterp. Technol.* **2020**, *1*, 6–14. [CrossRef]
30. Mubarok, M.Z.; Yudiarto, A. *Synthesis of Magnesium Oxide from Ferronickel Smelting Slag through Hydrochloric Acid Leaching—Precipitation and Calcination*; Springer: Berlin/Heidelberg, Germany, 2017.
31. Gao, F.; Huang, Z.; Li, H.; Li, X.; Wang, K.; Hamza, M.F.; Wei, Y.; Fujita, T. Recovery of magnesium from ferronickel slag to prepare hydrated magnesium sulfate by hydrometallurgy method. *J. Clean. Prod.* **2021**, *303*, 127049. [CrossRef]
32. Linchuan, L. The Study on the Comprehensive Utilization and Recovery of Valuable Components from Electric Furnace Nickel-Iron Slag. Master's Thesis, University of Science and Technology, Beijing, China, 2018; pp. 1–79.

33. Bergmann, J.; Friedel, P.; Kleeberg, R. BGMN—A New Fundamental Parameters Based Rietveld Program for Laboratory X-ray Sources, it's Use in Quantitative Analysis and Structure Investigations. *IUCr Comm. Powder Diffr. Newsl.* **1998**, *20*, 5–8.
34. Xu, H.; Liu, W.P.; Yang, X.Y.; Shi, X.C.; Chen, S.Y.; Yu, L.L. Preparation of high-purity light magnesia from sulfate salt subtype salt lake brine. *Zhongnan Daxue Xuebao (Ziran Kexue Ban)/J. Cent. South Univ. (Sci. Technol.)* **2011**, *42*, 2204–2208.
35. Huang, J.; Chen, M.; Chen, H.; Chen, S.; Sun, Q. Leaching behavior of copper from waste printed circuit boards with Brønsted acidic ionic liquid. *Waste Manag.* **2014**, *34*, 483–488. [CrossRef] [PubMed]
36. Khatoon, H.; Rai, J.P.N. Optimization studies on biodegradation of atrazine by Bacillus badius ABP6 strain using response surface methodology. *Biotechnol. Rep.* **2020**, *26*, e00459. [CrossRef] [PubMed]
37. Apostolidis, C.I.; Distin, P.A. The kinetics of the sulphuric acid leaching of nickel and magnesium from reduction roasted serpentine. *Hydrometallurgy* **1978**, *3*, 181–196. [CrossRef]
38. Hollagh, A.R.E.; Alamdari, E.K.; Moradkhani, D.; Salardini, A.A. Kinetic Analysis of Isothermal Leaching of Zinc from Zinc Plant Residue. *Int. J. Nonferrous Metall.* **2013**, *2*, 10–20. [CrossRef]
39. Lin, Q.; Gu, G.; Wang, H.; Zhu, R.; Liu, Y.; Fu, J. Preparation of manganese sulfate from low-grade manganese carbonate ores by sulfuric acid leaching. *Int. J. Miner. Metall. Mater.* **2016**, *23*, 491–500. [CrossRef]

minerals

MDPI

Article

Optimal Thermal Treatment for Effective Copper Recovery in Waste Printed Circuit Boards by Physical Separation: Influence of Temperature and Gas

Boram Kim [1,†], Seongsoo Han [2,†], Seungsoo Park [3,†], Seongmin Kim [4], Minuk Jung [3,4], Chul-Hyun Park [5], Ho-Seok Jeon [4,6], Dae-Weon Kim [1,*] and Yosep Han [4,6,*]

1 Center for Advanced Materials & Processing, Institute for Advanced Engineering (IAE), 175-28 Goan-ro 51, Yongin-si 17180, Korea; boramkim@iae.re.kr
2 Department of Chemical and Biomolecular Engineering, Korea Advanced Institute of Science and Technology (KAIST), 291, Daehak-ro, Yuseong-gu, Daejeon 34141, Korea; sshan12@kaist.ac.kr
3 Department of Earth Resources and Environmental Engineering, Hanyang University 222, Wangsimni-ro, Seongdong-gu, Seoul 04763, Korea; ssglenpark@hanyang.ac.kr (S.P.); mujung@kigam.re.kr (M.J.)
4 Resources Recovery Research Center, Mineral Resources Division, Korea Institute of Geoscience & Mineral Resources (KIGAM), 124, Gwahak-ro, Yuseong-gu, Daejeon 34132, Korea; smkim@kigam.re.kr (S.K.); hsjeon@kigam.re.kr (H.-S.J.)
5 Department of Energy and Resource Engineering, Chosun University, Gwagju 61452, Korea; chpark@chosun.ac.kr
6 Department of Resources Recycling, University of Science and Technology (UST), 217, Gajeong-ro, Yuseong-gu, Daejeon 34113, Korea
* Correspondence: mdsimul@iae.re.kr (D.-W.K.); yosep@kigam.re.kr (Y.H.); Tel.: +81-31-330-7468 (D.-W.K.); +82-42-868-3181 (Y.H.)
† These authors contributed equally to this work.

Citation: Kim, B.; Han, S.; Park, S.; Kim, S.; Jung, M.; Park, C.-H.; Jeon, H.-S.; Kim, D.-W.; Han, Y. Optimal Thermal Treatment for Effective Copper Recovery in Waste Printed Circuit Boards by Physical Separation: Influence of Temperature and Gas. *Minerals* **2021**, *11*, 1213. https://doi.org/10.3390/min11111213

Academic Editors: Ilhwan Park and Marthias Silwamba

Received: 6 October 2021
Accepted: 27 October 2021
Published: 30 October 2021

Publisher's Note: MDPI stays neutral with regard to jurisdictional claims in published maps and institutional affiliations.

Abstract: Printed circuit boards (PCBs) are difficult to recycle because of the layered structure of non-metal (i.e., epoxy resin, glass fiber) and copper. In this work, we conducted a systematic investigation to effectively recover copper from PCB. A thermal treatment was employed for improving the crushing performance of PCB and conducted by varying the temperature and the gas. Then, the mechanical strength, degree of liberation (DL), and copper separation efficiency of the heat-treated and untreated PCBs were investigated. After heat treatment under a 300 °C air atmosphere, the mechanical strength of PCB decreased from 386.36 to 24.26 MPa, and copper liberation improved from 9.3% to 100% in the size range of a coarser size fraction (>1400 μm). Accordingly, when electrostatic separations were performed under these conditions, a high-Cu-grade concentrate and high recovery could be obtained. The results show that the change in the physical properties of the PCBs leads to an improvement in the DL following thermal decomposition at 300 °C in air. Our study elucidates the physical properties of PCBs and the DL under various heat treatment conditions. Furthermore, it shows that the heat treatment condition of 300 °C in air is ideal for recovering copper from the PCB.

Keywords: printed circuit boards; heat treatment; strength; liberation; Cu recovery; physical separation

1. Introduction

The quantity of electronic waste (e-waste) generated has increased with the growing use of electronic devices, increasing by approximately 4–5% every year [1]. At least 53.6 million tons of e-waste was produced in 2019 [2]. Most e-waste has a vital module—the printed circuit board assembly (PCBA)—which is designed to exchange electronic signals between the parts of an electronic device. Therefore, an increase in e-waste naturally leads to an increase in waste PCBA.

PCBA comprises printed circuit boards (PCB) and electronic components (ECs), and it consists of approximately 40% metal, 30% plastic, and 30% ceramic [3–8]. These components are arranged in a layered structure, with Cu being in particularly high amounts [9–11].

Among the metals present in PCBA, Au and Cu are the two most valuable metals, and their grade is much larger than the grade expected in gold and copper ore [12–14]. Accordingly, a large number of studies have been conducted on developing technologies to recover valuable metals through the proper treatment of waste PCBA. Metal recovery from PCBA by recycling has been conducted by mechanical, metallurgical, and incineration methods [15].

Generally, the ECs attached to the PCB are fixed with a metallic material such as lead; they can easily be detached following constant thermal exposure and simple surface damage [11,16,17]. However, because copper contained in PCB mainly exists in a laminated structure with epoxy resin and glass fiber, the copper should effectively be separated [17]. Cu in PCBs largely exists in a high-purity metallic state and is not chemically bonded with other metals. Thus, an efficient physical process to liberate and separate Cu should precede the metallurgical process. Accordingly, diverse studies have been conducted to increased metal liberation and separation efficiency from PCB waste [18–20].

PCB must be pulverized into particles to effectively liberate and separate Cu [18,21,22]. According to a previous study, it should be comminuted to approximately 150 µm or less (e.g., liberation size) [6,23,24]. However, because the liberation size is small, the grinding of PCB causes Cu loss and also consumes a lot of energy [23,24]. In addition, because of the large number of fine particles (i.e., smaller liberation size), the separation efficiency in the electrostatic separation process is greatly reduced. Accordingly, several studies have been conducted to find more favorable liberation size with the pretreatment before comminution, and liberation enhancement by thermal treatment has been reported as the most suitable method (see Figure 1) [25,26].

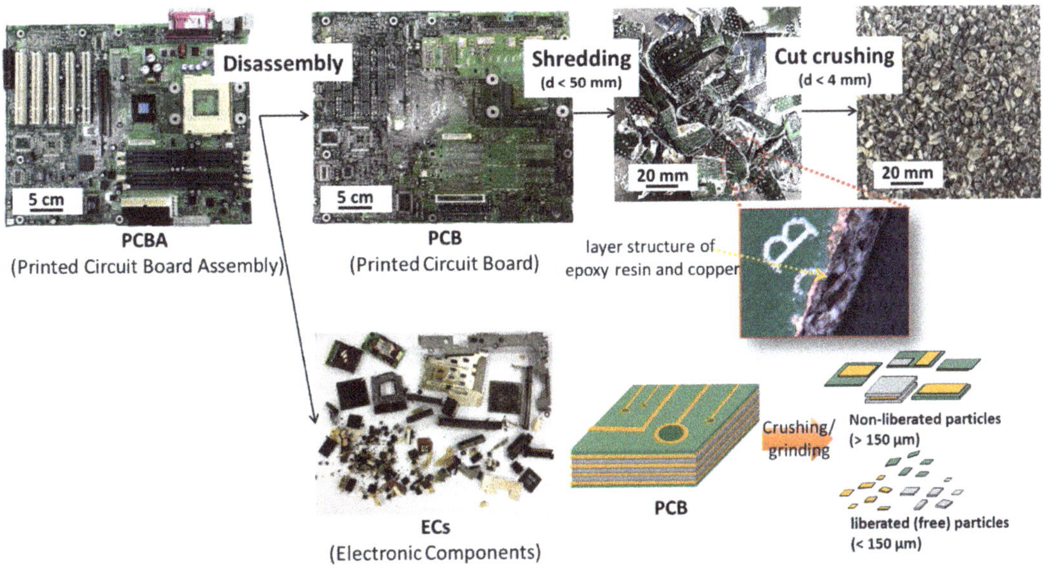

Figure 1. Flowsheet and photo images of PCBA used in this study. PCB and ECs were separated in PCBA, and the ground particles were prepared from PCB by shredding and cut crushing. Schematic of liberation of copper and epoxy resin according to particle size by the typical crushing/grinding method.

Jie et al. [23] studied PCB pyrolysis characteristics and mechanical strength changes under a nitrogen atmosphere at 300–700 °C. The pyrolysis affects the boundary of the layered structure and removes the adhesive between Cu and epoxy resin, facilitating the separation of metal from PCB. In addition, they analyzed the pyro-oil and pyro-gas that formed depending on the pyrolysis temperature and presented references on PCB pyrolysis according to the heat treatment temperature. Huang et al. suggested a microwave-based pyrolysis to recover metal from PCB. They implemented an acid-leaching process to skip the grinding process after thermal treatment [27]. Li et al. [28] focused on thermal analysis and kinetics of low-temperature PCB pyrolysis under nitrogen, air, and vacuum. Moreover, they calculated the activation energy of thermal decomposition and quantified the efficiency of the thermal decomposition process according to the heat treatment conditions. Li et al. [29] proposed heat treatment at 250 °C under vacuum by employing an existing physical treatment technique for high-efficiency metal recovery. They mentioned that the product strength decreased sharply with heat treatment, improving metal liberation and recovery.

Corona-induced electrostatic separation is the most efficient physical separation technique, where both metals and non-metals remain as a mixture after comminution [30–32]. This method has the advantage of a very high metal recovery performance and does not cause secondary pollution. Li et al. [33] proposed a metal recovery method consisting of crushing, screening, drying, and electrostatic separation. They found that a particle size of 0.6–1.2 mm was the most appropriate for the electrostatic separation of metals. Xue et al. [34] conducted the electrostatic separation of a mixture of Cu, epoxy resin, and glass fiber. They proposed a multistage separation for the effective separation of conductors, semiconductors, and nonconductors. Although a large volume of research has been conducted in the past/mentioned above, no study has focused on the relationship between strength, grindability, liberation properties, and the physical recovery of metal components of heat-treated PCB comminution products.

In this work, a systematic investigation was conducted on the physical separation process to obtain Cu-containing particles (i.e., Cu concentrates) separated from the thermally processed/treated PCB. First, we evaluated the mechanical strength of the PCB expanded by heat treatment. To confirm the optimum temperature for heat treatment, the particle size distributions and degree of liberation (DL) of the PCB under the same grinding conditions were analyzed. Finally, the electrostatic separation was attempted to demonstrate the superior separation efficiency of the PCB particles treated at the confirmed temperature than untreated PCB particles.

2. Materials and Methods

2.1. Printed Circuit Boards (PCBs)

PCBAs of waste computers were supplied by an e-waste recycling company (RTECH KOREA, Namyangju-si, Korea). From the supplied PCBAs, ECs were separated using a heating fan, and PCBs without ECs were used. As shown in Figure 2, the PCBs had a layer structure with a thickness of approximately 1.6 mm and contained a mixture of Cu and glass fiber-reinforced epoxy resin (FR-4). The Cu content in the PCBs was found to be 25.36 wt % using ICP-OES (iCAP 7000SERIES, Thermo Fisher Scientific, Waltham, MA, USA). To determine the conditions for heat treatment and thermal property changes, the thermal behavior of the PCBs was evaluated via thermogravimetric analysis (TGA, N-1000, Scinco, Seoul, Korea). In the TGA, the heating rate was set to 5 °C/min, and weight changes were observed while heating from 25 to 600 °C.

2.2. Thermal Treatment Conditions

An electrical furnace was used for thermal treatment. Harmful gas generated during the thermal treatment was minimized by connecting it to a precipitator equipped with a HEPA filter and activated charcoal filter. The PCBs were cut into 50 mm square samples. Twenty samples were placed in an electrical furnace. Heating was performed in air at a heating rate of 5 °C/min and temperatures of 200, 225, 250, 275, 300, and 325 °C. After reaching the target temperature, it was maintained for 30 min.

2.3. Evaluation of Flexible Strength and Degree of Liberation (DL)

To determine the changes in the physical properties of the PCB with thermal treatment, the expansibility and mechanical strength of the PCB were measured. Additionally, the comminution product particle size distribution and the DL of Cu were also evaluated.

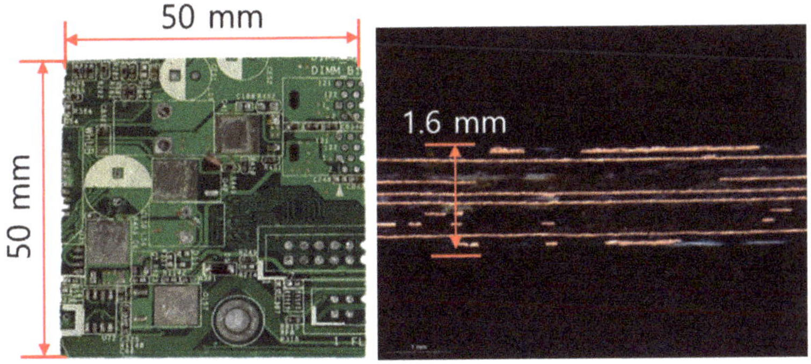

Figure 2. Front and side images of waste shredded PCB with EC disassembled for copper separation from waste PCB.

Changes in PCB thickness according to thermal treatment conditions were measured using an optical microscope (Leica DVM2500, Leica VZ80) to determine the expansibility of the board. Mechanical strength was measured using a universal strength analyzer (H25KS, Tinius Olsen). For strength measurement, the samples used were 50 mm × 15 mm. A load was applied to the central part of the sample, and the load value at the moment of destruction was calculated using Equation (1) to calculate the bending strength σ (MPa) [35].

$$\sigma = \frac{3Pl}{2bd^2} \tag{1}$$

where P represents the load at the moment of destruction (N), l is the length of the support span cycle (m), b is the specimen width (m), and d is the specimen thickness (m). Ten bending strength values were obtained for each set of thermal treatment conditions. The PCBs were heat-treated until the bending strength dropped sharply. Then, they were utilized for the comminution process.

A shredder (TOP-2515-SH, Topcrusher, Gunpo-si, Korea) and cut crusher (TOP-10-CC, Topcrusher, Gunpo-si, Korea) were used in the comminution experiment. The comminution product was separated by sieving into 10 size categories in the range 4000–250 μm to evaluate particle size distribution. The particle size distribution was obtained from the Gates–Gaudin–Schuhmann (GGS) model (Equation (2)) and plotted.

$$F\,(D) = \left(\frac{D}{D^*}\right)^{\alpha} \tag{2}$$

where D represents the particle size (μm); $F\,(D)$ represents the cumulative weight fraction undersize; α represents the distribution modulus; and D^* represents the size modulus (μm). From Equation (2), the smaller α is, the broader the distribution of particle size, whereas the larger the D^*, the larger will be the absolute size of a particle.

DL is defined as the proportion of effective components existing as liberated particles (free particles) among all the effective components. This is expressed in Equation (3).

$$DL_D = \frac{\mathrm{p}\,(g = 1|D)}{\int_{g=0}^{1} g\mathrm{p}\,(g|D).dg} \tag{3}$$

In this equation, DL_D is the DL of Cu in the particle size range D; g is the Cu grade; $\mathrm{p}\,(g|D)$ is the proportion of particles with Cu grade g in the particle size range D; and $\mathrm{p}\,(g = 1|D)$ is the proportion of particles with Cu grade 1 in the particle size range D.

To check the DL of Cu in the PCB comminution product, the particles were sampled in the 355–500 μm, 710–1000 μm, 1400–2000 μm, and 2800–4000 μm size ranges. The ground PCB samples were produced using the cold-mounting method. The cutting plane of the ground pieces was captured using an electron microscope, from which the color representative value of the Cu component was extracted to calculate the total area of the Cu particles [36].

To quantitatively examine the Cu particle distribution, which in turn depends on the comminution product particle size, the Cu grade was analyzed using ICP-OES (iCAP 7000SERIES, Thermo Fisher Scientific). For particle size ranges with high-grade Cu, the entire range of elements was analyzed to improve data reliability, and all impurities except Cu were excluded from the whole metal grade to obtain the Cu grade.

2.4. Electrostatic Separation Experiment

The PCB comminution product was separated into metallic (Cu) and non-metallic particles using a corona discharge electrostatic separator (laboratory-type corona discharge electrostatic separator, ES 01/04S, Eriez, Erie, PA, USA). A corona discharge electrostatic separator consists of a rotating disk electrode, a corona electrode, and an induction electrode. For metallic particles, their high conductivity causes fast electric charge extinction, and the particles fall from the rotating electrode due to inertia. On the other hand, nonmetallic particles have poor conductivity, and charge extinction is very slow. This keeps them attached to the rotating disk electrode until they are separated by a brush. The operational conditions of the electrostatic separator are listed in Table 1.

Table 1. Operating parameters and conditions of corona discharge electrostatic separation for recycling copper from waste PCB particles.

Operating Parameters	Operating Conditions
Supplied voltage (kV)	−30
Rotation speed (rpm)	30
Distance of induction electrode (m)	0.3
Degree of induction electrode (°)	20
Distance of corona electrode (m)	0.3
Degree of corona electrode (°)	60
Relative humidity (%)	35 ± 5

Specimens with (325 °C) and without thermal treatment were subjected to electrostatic separation. Particles sizes ranged from 250 to 4000 μm. In the first electrostatic separation, the comminution product was separated into three types: conductor (metallic particle), middling, and nonconductor (non-metallic particle). The middling was further sorted into conductors and nonconductors. The separated samples were examined for their Cu particle distribution status according to thermal treatment using ICP-OES (iCAP 7000SERIES, Thermo Fisher Scientific). For the separated conductors and nonconductors, we evaluated their yield with respect to the input of each particle size. The Cu grade was analyzed for feed, conductor, and nonconductor products to determine the effect of thermal treatment on electrostatic separation.

3. Results

3.1. Thermal Behavior of the PCB

Figure 3 shows the TGA-DTA of changes in PCB mass with changes in temperature up to 600 °C under air and nitrogen. Pyrolysis occurred in two phases. The first phase involved conversion of the moisture content within the substrate to vapor and the formation of CO_2 in the temperature range of 200–310 °C, resulting in a small decrease in the mass of the sample. The second phase was marked by the disintegration and carbonization of the epoxy resin at 310 °C or higher, causing a rapid reduction in the mass [26]. The temperature with the maximum mass reduction (T_m) was 302 °C. The organic matter was completely decomposed to show no further change in mass at 600 °C or higher. After completion of the reaction, the mass of the specimen was approximately 70% of its initial mass.

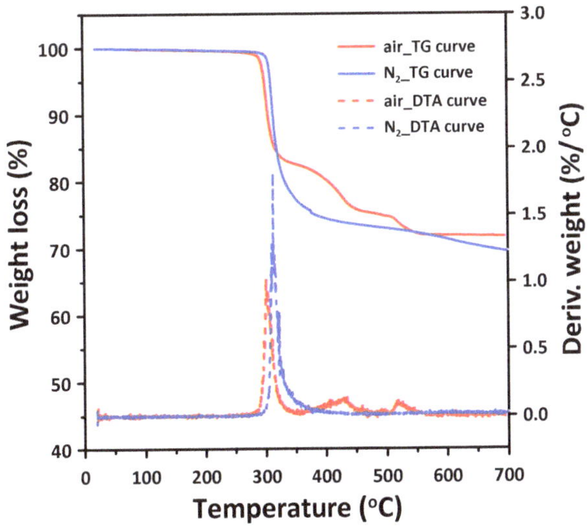

Figure 3. TGA-DTA curves of the PCBs according to the nitrogen and air conditions. Solid lines represent the TGA curve and dotted lines represent the DTA curve.

The purpose of pyrolysis was to minimize the emission of organic matter from the PCB while improving the degree of Cu liberation. We expected that hazardous gas generation would increase and Cu would be oxidized at 325 °C or higher; thus, we set 325 °C as the upper limit for thermal treatment [37,38].

Figure 4 shows the results of the bending strength test of the PCB used in this study according to different temperatures and gas conditions. The PCBs were thermally treated according to each set of conditions, and their bending strengths were measured. For each temperature condition, 10 samples were used to measure the mechanical strength, and their means and standard deviations were calculated. The average bending strength of the boards without any heat treatment was 386.36 MPa. The average bending strengths of PCB-A300 and PCB-A325 were 74.52 MPa and 24.26 MPa respectively. In general, the higher the thermal treatment temperature, the lower is the strength to converge to a certain value. For each thermal treatment atmosphere, the temperature at which the strength rapidly drops (300 °C) was found to be similar to T_m (302 °C), as observed from the DTG analysis results above.

The PCBs were thermally treated from 200 to 325 °C at 25 °C intervals, and their cross-sectional enlargement was observed via optical microscopy, as shown in Figure 4c–h. The board thickness hardly changed from 200 to 275 °C. The board expanded by approximately 11% (1.6 to 1.77 mm) at 300 °C, and the enlargement was approximately 28% (1.6 to 2.04 mm) at 325 °C. Such PCB cross-sectional expansion seemed to have been caused by the fixed glass fiber, but it was eventually released as the epoxy resin disintegrated during pyrolysis. Such delaminated samples are expected to have structural defects inside and experience a mechanical strength drop. To verify this, the bending strength of the PCB was measured according to the thermal treatment conditions. The PCB thermally treated at 300 °C in air and nitrogen are referred to as PCB-A300 and PCB-N300, respectively. PCB without any thermal treatment is referred to as a PCB-WT.

3.2. Comminution Behavior of Thermally Treated PCBs

Figure 5 shows the particle size distributions of the heat-treated PCB particles crushed by shredding and cut crushing according to the different temperatures and gas conditions. The PCB comminution product was sieved into 10 size categories ranging from 250 to 4000 µm, and their particle size distribution was identified. This model was applied to the GGS model. The higher the thermal treatment temperature, the lower the distribution modulus α, and the higher the size modulus D* because PCB-A300 and PCB-A325 were liberated with particle sizes larger than those of PCB-WT. In other words, the liberated FR-4 component was rapidly comminuted, owing to brittleness and high concentration in the size range of 250 μm or smaller, lowering the distribution modulus values. In contrast, the thin Cu plate liberated in the comminution process was ductile and produced with a larger particle size, resulting in higher size modulus values. The PCB-A300 and PCB-A325 also generated untied glass fibers that remained tangled during sieving and concentrated at particle sizes of 4000 μm or larger. Therefore, the maximum particle size also increased.

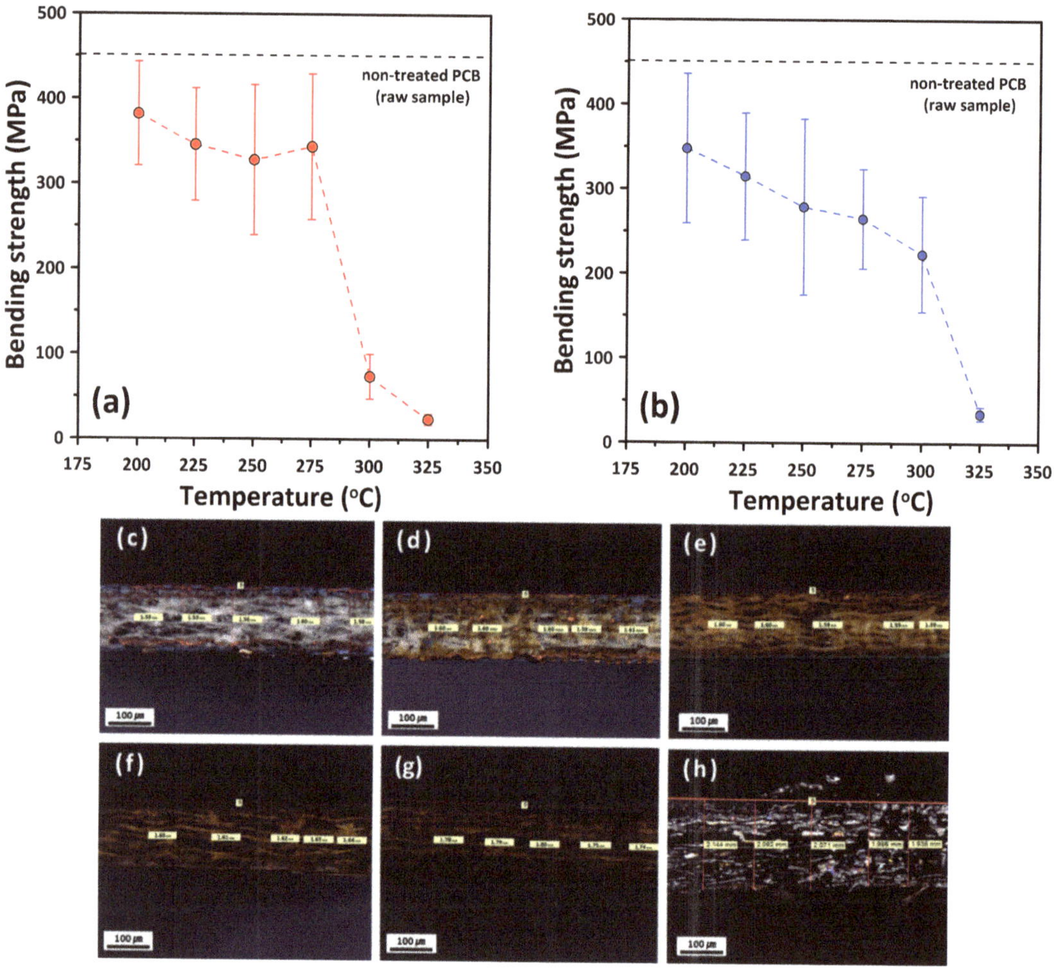

Figure 4. Bending strength results of PCB used in this study according the different temperatures and gas conditions: (**a**) air and (**b**) nitrogen. The black dotted line represents the nontreated PCB sample. The error bar denotes standard deviations. SEM images of the PCB cross-section treated from thermal process: (**c**) 275 °C in air, (**d**) 300 °C in air, (**e**) 325 °C in air, and (**f**) 275 °C in N_2, (**g**) 300 °C in N_2, and (**h**) 325 °C in N_2.

From the results of the size distribution of the crushed PCB particles according to the heat treatment, we determined that 300 °C is more suitable than a relatively high temperature (i.e., 325 °C). Therefore, among the comminution products of PCB-WT and PCB-A300, products ranging in size from 2800 to 4000 μm were used as test samples to show the degree of liberation (by image/optically). The particles were observed using an optical microscope, as shown in Figure 6. Of the PCB comminution products without thermal treatment, Cu was observed to remain in a layered structure together with FR-4. On the other hand, of the comminution products of PCB-A300, Cu was found as free particles with some irregular shapes, even in the 2800–4000 μm size range. This is because the liberated Cu particles were crumpled by shock inside the crusher, and it was indirectly verified that for this reason, the materials were separated from one another.

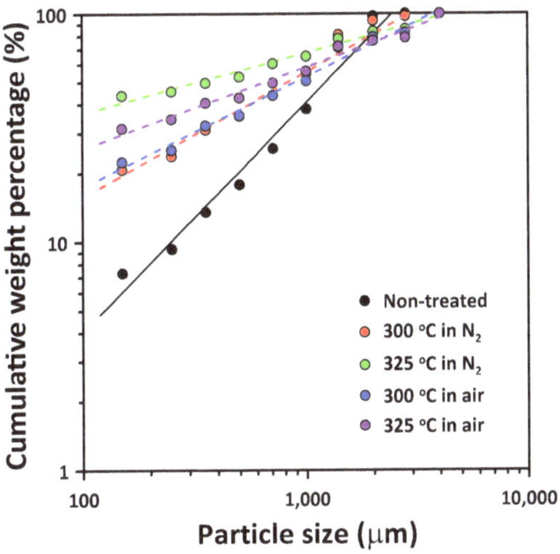

Figure 5. Particle size distributions of the thermally treated PCB particles crushed by shredding and cut crushing according to the different temperatures and gas conditions.

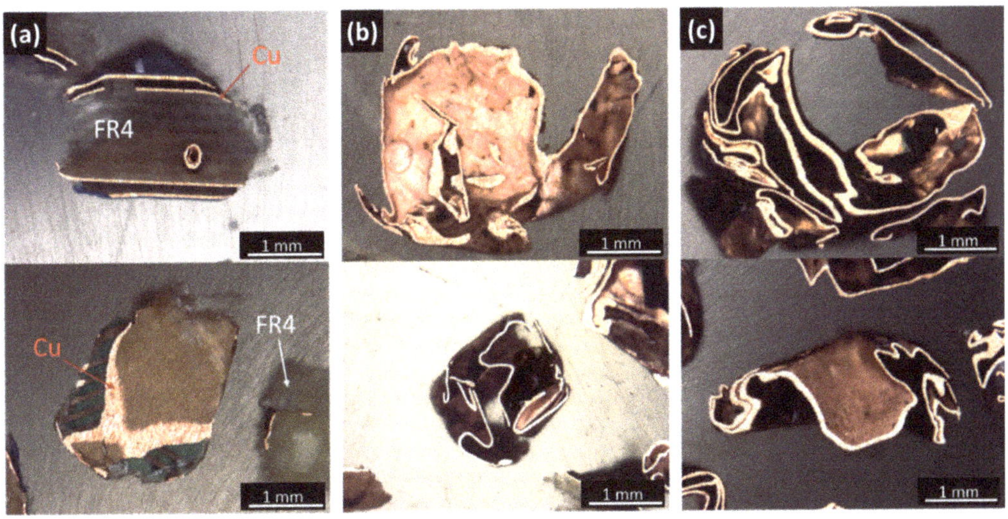

Figure 6. SEM images of the crushed PCB particles in the range of −4000 to +2800 μm: (**a**) nontreated raw, (**b**) 300 °C in air, and (**c**) 300 °C in nitrogen.

The images of PCB comminution products in the 355–500 μm, 710–1000 μm, 1400–2000 μm, and 2800–4000 μm size range were taken with a digital microscope, and the DL of Cu was assessed (see Figure 7). In the comminution products of PCB-WT, the DL of Cu was 0% in the range of 2800–4000 μm and less than 80% in the 355–500 μm range. On the other hand, PCB-A300 and PCB-N300 showed a significant increase in the DL of Cu in the comminution products. More specifically, for PCB-A300, the DL was 80.1% at 2800 μm or higher.

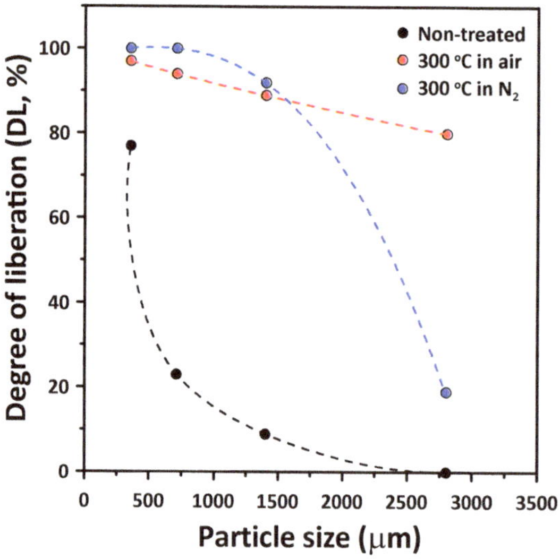

Figure 7. Degree of liberation (DL) of the thermally treated PCB particles at 300 °C in the various gas conditions (air and nitrogen).

For PCB-N300, the DL was lower at particle sizes greater than 1400 μm. From this, it is expected that the PCB-A300 sample will have excellent Cu separation performance by electrostatic separation compared to the PCB-N300 sample.

3.3. Cu Concentration by Electrostatic Separation

The above comprehensive results (e.g., particle size and liberation) indicate a high DL in the relatively large PCB-A300 particles. Therefore, the yield, Cu grade, and recovery evaluation through the electrostatic separation test were performed with PCB-A300 as well as with PCB-WT for comparison.

The comminution products of both PCB-WT and PCB-A300 in the 250–4000 μm particle size range were placed in an electrostatic separator. The yield of each product was calculated against the initial amount, as shown in Figure 8. PCB-WT showed no significant difference in yield between conductors and nonconductors. However, PCB-A300 exhibited a large difference in yield between conductor and nonconductor, and this difference was very apparent with particle size. At 500 μm or larger, the conductor ratio was very high, whereas for 500 μm or lower, the nonconductor ratio was high.

Figure 9 shows Cu grades of conductive particles according to the different heat treatment conditions. As seen in Figure 9a, for PCB-WT, the Cu grade of the conductor product increased with a decrease in the particle size and became the highest at 70 wt % in the 355–500 μm size range. For PCB-A300, the Cu grade was 70 wt % or more for all size ranges. The higher the particle size, the higher the Cu grade. In the particle size range of 710–1000 μm or more, the Cu grade was 90% or larger, and in the 2000–2800 μm range, the Cu grade was the highest at 99 wt %.

Figure 9b shows the recovery of PCB-A300 and PCB-WT. The PCB-WT first showed an increase in recovery with an increase in particle size, after which it fell sharply in the range of 2800–4000 μm. For PCB-A300, the recovery started at approximately 84 wt % in the 250–355 μm range; then, it gradually decreased until the 1400–2000 μm range, and then, it increased to 90 wt % or higher. Particles with a high DL were able to achieve a high recovery regardless of the particle size. Remarkably, the recovery increased compared to the recovery of non-thermally treated particles, which was low in the 2800–4000 μm size range.

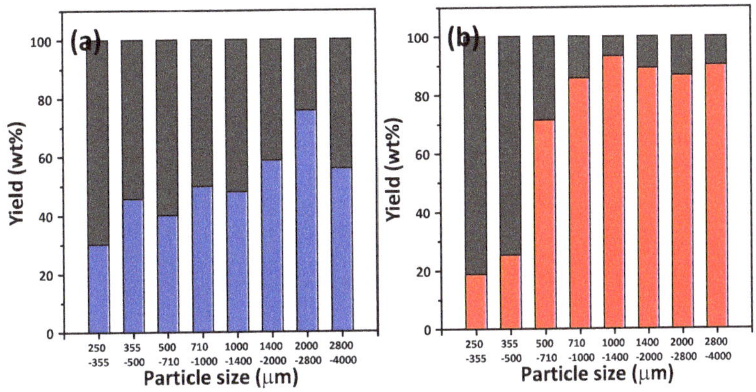

Figure 8. Yield of conductive and nonconductive particles after electrostatics separation for (**a**) the crushed PCB particles (PCB-WT) and (**b**) the crushed PCB particles treated at 300 °C in air (PCB-A300). Gray bar represents nonconductive yield, and red and blue represent conductive yield, respectively.

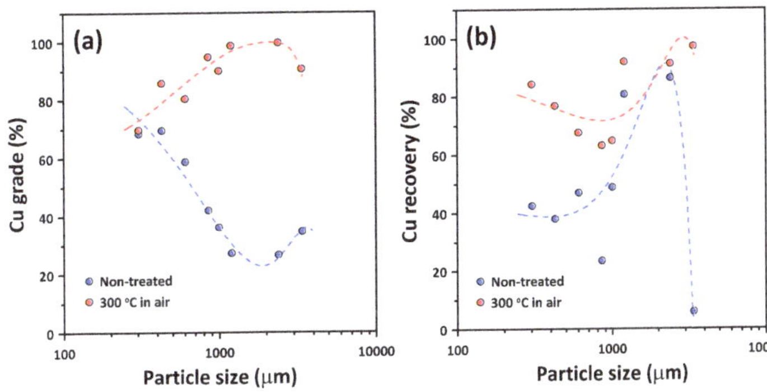

Figure 9. Cu (**a**) grade and (**b**) recovery of the crushed PCB particles (PCB-WT) and (**b**) the crushed PCB particles by electrostatic separation treated at 300 °C in air (PCB-A300).

4. Discussion

PCB layer separation according to thermal treatment can be explained by the differences in the coefficient of thermal expansion between the component materials of the PCB. The coefficients of thermal expansion of Cu, glass fiber, and epoxy resin that form the lamination structure of PCB are 16.7, 15–100, and 54 ppm/°C, respectively. When a PCB is heated, the differences in the thermal expansion ratios between the component materials lead to their expansion at mutually different expansion rates. In this process, internal shear stress is generated in the layers between them. When this internal stress exceeds the bonding strength between the board components, delamination occurs inside the board. If delamination does not occur, the components are still cooled to room temperature, and the internal stress remains in the PCB as residual stress to reduce the mechanical strength. This concept has already been proven both numerically and experimentally. Janáčová et al. [38] simulated the thermal expansion by employing the linear expansion ratios of epoxy resin and Cu that form PCB as well as the bond strength between the two. Through numerical analysis, they concluded that 350 °C was the optimal temperature for Cu layer separation. Li et al. [29] obtained SEM images of the cross-section of a PCB layer and found that micro delamination occurred after thermal treatment. They also found from the SEM micrographs of the epoxy resin interface that PCB thermal treatment helped epoxy resin to decrease the cohesive force between Cu and glass fiber that formed the laminar structure.

Materials consisting of conjugated substances, such as PCBs, have two markedly different properties. Such materials exhibit many unique characteristics during comminution. Ductility has the largest influence on comminution. In the grinding of a substance, its ductility determines the breakage rate (commonly called the specific rate of breakage in traditional mineral processing). Specifically, because Cu has a very high ductility, similar to any other metal components, it tends to be wrinkled rather than breakage during comminution. Thus, the particle size did not decrease easily. On the other hand, FR-4, a compound of epoxy resin and glass fiber, has a very low ductility (or has a very high brittleness, to put it differently). Therefore, once liberated, it was comminuted rapidly [39]. Park et al. [22] suggested that the comminution of objects consisting of two substances with mutually different ductilities shows the following behavior with a decrease in particle size.

(1) Only size reduction without liberation causes a grade change in each size fraction (e.g., Cu and FR-4 in the present study). (2) Liberation first increases the grade of the high-ductility substance (Cu in the present study) at coarser size fraction. (3) The high-ductility substance grade reaches its peak value and then decreases gradually at the fraction. (4) The high-ductility grade at the fraction of the substance reaches its minimum value and converges to a certain value. If the test particle sizes for moving from stage 1 to stage 2, stage 2 to stage 3, and stage 3 to stage 4 are D1, D2, and D3, respectively, the key of comminution preprocessing is to increase D1 and D2. In the present study, the PCBs without thermal treatment showed that D1—the particle size at which each component started to be liberated—was approximately 2000 μm; and D2, the particle size at which the Cu grade peaks, was approximately 800 μm. PCBs with potential layer separation through thermal treatment were rapidly liberated, even with a small impact. In the sieve separation with the size ranging between 250 and 4000 μm, D1 was not observed, and D2 was also indeterminate, suggesting that Cu liberation started from a particle size of 4000 μm or larger. The Cu and FR-4 liberated for large particle sizes had the maximum ductility difference, showing a large gap in each substance grade according to particle size. Given that it is difficult to physically separate the effective components from the ineffective ones in smaller particle groups [40,41], it is necessary to liberate the target component, Cu, from a large particle size range. In addition, for a particle size of 4000 μm or larger, a considerable amount of glass fibers remained tangled. Therefore, approximately 20.45% of the non-metal components could be removed by screening only. As shown above, under thermal treatment, Cu was liberated in particle sizes of a few millimeters only through crushing without fine grinding. This is a significant result in terms of energy savings and the reduction of metal loss by fine dust scattering.

Based on the electrostatic separation results, we could reaffirm the effect of thermal treatment on the liberation and separation efficiency enhancement. Compared with normal PCB comminution products, thermally treated PCB comminution products were found to have a marked contrast in yield between conductor and nonconductor for each particle size. This is evidently because of the Cu grade difference in the PCB comminution products depending on the particle size. Electrostatic separation enhancement can also be verified by the Cu grade in the conductor recovered via electrostatic separation under various conditions.

In the recovery studies, a reduction in Cu recovery for particles larger than 4000 μm of comminution products of PCB-NT was not expected. This can be explained as follows: In electrostatic separation, the selection efficiency is affected by the component through which a particle is attached to the rotation electrode. In other words, if a metal ingredient is attached to the electrode, the particles are collected in a conductor container. However, if a non-metal ingredient is attached, the particles are collected in a nonconductor container. If there is no delamination, only the nonconductor part of the particles is exposed, and then, the particles are collected in a nonconductor container. Therefore, many particles 4000 μm or larger were collected in the nonconductor container, resulting in decreased recovery. The PCB-A300 comminution products showed higher recovery than the comminution products of PCB-NTs for all sizes.

The collection of metal particles with large particle sizes is also beneficial in the metallurgy stage after physical separation. Specifically, it was reported that the increasing numbers of minute particles undermine the leaching efficiency because of increased particle–particle collision and severe attrition imposition. Furthermore, a problem is introduced in the filtration process following the leaching procedure [42–44]. Subsequently, the physical separation of coarse particles can be regarded as important for improving the efficiency of chemical processing after mechanical processing.

Therefore, after thermal treatment as a pretreatment step of comminution, the delamination of PCB increased the liberated particle size and drastically enhanced the physical separation efficiency. The thermal treatment temperature and conditions necessary for enhancing the liberation of metal/non-metal components in PCBs vary according to the origin of the PCB. Therefore, it is necessary to determine the properties of the corresponding samples. More studies are necessary to improve the physical separation process and purification process for Cu recovered in coarse size.

5. Conclusions

We implemented thermal treatment as a pretreatment to enhance Cu separation efficiency for waste PCBs from waste computers. We also investigated the changes in the properties of PCBs, DL, and separation efficiency, which depended upon the thermal treatment. The heat treatment of PCB was conducted at temperatures ranging from 200 to 325 °C under N_2 and air atmospheres. TGA revealed that the largest mass reduction (T_m) occurred at 302 °C under both N_2 and air. For PCB undergoing thermal treatment, delamination at 300 °C caused an increase in thickness. Moreover, at 300 °C, i.e., near the T_m, the bending strength decreased from approximately 386.36 to 74.52 MPa, and at 325 °C, it was approximately 24.26 MPa. The heat treatment has the effect of reducing the bonding strength between the glass fiber and the copper layer. This reduction in bonding force can play a role in improving the DL when crushing PCB. In this context, the PCB samples thermally treated at 300 °C were comminuted, which was followed by sieving to investigate the Cu liberation at various particle size ranges. The DL of copper was enhanced for all particle sizes. In particular, the DL increased significantly from 9.3% to 100% in the size range of 1400–2000 μm when the PCB was heat treated at 300 °C in air. Under these conditions, a high Cu grade concentrate with a high recovery could be obtained via electrostatic separation. As a result, this study elucidates the physical properties of PCBs and the DL under various heat treatment conditions. Heat treatment at 300 °C in air is found to be the most suitable for recovering copper from the PCB in terms of the DL.

Author Contributions: Conceptualization: B.K., S.H., S.P., D.-W.K., and Y.H.; methodology: B.K., S.H., S.P., M.J., S.K., H.-S.J., D.-W.K., and Y.H.; validation: B.K., S.H., S.P., M.J., S.K., and Y.H.; formal analysis: B.K., S.H., S.P.; investigation: B.K., S.H., S.P., M.J., C.-H.P., H.-S.J., and Y.H.; writing—original draft: B.K., S.H., S.P., S.K., M.J., and Y.H.; writing—review and editing: B.K., S.H., S.P., S.K., M.J., D.-W.K., and Y.H.; resources: C.-H.P., H.-S.J., D.-W.K., and Y.H.; supervision: D.-W.K., and Y.H. All authors have read and agreed to the published version of the manuscript.

Funding: This research was supported by the Basic Research Project (GP2020-013, 21-3212-1) of the Korea Institute of Geoscience and Mineral Resources (KIGAM) funded by the Ministry of Science, ICT, and Future Planning of Korea, and this work was also supported by the Korea Institute of Energy Technology Evaluation and Planning (KETEP) and the Ministry of Trade, Industry & Energy (MOTIE) of the Republic of Korea (No. 20206110100030).

Conflicts of Interest: The authors declare no conflict of interest.

References

1. Li, J.; Xu, T.; Liu, J.; Wen, J.; Gong, S. Bioleaching metals from waste electrical and electronic equipment (WEEE) by Aspergil-lus niger: A review. *Environ. Sci. Pollut. Res.* **2021**, *28*, 1–16. [CrossRef] [PubMed]
2. Cardamone, G.F.; Ardolino, F.; Arena, U. About the environmental sustainability of the European management of WEEE plastics. *Waste Manag.* **2021**, *126*, 119–132. [CrossRef] [PubMed]
3. Behnamfard, A.; Salarirad, M.M.; Veglio, F. Process development for recovery of copper and precious metals from waste printed circuit boards with emphasize on palladium and gold leaching and precipitation. *Waste Manag.* **2013**, *33*, 2354–2363. [CrossRef]
4. Fujita, T.; Ono, H.; Dodbiba, G.; Yamaguchi, K. Evaluation of a recycling process for printed circuit board by physical separation and heat treatment. *Waste Manag.* **2014**, *34*, 1264–1273. [CrossRef]
5. Ghosh, B.; Ghosh, M.; Parhi, P.; Mukherjee, P.; Mishra, B. Waste Printed Circuit Boards recycling: An extensive assessment of current status. *J. Clean. Prod.* **2015**, *94*, 5–19. [CrossRef]
6. Otsuki, A.; De La Mensbruge, L.; King, A.; Serranti, S.; Fiore, L.; Bonifazi, G. Non-destructive characterization of mechanically processed waste printed circuit boards-particle liberation analysis. *Waste Manag.* **2020**, *102*, 510–519. [CrossRef]
7. Park, Y.J.; Fray, D.J. Recovery of high purity precious metals from printed circuit boards. *J. Hazard. Mater.* **2009**, *164*, 1152–1158. [CrossRef]
8. Kim, S.; Park, S.; Han, S.; Han, Y.; Park, J. Silanol-rich ordered mesoporous silica modified thiol group for enhanced recovery performance of Au(III) in acidic leachate solution. *Chem. Eng. J.* **2018**, *351*, 1027–1037. [CrossRef]
9. Cui, J.; Zhang, L. Metallurgical recovery of metals from electronic waste: A review. *J. Hazard. Mater.* **2008**, *158*, 228–256. [CrossRef]
10. Hino, T.; Agawa, R.; Moriya, Y.; Nishida, M.; Tsugita, Y.; Araki, T. Techniques to separate metal from waste printed circuit boards from discarded personal computers. *J. Mater. Cycles Waste Manag.* **2009**, *11*, 42–54. [CrossRef]
11. Lee, J.; Kim, Y.; Lee, J.-C. Disassembly and physical separation of electric/electronic components layered in printed circuit boards (PCB). *J. Hazard. Mater.* **2012**, *241*, 387–394. [CrossRef] [PubMed]
12. Cucchiella, F.; D'Adamo, I.; Koh, S.L.; Rosa, P. Recycling of WEEEs: An economic assessment of present and future e-waste streams. *Renew. Sustain. Energy Rev.* **2015**, *51*, 263–272. [CrossRef]
13. Cilek, E.C.; Tuzci, G. Flotation behavior of native gold and gold-bearing sulfide minerals in a polymetallic gold ore. *Part. Sci. Technol.* **2021**, 1–9. [CrossRef]
14. Jameson, G.J.; Emer, C. Coarse chalcopyrite recovery in a universal froth flotation machine. *Miner. Eng.* **2019**, *134*, 118–133. [CrossRef]
15. Silvas, F.P.; Correa, M.M.J.; Caldas, M.P.; de Moraes, V.T.; Espinosa, D.C.; Tenório, J.A. Printed circuit board recycling: Physical processing and copper extraction by selective leaching. *Waste Manag.* **2015**, *46*, 503–510. [CrossRef] [PubMed]
16. Duan, H.; Hou, K.; Li, J.; Zhu, X. Examining the technology acceptance for dismantling of waste printed circuit boards in light of recycling and environmental concerns. *J. Environ. Manag.* **2010**, *92*, 392–399. [CrossRef]
17. Park, S.; Kim, S.; Han, Y.; Park, J. Apparatus for electronic component disassembly from printed circuit board assembly in e-wastes. *Int. J. Miner. Process.* **2015**, *144*, 11–15. [CrossRef]
18. Cui, J.; Forssberg, E. Mechanical recycling of waste electric and electronic equipment: A review. *J. Hazard. Mater.* **2003**, *99*, 243–263. [CrossRef]
19. Li, J.; Lu, H.; Guo, J.; Xu, Z.; Zhou, Y. Recycle Technology for Recovering Resources and Products from Waste Printed Circuit Boards. *Environ. Sci. Technol.* **2007**, *41*, 1995–2000. [CrossRef] [PubMed]
20. Veit, H.M.; De Pereira, C.C.; Bernardes, A.M. Using mechanical processing in recycling printed wiring boards. *JOM* **2002**, *54*, 45–47. [CrossRef]
21. Guo, C.; Wang, H.; Liang, W.; Fu, J.; Yi, X. Liberation characteristic and physical separation of printed circuit board (PCB). *Waste Manag.* **2011**, *31*, 2161–2166. [CrossRef] [PubMed]
22. Park, S.; Kim, S.; Han, S.; Kim, B.; Kim, B.; Han, Y.; Park, J. Liberation Characteristics Assessment for Copper Component in PCB Comminution Product by Image Analysis. *Mater. Trans.* **2018**, *59*, 1493–1500. [CrossRef]
23. Jie, G.; Ying-Shun, L.; Mai-Xi, L. Product characterization of waste printed circuit board by pyrolysis. *J. Anal. Appl. Pyrolysis* **2008**, *83*, 185–189. [CrossRef]
24. Zhao, Y.; Wen, X.; Li, B.; Tao, D. Recovery of copper from waste printed circuit boards. *Mining, Met. Explor.* **2004**, *21*, 99–102. [CrossRef]
25. Kang, K.D.; Ilankoon, I.M.S.K.; Dushyantha, N.; Chong, M.N. Assessment of Pre-Treatment Techniques for Coarse Printed Circuit Boards (PCBs) Recycling. *Minerals* **2021**, *11*, 1134. [CrossRef]
26. Verma, H.R.; Singh, K.K.; Mankhand, T.R. Liberation of metal clads of waste printed circuit boards by removal of halogenated epoxy resin substrate using dimethylacetamide. *Waste. Manag.* **2017**, *60*, 652–659. [CrossRef]
27. Huang, Y.F.; Pan, M.W.; Lo, S.L. Hydrometallurgical metal recovery from waste printed circuit boards pretreated by microwave pyrolysis. *Resour. Conserv. Recycl.* **2020**, *163*, 105090. [CrossRef]
28. Li, J.; Duan, H.; Yu, K.; Liu, L.; Wang, S. Characteristic of low-temperature pyrolysis of printed circuit boards subjected to various atmosphere. *Resour. Conserv. Recycl.* **2010**, *54*, 610–615. [CrossRef]
29. Li, J.; Duan, H.; Yu, K.; Wang, S. Interfacial and Mechanical Property Analysis of Waste Printed Circuit Boards Subject to Thermal Shock. *J. Air Waste Manag. Assoc.* **2010**, *60*, 229–236. [CrossRef]

30. Hadi, P.; Xu, M.; Lin, C.S.; Hui, C.-W.; McKay, G. Waste printed circuit board recycling techniques and product utilization. *J. Hazard. Mater.* **2014**, *283*, 234–243. [CrossRef]
31. Li, J.; Lu, H.; Liu, S.; Xu, Z. Optimizing the operating parameters of corona electrostatic separation for recycling waste scraped printed circuit boards by computer simulation of electric field. *J. Hazard. Mater.* **2008**, *153*, 269–275. [CrossRef] [PubMed]
32. Veit, H.M.; Diehl, T.R.; Salami, A.P.; Rodrigues, J.d.S.; Bernardes, A.M.; Tenório, J.A.S. Utilization of magnetic and electro-static separation in the recycling of printed circuit boards scrap. *Waste Manag.* **2005**, *25*, 67–74. [CrossRef]
33. Li, J.; Xu, Z.; Zhou, Y. Application of corona discharge and electrostatic force to separate metals and nonmetals from crushed particles of waste printed circuit boards. *J. Electrost.* **2007**, *65*, 233–238. [CrossRef]
34. Xue, M.; Yan, G.; Li, J.; Xu, Z. Electrostatic Separation for Recycling Conductors, Semiconductors, and Nonconductors from Electronic Waste. *Environ. Sci. Technol.* **2012**, *46*, 10556–10563. [CrossRef]
35. Han, Y.; Kim, S.B.; Kim, H. TiO_2-Coated Silica Foams by In-Situ Sol-Gel Reaction. *Mater. Trans.* **2011**, *52*, 2245–2249. [CrossRef]
36. Barontini, F.; Marsanich, K.; Petarca, L.; Cozzani, V. Thermal Degradation and Decomposition Products of Electronic Boards Containing BFRs. *Ind. Eng. Chem. Res.* **2005**, *44*, 4186–4199. [CrossRef]
37. Chiang, H.-L.; Lo, C.-C.; Ma, S.-Y. Characteristics of exhaust gas, liquid products, and residues of printed circuit boards using the pyrolysis process. *Environ. Sci. Pollut. Res.* **2009**, *17*, 624–633. [CrossRef]
38. Janáčová, D.; Kolomazník, K.; Vašek, V.; Křenek, J.; Uhlířová, M. Separation of printed circuit board by temperature shock. In Proceedings of the 5th WSEAS Conference on Heat Transfer, Thermal Engineering and Environment HTE'07, Athens, Greece, 25–27 August 2007.
39. Koyanaka, S.; Endoh, S.; Ohya, H. Effect of impact velocity control on selective grinding of waste printed circuit boards. *Adv. Powder Technol.* **2006**, *17*, 113–126. [CrossRef]
40. Richard, G.M.; Mario, M.; Javier, T.; Susana, T. Optimization of the recovery of plastics for recycling by density media separation cyclones. *Resour. Conserv. Recycl.* **2011**, *55*, 472–482. [CrossRef]
41. Tuncuk, A.; Stazi, V.; Akcil, A.; Yazici, E.Y.; Deveci, H. Aqueous metal recovery techniques from e-scrap: Hydrometallurgy in recycling. *Miner. Eng.* **2012**, *25*, 28–37. [CrossRef]
42. Huang, J.; Chen, M.; Chen, H.; Chen, S.; Sun, Q. Leaching behavior of copper from waste printed circuit boards with Brønsted acidic ionic liquid. *Waste Manag.* **2014**, *34*, 483–488. [CrossRef] [PubMed]
43. Kamberović, Ž.; Korać, M.; Ivšić, D.; Nikolić, V.; Ranitović, M. Hydrometallurgical process for extraction of metals from electronic waste-part I: Material characterization and process option selection. *Metall. Mater. Eng.* **2018**, *15*, 231–243. [CrossRef]
44. Jadhao, P.; Chauhan, G.; Pant, K.; Nigam, K. Greener approach for the extraction of copper metal from electronic waste. *Waste Manag.* **2015**, *57*, 102–112. [CrossRef] [PubMed]

minerals

MDPI

Article

Recovery of Rare Earth Metals (REMs) from Nickel Metal Hydride Batteries of Electric Vehicles

Manis Kumar Jha [1,*], Pankaj Kumar Choubey [1], Om Shankar Dinkar [1], Rekha Panda [1], Rajesh Kumar Jyothi [2], Kyoungkeun Yoo [3] and Ilhwan Park [4,*]

[1] Metal Extraction and Recycling Division, CSIR-National Metallurgical Laboratory, Jamshedpur 831007, India; impankaj.choubey@gmail.com (P.K.C.); omshankarojha@gmail.com (O.S.D.); rekhapanda1608@gmail.com (R.P.)
[2] Mineral Resources Research Division, Korea Institute of Geoscience and Mineral Resources, Daejeon 34132, Korea; rkumarphd@kigam.re.kr
[3] Department of Energy and Resources Engineering, Korea Maritime and Ocean University (KMOU), Busan 49112, Korea; kyoo@kmou.ac.kr
[4] Division of Sustainable Resources Engineering, Faculty of Engineering, Hokkaido University, Sapporo 060-8606, Japan
* Correspondence: mkjha@nmlindia.org (M.K.J.); i-park@eng.hokudai.ac.jp (I.P.)

Abstract: Nickel metal hydride (NiMH) batteries are extensively used in the manufacturing of portable electronic devices as well as electric vehicles due to their specific properties including high energy density, precise volume, resistance to overcharge, etc. These NiMH batteries contain significant amounts of rare earth metals (REMs) along with Co and Ni which are discarded due to illegal dumping and improper recycling practices. In view of their strategic, economic, and industrial importance, and to mitigate the demand and supply gap of REMs and the limited availability of natural resources, it is necessary to explore secondary resources of REMs. Therefore, the present paper reports a feasible hydrometallurgical process flowsheet for the recovery of REMs and valuable metals from spent NiMH batteries. More than 90% dissolution of REMs (Nd, Ce and La) was achieved using 2 M H_2SO_4 at 75 °C in 60 min in the presence of 10% H_2O_2 (v/v). From the obtained leach liquor, the REMs, such as Nd and Ce, were recovered using 10% PC88A diluted in kerosene at eq. pH 1.5 and O/A ratio 1/1 in two stages of counter current extraction. La of 99% purity was selectively precipitated from the leach liquor in the pH range of 1.5 to 2.0, leaving Cu, Ni and Co in the filtrate. Further, Cu and Ni were extracted with LIX 84 at equilibrium pH 2.5 and 5, leaving Co in the raffinate. The developed process flow sheet is feasible and has potential for industrial exploitation after scale-up/pilot trails.

Keywords: NiMH batteries; rare earth metals; leaching; solvent extraction; precipitation

check for updates

Citation: Jha, M.K.; Choubey, P.K.; Dinkar, O.S.; Panda, R.; Jyothi, R.K.; Yoo, K.; Park, I. Recovery of Rare Earth Metals (REMs) from Nickel Metal Hydride Batteries of Electric Vehicles. *Minerals* **2022**, *12*, 34. https://doi.org/10.3390/min12010034

Academic Editor: Kenneth N. Han

Received: 30 November 2021
Accepted: 21 December 2021
Published: 25 December 2021

1. Introduction

Rare earth metals (REMs) are called critical raw materials due to their economic, strategic and industrial importance [1]. The unavailability of the primary resources for these REMs and supply restrictions imposed by China, compelled the researchers to explore secondary resources for the extraction of REMs [2,3]. In this connection, e-waste is one of the fastest emerging areas for the generation of secondary resources containing REMs [4,5]; e-waste contains a several fold higher concentration of REMs in comparison to their availability in natural resources [6]. One of the most prominent secondary resources of e-waste is nickel metal hydride (NiMH) batteries, which are extensively used in portable electronic devices (e.g., notebook, computers, video recorders, mobile phones, etc.) as well as heavily in electric vehicles (EVs) due to their high energy density, specific volume, resistance to overcharge and no electrolyte consumption during charge and discharge [7,8]. Therefore, NiMH batteries are a potential secondary resource of rare earth elements as well as of cobalt and nickel. The proper recycling of NiMH batteries will not only add economic value but also protect the environment.

Several authors have reported various recycling methods for the recovery of REMs and other valuable non-ferrous metals from spent NiMH batteries. Three different paths (i.e., mechanical processing, pyrometallurgy and hydrometallurgy) have been employed for the recovery of metallic values from spent NiMH batteries [9]. Mechanical processing, usually consisting of grinding, sieving and bed elutriation, is carried out to recover valuable materials [10]. In pyrometallurgy, thermal separation of REMs, such as La, Ce, Nd and Pr in oxidized form, is carried out from spent NiMH batteries [11]. Among the three paths, hydrometallurgy is most suitable and appropriate for the refining and recovery of REMs from spent NiMH batteries [12–14].

Different hydrometallurgical processes, such as leaching [15–17], solvent extraction and precipitation [18–20], using various lixiviants, extractants and precipitants, respectively, have been reported for metal extraction from spent NiMH batteries. Yang et al. [20] reported the leaching of REMs using HCl from waste NiMH batteries. Other impurities were removed using ammonia solution and 99% rare earth oxides were recovered [20]. Bertuol et al. [18] reported the sulfuric acid leaching of REMs followed by their selective precipitation using NaOH at pH 1.2. Fernandes et al. [21] reported the chloride leaching of REMs from NiMH batteries followed by their selective extraction using PC88A. Innocenzi and Veglio [19] also studied the sulfuric acid leaching of REMs from spent NiMH batteries followed by their selective precipitation using NaOH at pH below 2. Zhang et al. [22] reported the chloride leaching of REMs followed by solvent extraction using D2EHPA. Xia et al. [23] reported recovery of REMs by a method of leaching using H_2SO_4, followed by solvent extraction using Cyanex 923. Finally, REM oxide was obtained using oxalic acid [23]. Two different extractants, D2EHPA diluted in kerosene [24] and Cyanex923 [25], have been used for the extraction of REMs from spent NiMH batteries. The method of chemical precipitation is most used for REM recovery owing to its ease of operation and low cost [20,26]. The use of anhydrous sodium sulfate [27] and sodium hydroxide [19] has been reported for the recovery of REMs from spent NiMH batteries. Pietrelli et al. [28] studied sulfuric acid leaching of REMs, Ni, Co, Cd and Fe from NiCd and NiMH batteries. Ahn et al. [29] reported the recovery of REMs in the form of their mixed oxides using a leaching and precipitation process. However, the selective extraction of Ce, La and Nd from the leach liquor of NiMH has not been discussed in this paper.

The literature review suggests that the precipitation process is not selective for the separation of REMs (e.g., La, Ce and Nd). A combined process, consisting of solvent extraction and precipitation, is needed to be developed to selectively extract the REMs. Therefore, the present study reports a novel hydrometallurgical process consisting of leaching, solvent extraction and precipitation for recovery of REMs, including valuable metals, such as Cu, Ni and Co, in a closed-loop system, considering the zero-waste discharge concept. REMs were first dissolved in sulfuric acid and thereafter Ce and Nd were extracted with PC88A. Subsequently, La was selectively precipitated in a range of pH 1.5 to 2. Further, Cu and Ni were selectively extracted using LIX 84 leaving Co in the raffinate. The developed process flow sheet is novel for the selective extraction of REMs.

2. Experimental

2.1. Materials

Discarded nickel metal hydride (NiMH) batteries of electric vehicles were collected locally to be used for the experimental purpose. Initially, discarded NiMH batteries were discharged using a saturated solution of sodium chloride. Then, NiMH batteries were pretreated by a crushing and flotation process to separate the black powder, metallic fraction and plastic materials, as shown in Figure 1. The black powder was dissolved in aqua regia to analyze the concentration of metals. The typical composition of black powder is shown in Table 1.

Figure 1. Pre-treatment of discarded electric vehicle batteries.

Table 1. Elemental composition by weight (wt. %) of the black power of an NiMH battery used in electric vehicles.

Elements	La	Ce	Nd	Cu	Ni	Co	Other
Contents	7.7	12	4.0	6.5	29	5.3	balance

2.2. Methodology

2.2.1. Leaching Procedure

Leaching experiments were carried out in a three-necked Pyrex glass reactor (1 L capacity) fitted with a condenser to condense back the gases evolved during the leaching studies (Figure 2). Hotplates with a temperature sensor and magnetic stirrer facilities were used during leaching experiments. The various concentration of sulfuric acid (0.5–2.0 M) was used as a leachant to dissolve the metals at different temperatures, varying from 25 to 75 °C. When the solution attained the desired temperature, a certain amount of black powder was poured into the solution to leach the metals. Further, the slurry was processed for filtration to obtain leach liquor and leached residue. The dissolved metals present in the leach liquor and remaining in the residue were analyzed using an atomic absorption spectrometer (AAS, AAnalyst 200, Perkin Elmer, Waltham, MA, USA) and inductively coupled plasma-optical emission spectroscopy (ICP-OES, VISTA-MPX, Varian Inc., Victoria, Australia).

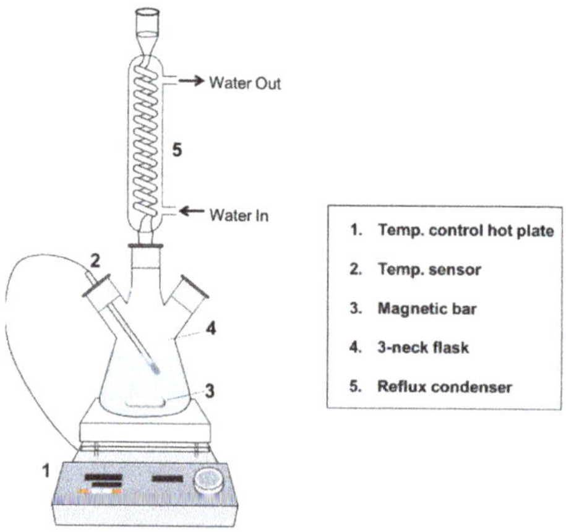

Figure 2. Experimental setup for leaching of REMs from discarded NiMH batteries.

2.2.2. Solvent Extraction Procedure

Liquid-liquid extraction studies were carried out in a beaker fitted with a magnetic stirrer using an equal volume of leach liquor and extractant (25 mL/25 mL). Two types of extractants were used, i.e., 2-ethylhexyl 2-ethylhexyphosphonic acid (PC88A, Figure 3a) for REMs and 2-hydroxy-5-nonylacetophenoneoxime (LIX 84, Figure 3b) for base metals. Kerosene and isodecanol were used as a diluent for extractant and a phase modifier, respectively. Ammonium hydroxide and dilute sulfuric acid were used to maintain the equilibrium pH (eq. pH) in the range of 1 to 5. When the solution attained equilibrium, the organic and aqueous phases were separated using a separating funnel and the concentrations of metal ions in the raffinate solution were analyzed using ICP-OES. The concentrations of metals in the organic phase were calculated.

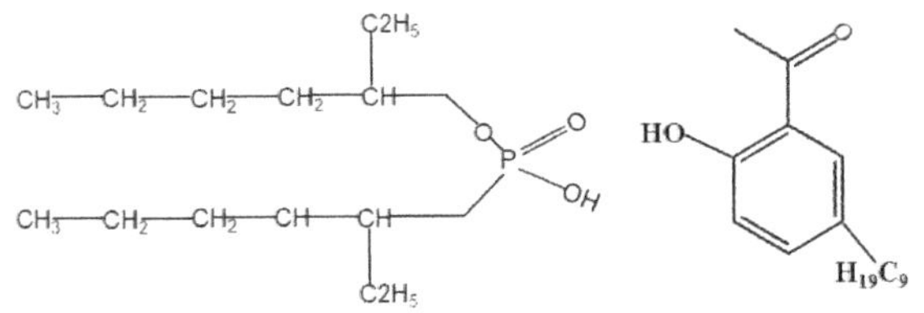

Figure 3. Structure of (**a**) PC88A and (**b**) LIX 84.

2.2.3. Precipitation Procedure

Precipitation studies were carried out in a beaker (capacity: 100 mL) under constant stirring speed at room temperature using ammonium hydroxide as a precipitant. During the experiments, solution pH was varied between 0.75 and 2.0 by adding ammonium hydroxide to precipitate La as lanthanum hydroxide. The solution was taken at different intervals of time to analyze the concentrations of metals in leach liquor after precipitation.

2.2.4. Characterization and Analysis of Samples

X-ray diffraction (XRD, D8, Bruker AXS Inc., Wisconsin, WI, USA) was used to analyze the phases of samples. Figure 4 shows that XRD patterns mainly contained the peak of oxides of rare earth metals, Cu and Ni in black powder of NiMH batteries. Further, AAS and ICP-OES were used to analyze the concentrations of metal ions in the leach liquor.

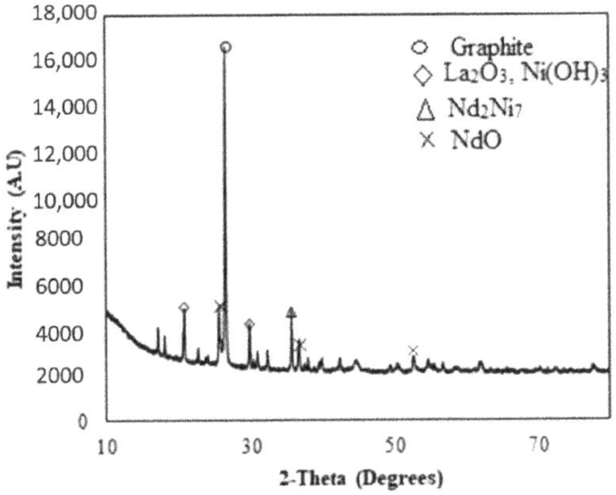

Figure 4. XRD pattern of black powder obtained from NiMH batteries.

3. Results and Discussion

Experiments were carried out using leaching followed by solvent extraction and precipitation to extract the REMs. The obtained results are discussed below:

3.1. Leaching Studies

To dissolve the metals and optimize the process parameters (i.e., acid concentration, time, temperature, pulp density, etc.), the leaching experiments were carried out varying the scale of the parameters.

3.1.1. Effect of Acid Concentration

To optimize the minimum acid requirement during the leaching experiments, studies were performed varying the sulfuric acid concentration in a range of 0.5 to 2.0 M H_2SO_4 at 75 °C for mixing time 75 min at pulp density 100 g/L in presence of 10% H_2O_2 (v/v).

The results (Figure 5a–f) showed that leaching of La, Ce, Nd, Cu, Ni and Co increased with increase in the concentration of acid, which happened due to the increase in acidic strength. It was found that 99.1% La, 96.1% Ce, 99.1% Nd, 99.9% Cu, 99.9% Ni and 99% Co leached in 2 M H_2SO_4 at 75 °C in 60 min. Further, the increase in acidic concentration had no significant effect on the enhancement of REM leaching efficiency. Therefore, 2 M H_2SO_4 was considered the optimum acid concentration for the dissolution of metals from black powder of spent electric vehicle batteries.

3.1.2. Effect of Temperature

To optimize the temperature for dissolution of REMs, experiments were carried out in a range of temperatures varying from 25 to 75 °C at a pulp density of 100 g/L in the presence of 10% H_2O_2 (v/v). Figure 6 shows that leaching of La, Ce, Nd, Cu, Ni and Co increased with increase in temperature due to the decrease in activation energy, resulting in the enhancement of metal leaching efficiency with rise in solution temperature [30]. It was found that 98.9% La, 98.5% Nd, 89.9% Ce, 99% Cu, 99% Ni and 95.9% Co leached at 75 °C. Thereafter, the increase in the solution temperature had no significant effect on the enhancement of metal leaching efficiency.

3.1.3. Effect of Pulp Density

Pulp density is an important parameter for the effective leaching of metals from solid. Therefore, to optimize the solid to liquid ratio (i.e., pulp density (S/L)), leaching experiments were carried out varying the pulp density from 50 to 200 g/L at temperature 75 °C, H_2O_2 10% (v/v) and mixing time 60 min for maximum dissolution of REMs. The results (Figure 7) showed that leaching of La, Ce, Nd, Cu, Ni and Co was found to decrease with increase in pulp density above 100 g/L. This may have been due to increase in the total number of moles of metallic constitutes (Nd, La Ce, Cu, Ni and Co) relative to available leachant molecules resulting in a decrease in the leaching efficiency of metals [31]. Therefore, 100 g/L pulp density has been chosen as the optimum condition for leaching of metals from black powder of NiMH batteries of electric vehicles.

3.2. Separation and Purification Studies

3.2.1. Solvent Extraction of Ce and Nd

Leach liquor containing 3.9 g/L Nd; 11.9 g/L Ce; 7.6 g/L La; 6.5 g/L Cu; 28.9 g/L Ni and 5.1 g/L Co was used for the solvent extraction studies using 10% PC88A to selectively separate the rare earth metals. It was found that extraction of Nd and Ce increased with increase in eq. pH, as shown in Figure 8. The maximum separation of Ce and Nd with respect to La, Cu, Ni and Co occurred at eq. pH~1.5. The separation factor (β) values of Nd and Ce were also calculated with respect to La and found to be 12.1 ($\beta_{Nd/La}$) and 56.33 ($\beta_{Ce/La}$) with respect to La. This indicates that selective extraction of Nd and Ce can be achieved by increasing the number of stages for extraction of Nd. Thus, a McCabe–Thiele plot was drawn to determine the number of stages required for the extraction of Nd from leach liquor. It was found that two counter-current stages were required to completely extract Nd from leach liquor, as shown in Figure 9.

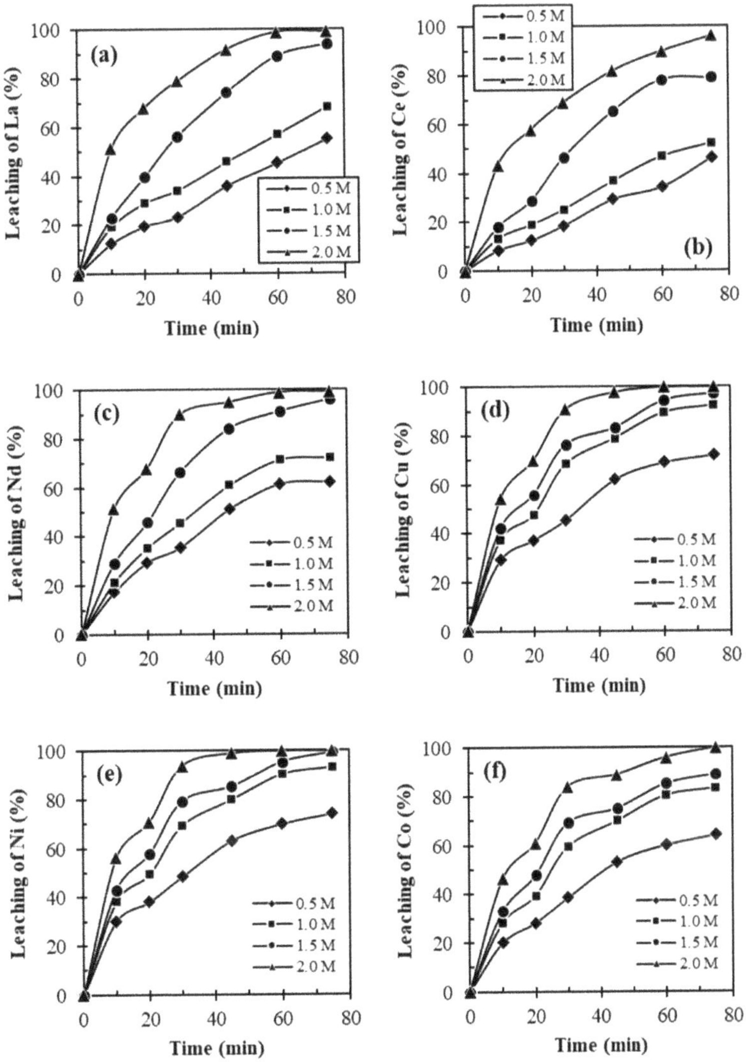

Figure 5. Effect of acid concentration on leaching of (**a**) La, (**b**) Ce, (**c**) Nd, (**d**) Cu, (**e**) Ni and (**f**) Co from black powder of spent NiMH battery of electric vehicle [Solid: black powder; liquid: 0.5–2.0 M H_2SO_4; additive: 10% H_2O_2 (*v*/*v*); temperature: 75 °C; pulp density 100 g/L; time: 75 min].

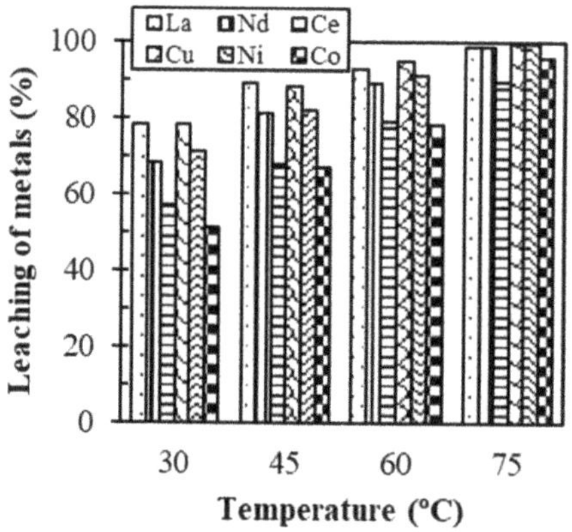

Figure 6. Effect of temperature on leaching of rare earth elements (La, Ce and Nd) and valuable metals (Cu, Ni and Co) [Solid: black powder; liquid: 2 M H_2SO_4; additive: 10% H_2O_2 (v/v); pulp density 100 g/L; time: 60 min].

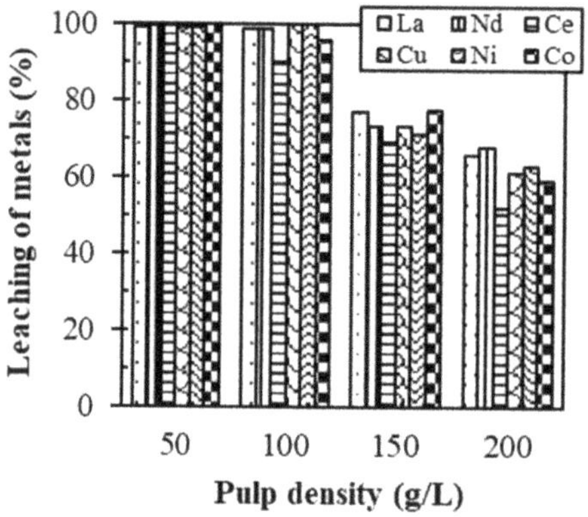

Figure 7. Effect of pulp density on the leaching of rare earth elements (La, Ce and Nd) and valuable metals (Cu, Ni and Co) [Solid: black powder; liquid: 2 M H_2SO_4; additive: 10% H_2O_2 (v/v); time: 60 min; temperature: 75 °C].

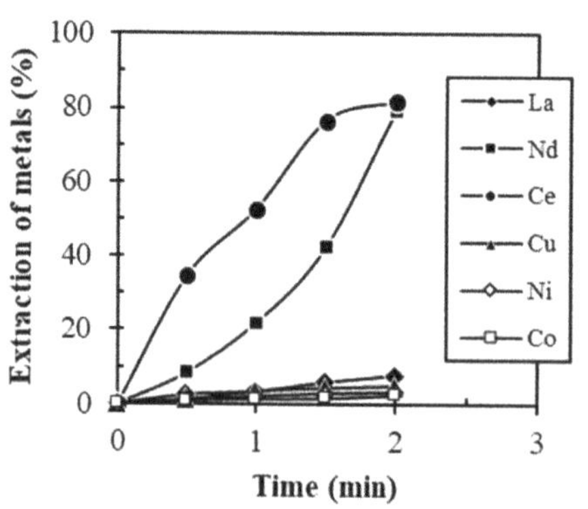

Figure 8. Solvent extraction of Ce and Nd from leach liquor [Metal concentration in leach liquor (g/L): Nd: 3.9; Ce: 11.9; La: 7.6; Cu: 6.5; Ni: 28.9; Co:5.1; extractant: 10% PC88A diluted in kerosene; time: 10 min; O/A ratio: 1/1].

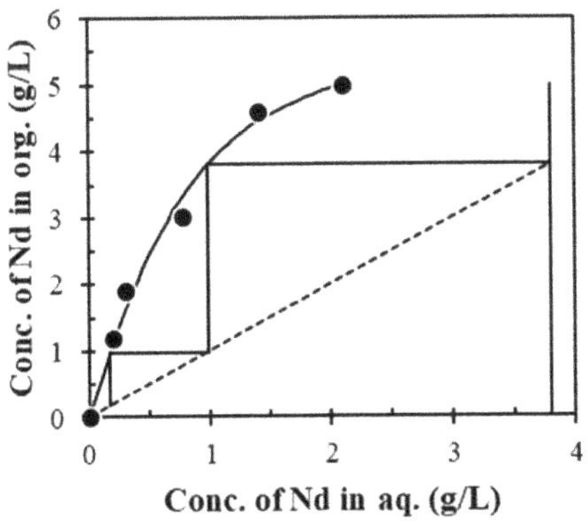

Figure 9. McCabe–Thiele plot for extraction of Nd from leach liquor [Leach liquor: 3.9 g/L Nd; extractant: 10% PC88A diluted in kerosene; time: 10 min; O/A ratio: 1/1].

3.2.2. Precipitation of La

The solution obtained after extraction of Nd and Ce that contained 7.6 g/L La; 6.5 g/L Cu; 28.9 g/L Ni and 5.1 g/L Co was used for recovery of La by precipitation process. It was found that precipitation of La increased with increase in pH as shown in Figure 10. The complete precipitation of La occurred at pH ~1.5 leaving Cu, Ni and Co remaining in the filtrate. Further, precipitation of La was validated with an Eh-pH diagram, which indicated that La precipitation occurred at or above pH 1.5 as $La(OH)_3$ [31], while Cu, Ni and Co form hydroxide above pH 5. Therefore, selective precipitation of La occurred between pH 1.5 and 2.0. Further, Cu, Ni and Co were separated by solvent extraction, as discussed below.

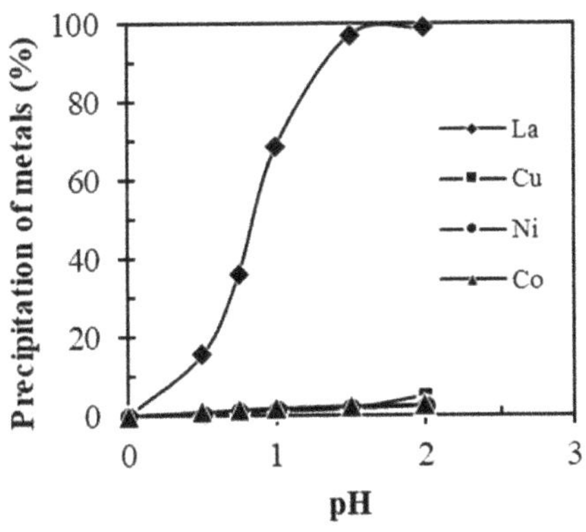

Figure 10. Precipitation of La from leach liquor [Metal concentration in leach liquor (g/L): La: 7.6; Cu: 6.5; Ni: 28.9; Co: 5.1; precipitant: ammonium hydroxide].

3.2.3. Extraction of Cu, Ni and Co

After the extraction of La, Ce and Nd, the obtained solution containing 6.5 g/L Cu, 28.9 g/L Ni and 5.1 g/L Co was used for selective extraction of Cu, Ni and Co using solvent extraction process. Solvent extraction experiments were carried out with LIX 84 varying eq. pH between 1.0 and 5.0 to extract the Cu, Ni and Co separately from leach liquor. The results (Figure 11) showed that 99.1% extraction of Cu occurred with 10% LIX 84 in a range of eq. pH 2.0 to 2.5 while Ni and Co remained in the raffinate. Further, 98.6% Ni was selectively extracted with 30% LIX 84 between eq. pH 4.5 and 5.0 at O/A ratio 1/1 in 10 min, leaving Co in the raffinate. This indicates that LIX 84 makes complexes with Cu and

Ni at different eq. pH resulting in the selective extraction of Cu and Ni. The metal-loaded organic was stripped using 10% H_2SO_4 to obtain a pure solution of Cu and Ni. Finally, a process flow sheet was developed for the extraction of La, Ce, Nd, Cu, Ni and Co, as shown in Figure 12. The metals/salts of Nd, Ce and La could be used in various applications, such as making magnets for hard disks, cathode ray tubes and phosphor powder, and pyrophoric alloys, respectively. In addition, the recovered REMs, Cu, Ni and Co can be re-used in the manufacture of rechargeable batteries. Based on the obtained results and pure metals/salts, the recovered materials of discarded NiMH can be used in different industries after scale-up/pilot trial.

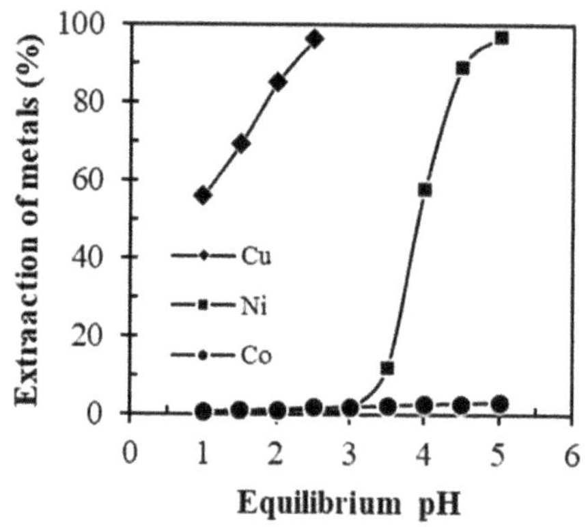

Figure 11. Solvent extraction of Cu, Ni and Co from leach liquor (Metals concentration in leach liquor (g/L): Cu: 6.5; Ni: 28.9; Co: 5.1; extractant: 10% & 30% LIX 84; time: 10 min; O/A ratio: 1/1).

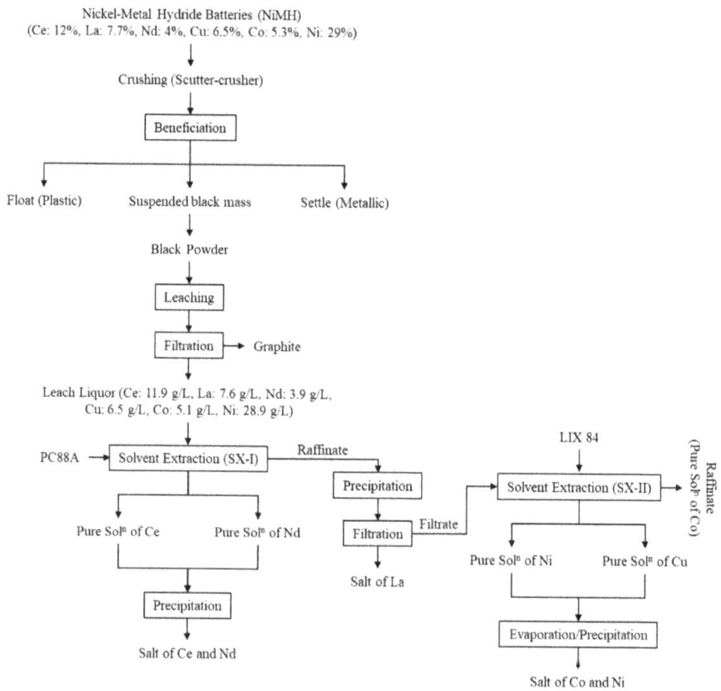

Figure 12. Developed process flowsheet for extraction of rare metals, Cu, Ni and Co from discarded NiMH batteries of electric vehicles.

4. Conclusions

Based on laboratory scale studies, the following conclusions have been drawn for the recovery of rare and valuable metals from NiMH batteries of electric vehicles:

(1) It was found that more than 98.9% La, 98.5% Nd, 89.9% Ce, 99% Cu, 99% Ni and 95.9% Co were leached in 2 M H_2SO_4 at 75 °C in 60 min in the presence of 10% H_2O_2 (v/v).
(2) Further, Nd and Ce were extracted using 10% PC 88A at equilibrium pH 1.5. A McCabe–Thiele plot showed that 99% Nd was extracted with PC 88A at eq. pH ~1.5 at O/A ratio 1/1 in two counter-current stages.
(3) After the extraction of Nd and Ce, 99% La was selectively precipitated with sodium hydroxide at room temperature between pH 1.5 and 2.0, leaving Cu, Ni and Co in the raffinate.
(4) Subsequently, Cu and Ni were selectively extracted using LIX 84 at eq. pH ~2.5 and ~5, respectively, leaving Co in the raffinate.

Author Contributions: Methodology, P.K.C. and O.S.D.; writing—original draft preparation, R.P. and P.K.C.; writing—review and editing, M.K.J., K.Y. and I.P.; supervision, M.K.J. and R.K.J.; project administration, M.K.J. and I.P.; funding acquisition, I.P. All authors have read and agreed to the published version of the manuscript.

Funding: This study was partially supported by the Japan Society for the Promotion of Science (JSPS) Grant-in-Aid for Early-Career Scientists (JP20K15214). The work has been supported by CSIR-NML under Urban Ore Recycling Centre (FTC-0014/MLP-3116) and various Indo-Korean long term collaboration programs.

Data Availability Statement: The data presented in this study are available on request from the corresponding author.

Acknowledgments: The authors are grateful to the Director CSIR-National Metallurgical Laboratory, Jamshedpur, India for giving permission to publish the paper and work done under CSIR-NML projects and international collaborations.

Conflicts of Interest: The authors declare no conflict of interest.

References

1. Ahn, N.K.; Swain, B.; Shim, H.W.; Kim, D.W. Recovery of rare earth oxide from waste NiMH batteries by simple wet chemical valorization process. *Metals* **2019**, *9*, 1151. [CrossRef]
2. Kumari, A.; Parween, R.; Chakravarty, S.; Parmar, K.; Patha, D.D.; Lee, J.C.; Jha, M.K. Novel approach to recover rare earth metals (REMs) from Indian coal bottom ash. *Hydrometallurgy* **2019**, *187*, 1–7. [CrossRef]
3. Park, I.; Kanazawa, Y.; Sato, N.; Galtchandmani, P.; Jha, M.K.; Tabelin, C.B.; Jeon, S.; Ito, M.; Hiroyoshi, N. Beneficiation of Low-Grade Rare Earth Ore from Khalzan Buregtei Deposit (Mongolia) by Magnetic Separation. *Minerals* **2021**, *11*, 1432. [CrossRef]
4. Cui, J.; Forssberg, E. Mechanical recycling of waste electric and electronic equipment: A review. *J. Hazard Mater.* **2003**, *99*, 243–263. [CrossRef]
5. Choubey, P.K.; Singh, N.; Panda, R.; Jyothi, R.K.; Yoo, K.; Park, I.; Jha, M.K. Development of Hydrometallurgical Process for Recovery of Rare Earth Metals (Nd, Pr, and Dy) from Nd-Fe-B Magnets. *Minerals* **1987**, *11*, 1987. [CrossRef]
6. Tan, Q.Y.; Li, J.H.; Zeng, X.L. Rare Earth Elements Recovery from Waste Fluorescent Lamps: A Review. *Crit. Rev. Environ. Sci. Technol.* **2015**, *45*, 749–776. [CrossRef]
7. Ying, T.K.; Gao, X.P.; Hu, W.K.; Wu, F.; Noreus, D. Studies on rechargeable NiMH batteries. *Int. J. Hydrogen Energy* **2005**, *31*, 525–530. [CrossRef]
8. Jung, D.Y.; Lee, B.H.; Kim, S.W. Development of battery management system for nickel-metal hydride batteries in electric vehicle applications. *J. Power Sources* **2002**, *109*, 1–10. [CrossRef]
9. Zhi, H.; Ni, S.; Su, X.; Xie, W.; Zhang, H.; Sun, X. Separation and recovery of rare earth from waste nickel-metal hydride batteries by phosphate based extraction-precipitation. *J. Rare Earths* **2021**. in Press. [CrossRef]
10. Tanabe, E.H.; Schlemmer, D.F.; Aguiar, M.L.; Dotto, G.L.; Bertuol, D.A. Recovery of valuable materials from spent NiMH batteries using spouted bed elutriation. *J. Environ. Manag.* **2016**, *171*, 177–183. [CrossRef]
11. Samane, M.; Khayyam, N.R.; Rumana, H.; Mohammad, A.; Veena, S. Recovery of rare earth (i.e., La, Ce, Nd, and Pr) oxides from end-of-life Ni-MH battery via thermal isolation. *ACS Sustain. Chem. Eng.* **2018**, *6*, 11811–11818.
12. Innocenzi, V.; Ippolito, N.M.; Michelis, I.D.; Prisciandaro, M.; Medici, F.; Veglio, F. A review of the processes and lab-scale techniques for the treatment of spent rechargeable NiMH batteries. *J. Power Sources* **2017**, *362*, 202–218. [CrossRef]

13. Jha, M.K.; Kumari, A.; Panda, R.; Kumar, J.R.; Yoo, K.; Lee, J.Y. Review on hydrometallurgical recovery of rare earth metals. *Hydrometallurgy* **2016**, *165*, 2–26. [CrossRef]
14. Li, L.Y.; Xu, S.M.; Ju, Z.Y.; Wu, F. Recovery of Ni, Co and rare earths from spent Ni-metal hydride batteries and preparation of spherical $Ni(OH)_2$. *Hydrometallurgy* **2009**, *100*, 41–46. [CrossRef]
15. Al-Thyabat, S.; Nakamura, T.; Shibata, E.; Iizuka, A. Adaptation of minerals processing operations for lithium-ion (LiBs) and nickel metal hydride (NiMH) batteries recycling: Critical review. *Miner Eng.* **2013**, *45*, 4–17. [CrossRef]
16. Bernardes, A.M.; Espinosa, D.C.R.; Tenorio, J.A.S. Recycling of batteries: A review of current processes and technologies. *J. Power Sources* **2004**, *130*, 291–298. [CrossRef]
17. Korkmaz, K.; Alemrajabi, M.; Rasmuson, A.; Forsberg, K. Recoveries of valuable metals from spent nickel metal hydride vehicle batteries via sulfation, selective roasting, and water leaching. *J. Sustain.* **2018**, *4*, 313–325. [CrossRef]
18. Bertuol, D.A.; Bernardes, A.M.; Tenorio, J.A.S. Spent NiMH batteries—the role of selective precipitation in the recovery of valuable metals. *J. Power Sources* **2009**, *193*, 914–923. [CrossRef]
19. Innocenzi, V.; Veglio, F. Recovery of rare earths and base metals from spent nickel-metal hydride batteries by sequential sulphuric acid leaching and selective precipitations. *J. Power Sources* **2012**, *211*, 184–191. [CrossRef]
20. Yang, X.; Zhang, J.; Fang, X. Rare earth element recycling from waste nickel-metal hydride batteries. *J. Hazard Mater.* **2014**, *279*, 384–388. [CrossRef]
21. Fernandes, A.; Afonso, J.C.; Dutra, A.J.B. Separation of nickel(II), cobalt(II) and lanthanides from spent Ni-MH batteries by hydrochloric acid leaching, solvent extraction and precipitation. *Hydrometallurgy* **2013**, *133*, 37–43. [CrossRef]
22. Zhang, P.; Yokoyama, T.; Itabashi, O.; Wakui, Y.; Suzuki, T.M.; Inoue, K. Hydrometallurgical process for recovery of metal values from spent nickel-metal hydride secondary batteries. *Hydrometallurgy* **1998**, *50*, 61–75. [CrossRef]
23. Xia, Y.; Xiao, L.; Tian, J.; Li, Z.; Zeng, L. Recovery of rare earths from acid leach solutions of spent nickel-metal hydride batteries using solvent extraction. *J. Rare Earths* **2015**, *33*, 1348–1354. [CrossRef]
24. Zhang, P.W.; Yokoyama, T.; Itabashi, O.; Wakui, Y.; Suzuki, T.M.; Inoue, K. Recovery of metal values from spent nickel-metal hydride rechargeable batteries. *J. Power Sources* **1999**, *77*, 116–122. [CrossRef]
25. Larsson, K.; Ekberg, C.; Jensen, A.Q. Using Cyanex 923 for selective extraction in a high concentration chloride medium on nickel metal hydride battery waste. *Hydrometallurgy* **2012**, *129–130*, 35–42. [CrossRef]
26. Zhang, W.C.; Noble, A.; Ji, B.; Li, Q. Effects of contaminant metal ions on precipitation recovery of rare earth elements using oxalic acid. *J. Rare Earths* **2020**, in press. [CrossRef]
27. Mei, G.J.; Xia, Y.; Shi, W.; Liu, B. Recovery of rare earth from spent MH-Ni battery negative electrode. *Environ. Prot. Chem. Ind.* **2008**, *28*, 70–73.
28. Pietrelli, L.; Bellomo, B.; Fontana, D.; Montereali, M. Characterization and leaching of NiCd and NiMH spent batteries for the recovery of metals. *Waste Manag.* **2005**, *25*, 221–226. [CrossRef] [PubMed]
29. Ahn, N.K.; Shim, H.-W.; Kim, D.-W.; Swain, B. Valorization of waste NiMH battery through recovery of critical rare earth metal: A simple recycling process for the circular economy. *Waste Manag.* **2020**, *104*, 254–261. [CrossRef]
30. Kumari, A.; Jha, M.K.; Pathak, D.D. An innovative environmental process for the treatment of scrap Nd-Fe-B magnets. *J. Environ. Manag.* **2020**, *273*, 111063. [CrossRef]
31. Kumari, A.; Jha, M.K.; Yoo, K.; Panda, R.; Lee, J.Y.; Kumar, R.J.; Pathak, D.D. Advanced process to dephosphorize monazite for effective leaching of rare earth metals (REMs). *Hydrometallurgy* **2019**, *187*, 203–211. [CrossRef]

minerals

Article

Carbothermic Reduction Roasting of Cathode Active Materials Using Activated Carbon and Graphite to Enhance the Sulfuric-Acid-Leaching Efficiency of Nickel and Cobalt

Youngjin Ahn [1], Wonbeom Koo [1], Kyoungkeun Yoo [1,*] and Richard Diaz Alorro [2]

1 Department of Energy and Resources Engineering, National Korea Maritime and Ocean University (KMOU), Busan 49112, Korea
2 Western Australian School of Mines: Minerals, Energy and Chemical Engineering, Faculty of Science and Engineering, Curtin University, Kalgoorlie, WA 6430, Australia
* Correspondence: kyoo@kmou.ac.kr

Abstract: Carbothermic reduction-roasting tests of NCM (nickel, cobalt, and manganese) cathode active materials with carbon sources such as activated carbon and graphite followed by sulfuric acid leaching were performed to investigate the effects of roasting temperature, molar mixing ratio of cathode active materials and carbon sources, and type of cathode active materials. When the virgin NCM622 materials were roasted with activated carbon, the peaks of Ni and Co metals were observed in the XRD data. The leaching efficiencies of Li, Ni, Co, and Mn increased to over 99.9% within 120 min in all samples roasted at 600 °C–900 °C, but, at the beginning of leaching, the leaching efficiencies increased more slowly with increasing roasting temperature. The leaching efficiencies of Ni and Co decreased with decreasing the molar mixing ratio of active cathode materials and carbon sources, but the leaching efficiencies were more than 99.9% in all ratios. These results indicate that roasting can enhance the leaching of cathode active materials and improve the conventional leaching process using hydrogen peroxide.

Keywords: lithium ion battery; recycling; carbothermic reduction; activated carbon; graphite

check for updates

Citation: Ahn, Y.; Koo, W.; Yoo, K.; Alorro, R.D. Carbothermic Reduction Roasting of Cathode Active Materials Using Activated Carbon and Graphite to Enhance the Sulfuric-Acid-Leaching Efficiency of Nickel and Cobalt. *Minerals* **2022**, *12*, 1021. https://doi.org/10.3390/min12081021

Academic Editor: Kenneth N. Han

Received: 6 July 2022
Accepted: 9 August 2022
Published: 14 August 2022

1. Introduction

The Korean government announced in October 2020 that it would achieve carbon neutrality by 2050, and that recycling is a key factor for the goal [1]. Since electric vehicles are also an important way toward carbon neutrality, there has been increasing demand for lithium ion batteries (LIB) [2–4]. In 2021, the EU announced a new EU regulatory framework, which includes a recycled-content declaration requirement [5]. This regulation would apply to industrial batteries, EV batteries, and automotive batteries containing cobalt, lead, lithium, or nickel in active materials. Mandatory minimum levels of recycled content would be set to 12% cobalt, 85% lead, 4% lithium, and 4% nickel in 2030, increasing to 20% cobalt, 10% lithium, and 12% nickel in 2035 [5].

Various cathode active materials have been used in lithium ion batteries such as LCO (lithium cobalt oxide), NCM (lithium nickel cobalt manganese oxide), NCA (lithium nickel cobalt aluminum oxide), and LFP (lithium iron phosphate) [6]. Nickel and cobalt are recognized as target metals in LIB recycling because Ni-rich cathode materials have been used for high-end electric vehicles [7]. SungEel HiTech, a representative LIB recycler in Korea, runs an LIB-recycling process consisting of pretreatment, leaching, solvent extraction, and evaporation/electrowinning, where H_2SO_4 with H_2O_2 is used as a leaching medium for the leaching process [8,9]. The H_2O_2 in this recycling process is used as a reductant because the oxidation state of metals is higher than +3 in cathode active materials [10,11]. Although the company recovers valuable metals successfully from spent LIB using the process, it is difficult to operate the leaching process continuously because the amount of H_2O_2 added varies depending on waste LIB.

Flotation processes as pretreatments have been examined to separate graphite and cathode materials [12] or to recover Cu and Al from ground-LIB particles [13]. Although graphite was recovered, it is difficult to reuse graphite as anode materials. Because carbon such as graphite can act as a reductant, it is better to use graphite components as reductants than to recover graphite in LIB-recycling processes, in which graphite could enhance the dissolution of metal in cathode materials by reducing the oxidation state of metals from +3 to +2 or less in leaching processes. Therefore, a pretreatment process using roasting with a carbon source as a reductant was proposed [14–21]. In these conventional studies, it was reported that the metals in the cathode materials were reduced to monoxide or metal [14,15] with graphite or with graphite [14,15,17–21] and CO_2 [16]. Ammonia or electrochemical leaching was tested with NCM powder [20,21], but the use of sulfuric acid is economical and advantageous for the recovery of Ni and Co as sulfate [22–24]. In other conventional studies [18,20], water leaching for the selective leaching of Li was performed before sulfuric acid or ammonia leaching, but the leaching efficiency of Li was not more than 95%.

Conventional LIB recyclers have proceeded with various types of LIB or cathode materials, which can contain anode materials such as graphite. Therefore, when spent LIB without graphite is recycled, other carbon sources should be added. In the present study, carbothermic reduction roasting processes with carbon sources such as activated carbon and graphite were investigated as a pretreatment process, in which activated carbon was selected because it can be obtained easily and economically. The effects of roasting temperature, molar mixing ratio of cathode active materials and carbon sources, and type of cathode active materials were examined to investigate the leaching behaviors of Li, Ni, Co, and Mn from the cathode active materials after roasting.

2. Materials and Methods

Three kinds of cathode active materials (NCM), such as virgin NCM622, waste cathode powder, and black powder were used in this study, in which the waste cathode powder was generated from battery-material manufacturing processes and the black powder was generated from an LIB recycling process using spent LIB. All samples were obtained from an LIB manufacturer or recycler in Korea. The metal contents of Li, Ni, Co, and Mn are given in Table 1, which were measured using an optima 8300 ICP-OES (inductively coupled plasma optical emission spectrometer, PerkinElmer Inc., Waltham, MA, USA) after acid-digestion with aqua regia, and, in the case of black powder, graphite components were removed by roasting at 950 °C. The content of graphite was 25.6%, calculated from the roasting test.

Table 1. Metal composition of samples used in this study (mass%).

	Li	Ni	Co	Mn	Others
Virgin material	9.4	54.5	19.4	16.6	-
Waste material	9.2	54.3	19.1	17.4	-
Black powder [1]	9.6	44.7	23.2	21.4	1.1 [2]

[1] The metal contents of black powder were calculated excluding the graphite component. [2] Other metals indicate Cu and Al.

The Gibbs free energies of metal oxides such as Li_2O, MnO, Mn_2O_3, NiO, CoO, and Co_3O_4, where Li, Mn, Ni, and Co are main components of NCM, were investigated for temperature using Outokumpu HSC Chemistry 5.11 to understand the behaviors of metal oxides by carbothermic reduction. The NCM samples were mixed with activated carbon (98%, Junsei Chemical, Co. Ltd., Tokyo, Japan) or graphite powder (reagent-grade, Kanto Chemical, Co. Inc., Tokyo, Japan) for roasting tests, and the molar ratio was adjusted from 1:1 to 1:10. The mole of active cathode material was calculated based on the chemical formula, Li $(Ni_{0.6}Co_{0.2}Mn_{0.2})$ O_2, where its molecular weight was calculated to be 96.93 g/mol. The reduction of Li, Ni, Co, and Mn in the cathode material needs

4 electrons, which could be provided from the oxidation of C to CO_2 (C^{4+}). The mixtures were isothermally reduced in a horizontal tube furnace (ID 60 x L 1000). An alumina boat loaded with 10 g of the mixture was placed in the middle of the tube, and the total gas-flow rate was fixed at 0.7 L/min Ar. The temperature range was from 600 to 900 °C, and the residence time was fixed at 60 min.

Before the leaching tests, a 1 mol/L H_2SO_4 solution was prepared with sulfuric acid (95%, Junsei Chemical Co., Ltd., Tokyo, Japan). The leaching tests of the cathode samples before and after roasting were performed in a 500 mL, four-necked Pyrex glass reactor using a heating mantle to maintain the temperature. The reactor was fitted with a reflux condenser and an agitator. The reflux condenser was inserted into one port to avoid solution loss at high temperatures. In a typical run, 200 mL of 1 mol/L sulfuric acid solution was poured into the reactor with an agitation speed of 400 rpm. After the temperature of solution reached the thermal equilibrium at 90 °C, 5 g of the roasted sample was added to the reactor in all the experiments. During the leaching tests, 3 mL of the solution was withdrawn periodically at a desired time interval (5–120 min) by a syringe. The sample was filtered using a 0.45 μm membrane filter, and the obtained filtrate was diluted with 5% HNO_3 solution. The leach residue was digested with aqua regia, and the solution was diluted with distilled water. All experimental procedures are summarized in Figure 1.

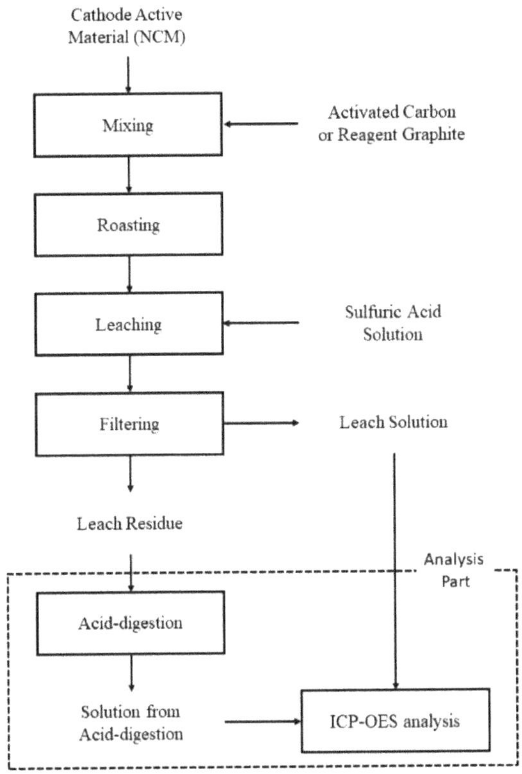

Figure 1. Flowchart of the experimental procedures in this study.

The phase change was characterized by X-ray diffraction (XRD, D/Max 2500, Rigaku, Tokyo, Japan) and by scanning-electron microscopy (SEM, MIRA-3, Tescan Co., Brno, Czech Republic). The concentration of Li, Ni, Co, and Mn was measured using an optima 8300 ICP-

OES (inductively coupled plasma optical emission spectrometer, PerkinElmer Inc., Waltham, MA, USA). The leaching efficiencies of metals were calculated using the following equation:

$$\text{The leaching efficiency (\%)} = M_{solution}/(M_{solution} + M_{residue}), \quad (1)$$

where $M_{solution}$ and $M_{residue}$ represent the mass of metals in the leach solution and leach residue, respectively.

3. Results and Discussion

The NCM cathode active materials are found to be oxides, although the oxidation state of metals is higher than +3 [10,11], and the crystal structure of cathode is transformed from a hexagonal crystal system to a monoclinic crystal system during repeated charging and discharging. However, there has been little information about the Gibbs free energy of active materials. Therefore, the carbothermic reduction processes of metal oxides such as Li_2O, MnO, Mn_2O_3, NiO, CoO, and Co_3O_4 were expressed to predict the reduction process, as a route represented in Figure 2, which was created using the software of Outokumpu HSC Chemistry 5.11. As shown in Figure 2, the Gibbs free energies decreased with increasing temperature; 1350 °C or higher is required to reduce Li_2O or MnO to Li or Mn metal, whereas CoO or NiO can be reduced to Co or Ni metal at 600 °C or lower.

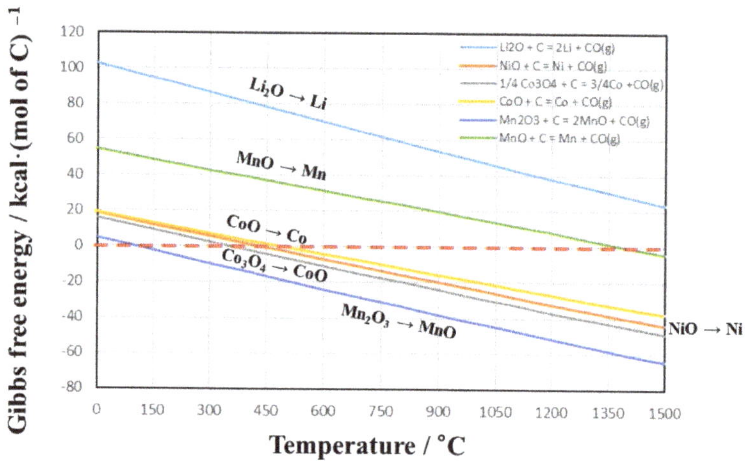

Figure 2. Gibbs free energies of metal oxides with temperature.

The virgin NCM622 samples were mixed with activated carbon in a molar ratio of 1:10 and then roasted at 600 °C–900 °C under Ar atmosphere. Figure 3 shows the XRD results of the virgin NCM622 and roasted products. As shown in Figure 3, the peaks of the NCM cathode active material were observed before roasting, and the peaks of the metal or oxide forms of Ni, Co, and Mn were shown when increasing the roasting temperature to 800 °C. When increasing the temperature further to 900 °C, only metal forms were observed. These results indicate that the lattice of anode material was collapsed by roasting and that parts of nickel and cobalt oxides were reduced to metal. Figure 4 shows SEM images of samples roasted at 700 °C, where Ni, Co, and Mn existed together, and each metal component did not grow as an independent granular.

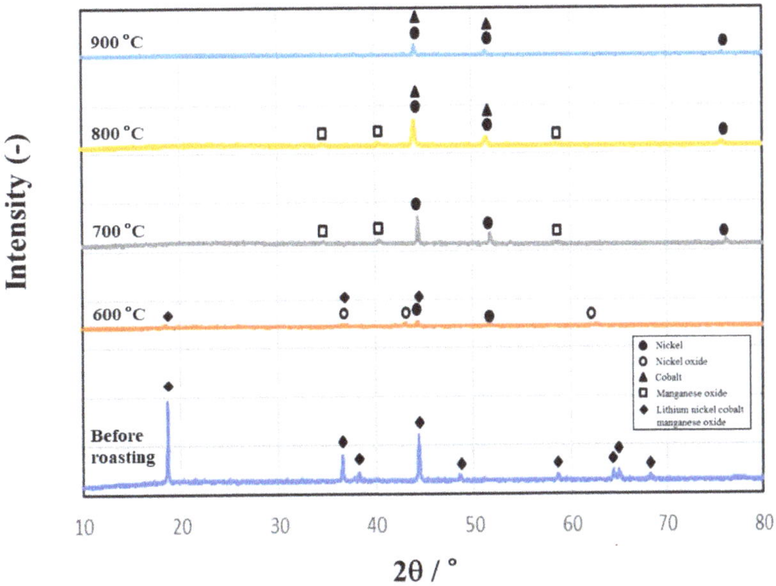

Figure 3. XRD data of virgin NCM622 active cathode material and samples roasted at 600 °C–900 °C.

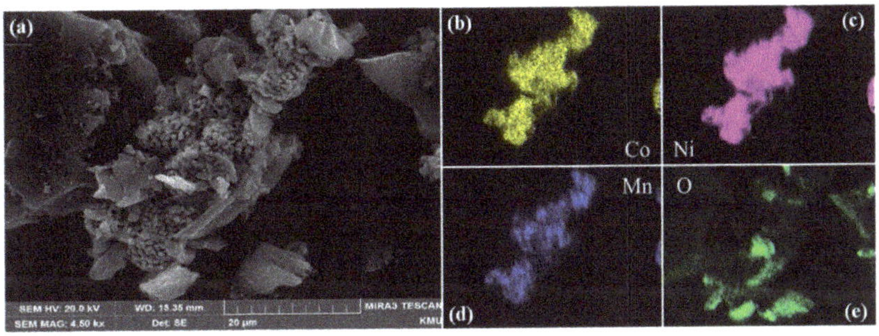

Figure 4. SEM images of (**a**) sample roasted at 700 °C and EDS results of (**b**) Co, (**c**) Ni, (**d**) Mn, and (**e**) O.

Figure 5 shows the leaching behaviors of Li, Ni, Co, and Mn from the samples roasted at 600 °C–900 °C in a 1 mol/L sulfuric acid solution at 90 °C and at 400 rpm with 2.5% pulp density (5g/200 mL). The leaching efficiencies increased rapidly within 15 min, and all leaching efficiencies reached over 99% at 60 min. In the cases of Li and Mn, the leaching efficiency increased from 73.6% and 76.7% at 600 °C to 99.2% and 97.2% at 800 °C, respectively, at 10 min of leaching time. Figure 6 shows the leaching efficiencies of Ni and Co at 10 min of leaching time with the roasted temperatures. The leaching efficiencies of Ni and Co decreased from 94.8% and 96.4% in the sample roasted at 600 °C, to 88.1% and 86.8% at 900 °C, respectively. At the beginning of leaching at approximately 10 min, the leaching efficiencies of Ni and Co decreased with increasing roasting temperature, but all leaching efficiencies increased over 99% within 60 min.

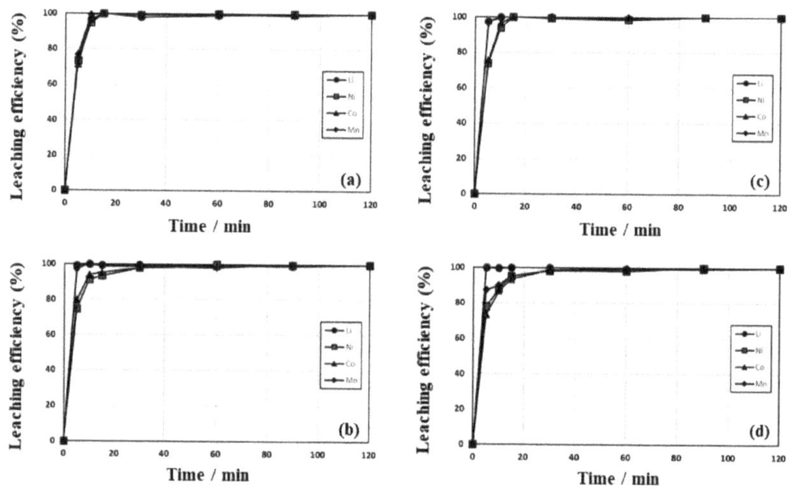

Figure 5. The leaching behaviors of Li, Ni, Co, and Mn from samples roasted at 600 °C (**a**), 700 °C (**b**), 800 °C (**c**), and 900 °C (**d**).

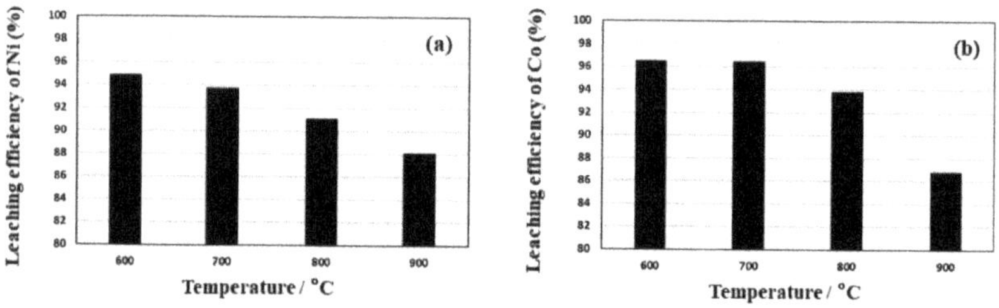

Figure 6. Leaching efficiencies of Ni (**a**) and Co (**b**) from samples roasted at 600 °C–900 °C.

The effects of carbothermic reduction roasting with activated carbon on the leaching of Ni and Co are shown in Figure 7. The NCM622 sample was roasted at 700 °C with a 1:10 molar ratio for 1 h, and then, the roasted product was leached in a 1 mol/L H_2SO_4 solution at 90 °C and at 400 rpm with 2.5% pulp density for 2 h. In the case of NCM622 leaching without roasting, the leaching efficiencies of Ni and Co were 60.5% and 56.8%, respectively, while the leaching efficiencies increased over 99% by roasting NCM622 with activated carbon. These results suggest that roasting with carbon sources enhances the leaching of cathode active materials. In the same leaching tests, the leaching efficiencies of Li and Mn increased over 99% within 2 h regardless of roasting. Figure 8 shows the effects of various mixing ratios of NCM622 and activated carbon on the leaching efficiencies of Ni, Co, Mn, and Li. Under most ratios, although the leaching efficiencies indicate rates above 99% regardless of the ratio, the leaching efficiencies of Ni and Co were slightly lower at the ratios of 1 and 2.

The effects of reductants such as activated carbon and graphite on the leaching were investigated after roasting at 700 °C with a 1:10 molar ratio for 1 h followed by leaching in a 1 mol/L sulfuric acid solution at 90 °C and at 400 rpm with 2.5% pulp density. Generally, because graphite was used as an anode material, the effects of roasting with activated carbon and graphite on the leaching were compared. In this study, the difference in leaching efficiencies was not observed when using activated carbon and graphite as reductants, as shown in Figure 9, but the carbothermic reduction with carbon sources enhanced the leaching efficiencies.

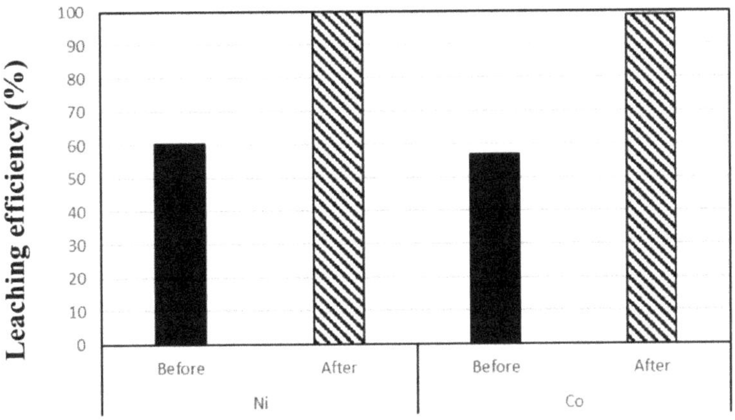

Figure 7. The leaching efficiencies of Ni and Co from samples before and after roasting.

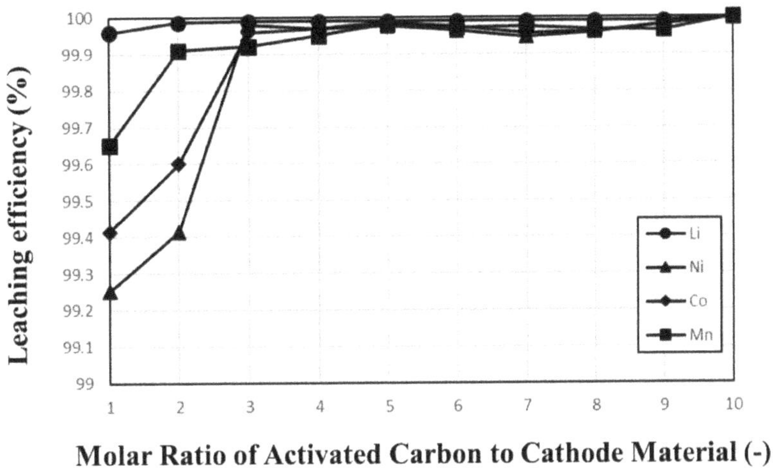

Figure 8. The leaching behaviors of Li, Ni, Co, and Mn with molar mixing ratio of NCM622 and activated carbon.

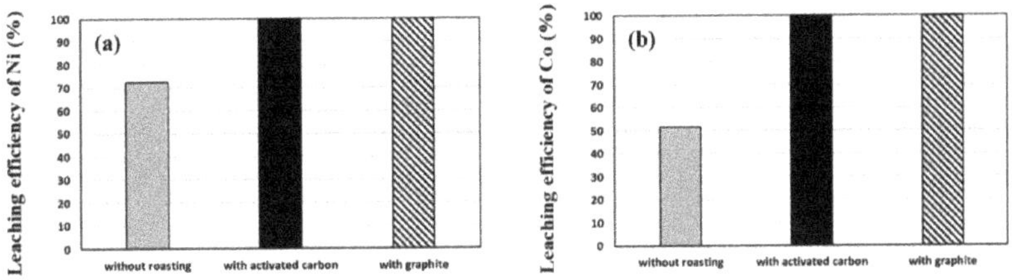

Figure 9. The leaching behaviors of Ni (**a**) and Co (**b**) from the waste cathode powder before and after roasting at 700 °C with activated carbon or graphite.

Currently, in Korea, there are very few end-of-life electric vehicles, and recycling companies in Korea have treated lithium-ion batteries generated from manufacturing processes. Therefore, leaching tests were performed with scrapped active cathode material obtained from manufacturing processes, and the samples were roasted at 700 °C with a 1:10 molar ratio for 1 h followed by leaching in a 1 mol/L sulfuric acid solution at 90 °C and at 400 rpm with 2.5% pulp density. The leaching behaviors of Ni and Co are shown in Figure 10, where the leaching results of Li and Mn are not shown because the leaching efficiencies increased rapidly over 99.9% within 10 min. Although the leaching efficiency of Ni increased more slowly than that of Co, the leaching efficiencies increased over 99.9% within 30 min, as shown in Figure 10a. The effects of the molar ratio of the graphite-to-cathode material were also investigated using the scrapped active cathode material, and, as shown in Figure 10b, the leaching efficiency of Ni increased by increasing the molar ratio even though the efficiency was over 99.6% when the ratio was 1:1, whereas the leaching efficiency of Co remained over 99.9%.

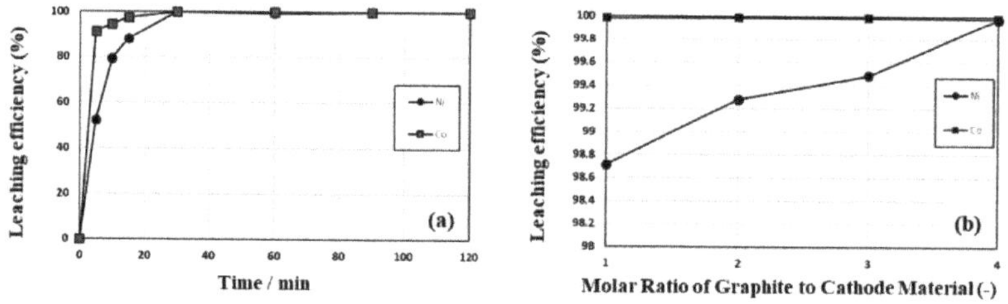

Figure 10. The leaching of Ni (**a**) and Co (**b**) from the waste cathode powder after roasting at 700 °C with graphite followed by 1 mol/L H_2SO_4 leaching.

Generally, in recycling processes of LIB, pretreatments such as dismantling, crushing, and screening have been performed before leaching processes, and the product obtained from the pretreatment processes contains active cathode and anode materials, which is called black powder. Figure 11 indicates the leaching behaviors of Ni and Co from the black powder with and without roasting, which was performed at 700 °C with a 1:4 molar ratio for 1 h. The reagent-grade graphite was added to the black powder to reach the molar ratio. The leaching test was conducted in a 1 mol/L sulfuric acid solution at 90 °C and at 400 rpm with 2.5% pulp density. The leaching efficiencies increased rapidly within 30 min, and then gradually. When the black powder was roasted, the leaching efficiencies increased to over 99.9% while the leaching efficiencies of Ni and Co increased to 77.6% and 86.5%, respectively, using the black powder without roasting.

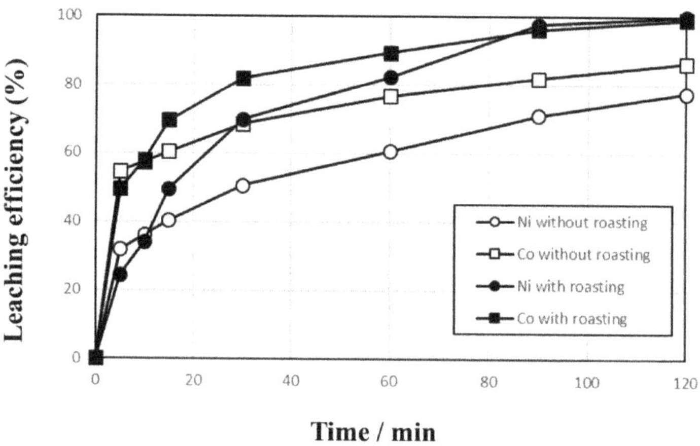

Figure 11. Leaching behaviors of Ni and Co using the black powder before and after roasting with graphite at 700 °C followed by H_2SO_4 leaching.

As discussed above, in the conventional leaching process for active cathode material or black powder, H_2O_2 has been used as a reductant. It was difficult to convert the conventional leaching process to a continuous process by using H_2O_2. Furthermore, hydrogen peroxide has been found to be unstable and expensive [25–29]. Therefore, pretreatment using roasting with carbon sources could be substituted for the conventional process using H_2O_2 and to enhance the efficiency of the leaching process for LIB recycling.

4. Conclusions

Three kinds of NCM cathode active materials were roasted with two carbon sources, activated carbon and graphite, and then, the leaching behaviors of Li, Ni, Co, and Mn were investigated in sulfuric-acid-leaching tests.

When the virgin NCM622 materials were roasted with activated carbon, the peaks of Ni and Co metals were observed in the XRD data. The leaching efficiencies of Li, Ni, Co, and Mn increased to over 99.9% within 120 min in all samples roasted at 600 °C–900 °C, but, at the beginning of leaching, the leaching efficiencies increased more slowly by increasing the roasting temperature. The lower the molar mixing ratio of cathode active materials and carbon sources, the lower the leaching efficiencies of Ni and Co, but the leaching efficiencies of Ni and Co were more than 99.9% in all ratios. These results indicate that carbothermic reduction roasting with carbon sources can enhance the leaching efficiencies of Ni and Co, regardless of the NCM cathode active materials such as virgin, scrapped, and black powder. Furthermore, activated carbon could be used instead of graphite in NCM cathode active materials without anode materials such as graphite.

Author Contributions: Methodology, Y.A., W.K. and K.Y.; writing—original draft preparation, Y.A. and K.Y.; writing—review and editing, K.Y. and R.D.A.; project administration, K.Y. All authors have read and agreed to the published version of the manuscript.

Funding: This work was supported by the Technology Innovation Program (or Industrial Strategic Technology Development Program—Development of Material Component Technology) (20011176, Development of Advanced Technology in Hydrometallurgy for High Added Value of Resources Recovery) funded by the Ministry of Trade, Industry, and Energy (MOTIE, Korea).

Data Availability Statement: Not applicable.

Conflicts of Interest: The authors declare no conflict of interest.

References

1. Kim, J.S.; Jun, Y.S.; Jun, J.H.; Cho, J.Y. Transition from Linear Economy to Circular Economy. *Resour. Recycl.* **2021**, *30*, 3–17. [CrossRef]
2. Jung, Y.J.; Park, S.C.; Kim, Y.H.; Yoo, B.Y.; Lee, M.S.; Son, S.H. A Study on Optimization of Nitric Acid Leaching and Roasting Process for Selective Lithium Leaching of Spent Batteries Cell Powder. *Resour. Recycl.* **2021**, *30*, 43–52. [CrossRef]
3. Nguyen, T.T.H.; Tran, T.T.; Lee, M.S. A Modified Process for the Separation of Fe(III) and Cu(II) from the Sulfuric Acid Leaching Solution of Metallic Alloys of Reduction Smelted Spent Lithium-ion Batteries. *Resour. Recycl.* **2022**, *31*, 12–20. [CrossRef]
4. Moon, H.S.; Song, S.J.; Tran, T.T.; Lee, M.S. Separation of Co(II), Ni(II), and Cu(II) from Sulfuric Acid Solution by Solvent Extraction. *Resour. Recycl.* **2022**, *31*, 21–28. [CrossRef]
5. European Parliamentary Research Service. New EU Regulatory Framework for Batteries. Available online: https://www.europarl.europa.eu/RegData/etudes/BRIE/2021/689337/EPRS_BRI(2021)689337_EN.pdf (accessed on 6 March 2022).
6. Nitta, N.; Wu, F.; Lee, J.T.; Yushin, G. Li-ion battery materials: Present and future. *Mater. Today* **2015**, *18*, 252–264. [CrossRef]
7. Kim, U.H.; Park, J.H.; Aishova, A.; Ribas, R.M.; Monteiro, R.S.; Griffith, K.J.; Yoon, C.S.; Sun, Y.K. Microstructure Engineered Ni-Rich Layered Cathode for Electric Vehicle Batteries. *Adv. Energy Mater.* **2021**, *11*, 2100884. [CrossRef]
8. Kang, J.; Senanayake, G.; Sohn, J.; Shin, S.M. Recovery of cobalt sulfate from spent lithium ion batteries by reductive leaching and solvent extraction with Cyanex 272. *Hydrometallurgy* **2010**, *100*, 168–171. [CrossRef]
9. Joo, S.H.; Shin, D.J.; Oh, C.; Wang, J.P.; Senanayake, G.; Shin, S.M. Selective extraction and separation of nickel from cobalt, manganese and lithium in pre-treated leach liquors of ternary cathode material of spent lithium-ion batteries using synergism caused by Versatic 10 acid and LIX 84-I. *Hydrometallurgy* **2016**, *159*, 65–74. [CrossRef]
10. Jamil, S.; Yousaf, A.B.; Yoon, S.H.; Han, D.S.; Yang, L.; Kasak, P.; Wang, X. Dual cationic modified high Ni-low co layered oxide cathode with a heteroepitaxial interface for high energy-density lithium-ion batteries. *Chem. Eng. J.* **2021**, *416*, 129118. [CrossRef]
11. Ryu, H.H.; Sun, H.H.; Myung, S.T.; Yoon, C.S.; Sun, Y.K. Reducing cobalt from lithium-ion batteries for the electric vehicle era. *Energy Environ. Sci.* **2021**, *14*, 844–852. [CrossRef]
12. Ruismäki, R.; Rinne, T.; Dańczak, A.; Taskinen, P.; Serna-Guerrero, R.; Jokilaakso, A. Integrating flotation and pyrometallurgy for recovering graphite and valuable metals from battery scrap. *Metals* **2020**, *10*, 680. [CrossRef]

of the 17 materials/metals identified as critical for the clean energy transition to succeed [4]. Unfortunately, high-grade, primary metal resources have become scarce, so alternative sources such as submarine deposits and wastes are currently being explored [5,6].

Zinc plant leach residues (ZPLRs), especially those that were produced using old technologies, are regarded as environmental nuisances and hazardous wastes due to their high amounts of leachable residual hazardous elements such as Pb, Zn, cadmium (Cd), copper (Cu), and arsenic (As) [7–10]. Lead, Cd, and As can cause various illnesses that affect the central nervous system, skin, lungs and kidneys, even in minute amounts, while Cu and Zn are essential micronutrients that are toxic at high concentrations [11–13]. Aside from Zn and Pb, ZPLRs can also contain other critical metals such as cobalt (Co), indium (In), gallium (Ga), and germanium (Ge) [14–16]. The extraction of these metals from ZPLRs serves two purposes: (1) exploitation for economic benefits and (2) the detoxification and clean-up of ZPLRs-impacted sites.

Metal extraction from ZPLRs by hydrometallurgical processes is preferred because they are less energy-intensive (especially for low-grade metallurgical wastes such as ZPLRs) compared to their counterpart, pyrometallurgical processes. Most hydrometallurgical techniques involve the use of strong acids to extract metals of interest. Because these acids are nonselective, they dissolve unwanted elements, the majority of which interferes with succeeding recovery processes, so purification processes (e.g., solvent extraction) are required [17–20].

The leaching of ZPLRs using alkaline lixiviants achieves the selective solubilization of amphoteric elements—Al, Pb, and Zn—leaving iron (Fe), calcium (Ca), and magnesium (Mg) host minerals that constitute a large percentage of ZPLRs undissolved. The dissolution of Pb and Zn under alkaline conditions is due to the formation of complexes with hydroxyl ions (OH^-) [21]. In weak to moderately strong alkaline solutions (i.e., pH 6–12), Pb and Zn dissolve as $Pb(OH)_3{}^-$ with small amounts of $Pb(OH)_4{}^{2-}$ and $Zn(OH)_3{}^-$ with small amounts of $Zn(OH)_4{}^{2-}$, respectively. In a strong alkaline solution (i.e., pH > 12), the dominant species are $Pb(OH)_4{}^{2-}$ for Pb and $Zn(OH)_4{}^{2-}$ for Zn. Many researchers have investigated and successfully extracted Pb and Zn from ZPLRs using alkaline solutions [14,22–24]. The alkaline extractive processes studied are as follows: leaching → solid–liquid separation → metal recovery stages. However, solid–liquid separation by filtration, especially for strong alkaline, is difficult [25]. Thus, some dissolved Pb and Zn from ZPLRs remain in residues if thorough filtration and the washing of leaching residues are not carried out. The residual metals in produced residues are economic losses and at the same time render the produced residues hazardous.

The authors previously developed concurrent-extraction cementation (CEC)—a new metals recovery technique that extracts metals and captures/sequesters them by cementation before solid–liquid separation. Cementation or reductive precipitation is an electrochemical process whereby zero-valent metals or alloys are used to selectively recover redox-sensitive dissolved metals from solution [26,27]. The CEC technique eliminates the need for thorough filtration and extensive washing to remove residual toxic elements in the leaching residues [28,29]. These previous studies, however, were conducted in acidic solutions and Zn could not be cemented by Al metal powder from the leaching pulp or filtered solution because of the competitive effects of proton reduction on cementation [30]. This study, therefore, investigates the CEC of dissolved Pb and Zn from ZPLRs in alkaline (NaOH) leaching pulp using Al metal powder as the cementation agent.

2. Materials and Methods

2.1. Materials

Zinc plant leach residues (ZPLRs) from a historic Pb-Zn mine dumpsite in Kabwe, Zambia were used in this study. The total amounts of Pb and Zn in the ZPLRs were around 6.19 wt% and 2.53 wt%, respectively. The major crystalline minerals of Pb were anglesite ($PbSO_4$), cerussite ($PbCO_3$), and esperite ($PbCa_2Zn_3(SiO_4)_3$). Meanwhile, only one crystalline mineral for Zn, zinkosite ($ZnSO_4$), was detected. Ultra-pure Al metal powder with a median particle size (D_{50}) of 126.8 μm (>99.99%, +50–150 μm, Wako Pure Chemical Industries, Ltd., Osaka, Japan) was used to cement dissolved Pb and Zn from ZPLRs. Detailed characterizations of Al powder and ZPLRs are reported elsewhere [28,31]. Reagent grade NaOH (Wako Pure Chemical Industries, Ltd., Osaka, Japan) was used to prepare the alkaline leaching solutions of different concentrations by dissolving and diluting in deionized (DI) water (18 MΩ·cm, Milli-Q® Integral Water Purification System, Merck Millipore, Burlington, VT, USA).

2.2. Methods

All batch experiments were done using 200-mL Erlenmeyer flasks, and the volume of leaching solutions of different concentrations (i.e., 0–6 M) was fixed at 50 mL. A leaching solution of known volume was added in a flask before the addition of ZPLRs to obtain a predetermined solid-to-liquid ratio (S/L). In the case of CEC experiments, 0.25 g of Al powder was added together with ZPLRs. The pulp was then shaken in a temperature-controlled water bath shaker maintained at 25 °C at a shaking speed of 120 strokes/min and a shaking amplitude of 40 mm. After shaking for preplanned durations, the pulp was carefully collected and filtered through 0.20 μm syringe-driven membrane filters. The filtrate was analyzed for Pb and Zn using an inductively-coupled plasma-atomic emission spectrometer (ICP-AES) (ICPE-9820, Shimadzu Corporation, Kyoto, Japan) (margin of error = ±2%). For the CEC experiments, one more step was added to remove cemented and agglomerated Pb and Zn by sieving using a stainless-steel sieve with an aperture size of 150 μm. The cementation products (i.e., +150 μm) were dried in a vacuum oven, digested by aqua regia in a microwave-assisted acid digestion system (Ethos Advanced Microwave Lab station, Milestone Inc., Sorisole, Italy), and the leachates were analyzed for Pb, Zn, and Fe by ICP-AES. Additionally, the cementation products were analyzed by a scanning electron microscope with an energy-dispersive X-ray spectrometer (SEM-EDX) (JSM-IT200, JEOL Ltd., Tokyo, Japan).

To calculate the Pb and Zn removal efficiencies (η_{Me}) from ZPLRs with and without the addition of Al powder, Equations (1) and (2), respectively, were used.

$$\eta_{Me} = \frac{(V \times C_{Me}) + (W_{cme} \times M_{cme})}{W_S \times M_s} \times 100 \tag{1}$$

$$\eta_{Me} = \frac{V \times C_{Me}}{W_S \times M_s} \times 100 \tag{2}$$

where C_{Me} is the concentration (g/L) of Pb and Zn, V is the volume (L) of the leaching solution, W_S is the weight % of either Pb and Zn in ZPLRs, M_s is the mass (g) of the leached ZPLRs, M_{cme} is the mass (g) of cemented and agglomerated particles, and W_{cme} is the weight % of Pb and Zn in cemented and agglomerated particles calculated based on the digested fraction of M_{cme} in aqua regia and analysis of the solution by ICP-AES.

3. Results and Discussions

3.1. Leaching of ZPLRs in NaOH without the Addition of Al Powder

The effects of leaching time, NaOH concentration, and S/L on Pb and Zn removal efficiencies from ZPLRs were investigated by batch leaching experiments without the addition of Al powder.

The leaching duration effects on Pb and Zn removal was investigated using 3 M NaOH, a 2.5 g/50 mL S/L ratio, and a temperature of 25 °C. The results show that the removal efficiencies for Pb and Zn increased with time up to 15 min, beyond which they changed only insignificantly (Figure 1a). At 15 min of leaching time, the Pb removal efficiency was 60.4% and remained the same even when the leaching time was increased to 120 min (i.e., 59.6%). Similarly, the removal efficiency for Zn was around 28% for 15 min of leaching and 25% when the leaching time was prolonged to 120 min. The Pb and Zn removal efficiencies corroborated and correlated with the water-soluble, exchangeable, and carbonate phases of Pb and Zn approximated by sequential extraction (experimental method and detailed discussion reported by the authors elsewhere [31]) (Figure 2). It is thermodynamically difficult to dissolve Pb and Zn bound to relatively stable phases (e.g., Fe/Mn oxyhydroxide, Fe oxide, and sulfides/organic) in NaOH leaching solution [32,33].

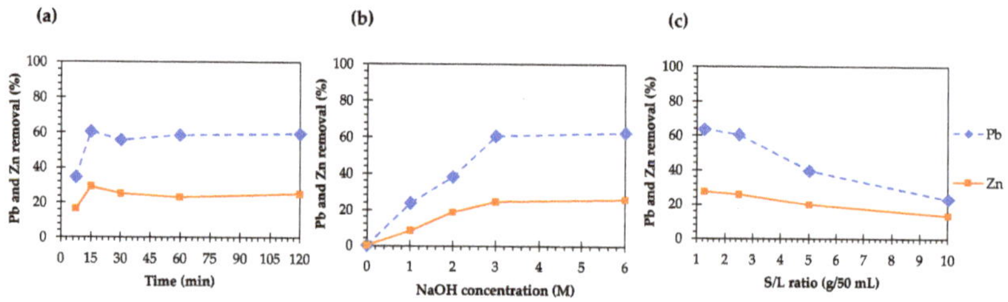

Figure 1. Pb and Zn removal efficiencies from ZPLRs: (**a**) Effects of leaching time when 2.5 g ZPLRs was leached in 50 mL of concentration 3 M NaOH and shaken at 120 strokes per minute in the water bath at 25 °C; (**b**) effects of NaOH concentration when 2.5 g ZPLRs was leached in 50 mL of different NaOH concentration and shaken at 120 strokes per minute in the water bath at 25 °C; and (**c**) effects of S/L ratio when various amounts ZPLRs were leached in 50 mL of 3M NaOH concentration and shaken at 120 strokes per minute in the water bath at 25 °C.

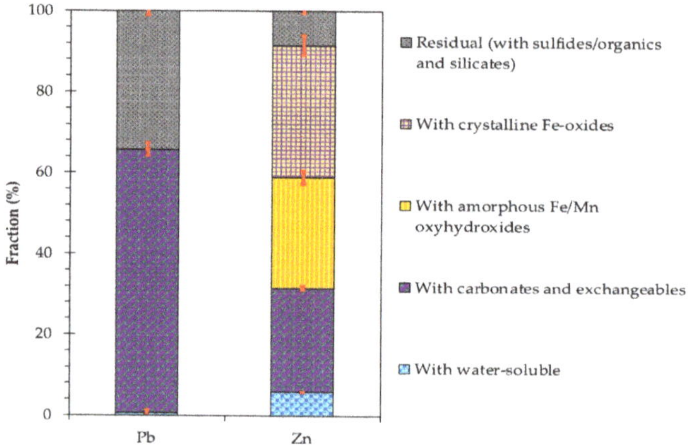

Figure 2. Phase partitioning by sequential extraction of Pb and Zn for ZPLRs (reprinted with permission from Silwamba et al., [31] copyright (2020) Elsevier) (for interpretation of the references to color in this figure legend, the reader is referred to the web version of this article).

To investigate the effects of the NaOH concentration on Pb and Zn removal efficiencies, NaOH in solution was varied from 0 to 6 M, the S/L maintained at 2.5 g/50 mL, and the temperature was maintained at 25 °C for 30 min of leaching time. The removal efficiencies of Pb and Zn increased with higher NaOH concentrations up to 3 M (i.e., Pb and Zn removal of around 60% and 25%), after which, the change in the removal efficiencies of Pb and Zn became insignificant (Figure 1b). The reactions between Pb- and Zn-hosting minerals and NaOH in the solution can be described by Equations (3)–(6) [22,32].

$$PbSO_4 + 3NaOH \rightarrow Pb(OH)_3^- + SO_4^{2-} + 3Na^+ \quad (3)$$

$$2PbO + 2NaOH + H_2O \rightarrow 2Pb(OH)_3^- + 2Na^+ \quad (4)$$

$$ZnSO_4 + 3NaOH \rightarrow Zn(OH)_3^- + SO_4^{2-} + 3Na^+ \quad (5)$$

$$2ZnO + 2NaOH + H_2O \rightarrow 2Zn(OH)_3^- + 2Na^+ \quad (6)$$

In weak to moderately strong alkaline solutions (pH 6–12), the dominant species for Pb and Zn are $Pb(OH)_3^-$ and $Zn(OH)_3^-$. When the NaOH concentration increases (strong alkaline solution, pH > 12) the equilibrium shifts and the more soluble Pb and Zn hydroxyl complexes $Pb(OH)_4^{2-}$ and $Zn(OH)_4^{2-}$, respectively, become more dominant [21]. This explains why Pb and Zn removal efficiencies increased at higher concentrations of NaOH (i.e., 3 M NaOH). Increasing the NaOH concentration beyond 3 M did not improve the Pb and Zn removal efficiencies because almost all the easily extractable Pb and Zn (as determined by sequential extraction) from the ZPLRs were already exhausted.

The S/L is another parameter that affects Pb and Zn removal efficiencies from ZPLRs due to changes in the ratio of hydroxyl concertation to Pb and Zn. To investigate the effects of the S/L ratio on Pb and Zn removal efficiencies, leaching experiments were carried out by varying the amounts of ZPLRs (i.e., 1–10 g) added in 50 mL of 3 M NaOH solution and shaking for 30 min in the water bath at 25 °C. The results show that Pb and Zn removal efficiencies decrease with increasing amounts of ZPLRs in a 50 mL of 3 M NaOH (Figure 1c). The Pb removal efficiency decreased from 62.5% for 1 g to 22.7% for 10 g of ZPLRs. Similarly, the Zn removal efficiency was negatively affected by the S/L. The Zn removal efficiency decreased from 27.1% for 1 g to 13.3% for 10 g of ZPLRs. This decrease in Pb and Zn removal efficiencies with an increase in the S/L can be attributed to the limited hydroxyl ions available to extract Pb and Zn [34], as highlighted above.

3.2. *Concurrent Cementation of Dissolved Pb and Zn in Leaching Pulp of ZPLRs*

The concurrent cementation of dissolved Pb and Zn in leaching pulp was conducted using Al metal powder. Al is not only environmentally friendly as a cementation agent, but it also has a very low standard electrode potential (i.e., −2.35 V vs. NHE in basic solution), which makes it a thermodynamically good candidate for the cementation of dissolved Pb and Zn [35–37]. The oxide layer (Al_2O_3) which covers and insulates Al and suppresses the transfer of electrons dissolves at a high pH [30,38]. As previously discussed, Pb and Zn oxides equally dissolve and are complexed with hydroxide, as shown in Figure 3.

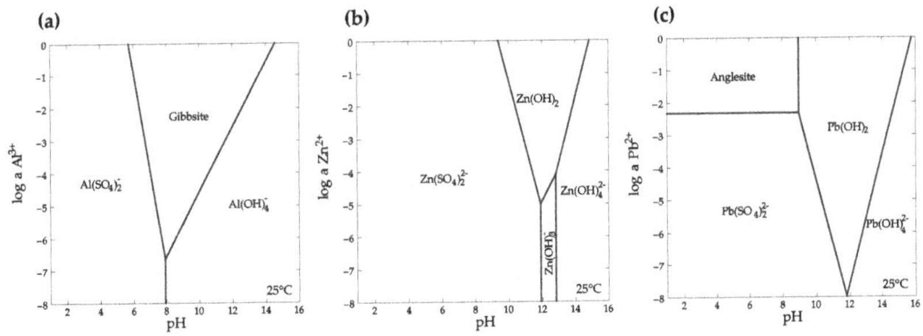

Figure 3. Log activity-pH predominant diagram for (**a**) 0.1 mM Al^{3+} and 0.1 mM SO_4^{2-}, (**b**) 0.1 mM Zn^{2+} and 0.1 mM SO_4^{2-}, and (**c**) 0.1 mM Pb^{2+} and 0.1 mM SO_4^{2-} at 25 °C and 1.013 bar created using the Geochemist's Workbench® with MINTEQ database [39] (for interpretation of the references to color in this figure legend, the reader is referred to the web version of this article).

3.2.1. Effects of Time and NaOH Concentration on Cementation and Distribution of Pb in Leaching Pulp

When 0.25 g of Al powder was added during the leaching of ZPLRs in 3 M NaOH solution, the overall chemical reaction that was thermodynamically expected is expressed by Equation (9), and the net reaction of anodic and cathodic half-reactions are expressed by Equations (7) and (8).

$$Al + 4OH^- \rightarrow Al(OH)_4^- + 3e^- \quad -2.35\ V \tag{7}$$

$$Pb + 4OH^- \rightarrow Pb(OH)_4^{2-} + 2e^- \quad -0.54\ V \tag{8}$$

$$2Al + 3Pb(OH)_4^{2-} \rightarrow 2Al(OH)_4^- + 4OH^- + 3Pb \quad (9)$$

Equation (9) is the cementation reaction of dissolved Pb by added Al powder. The standard Gibbs free energy change, ΔG^0 (i.e., $\Delta G^0 = -nF\Delta E^0$, n is number of electrons transferred, F is Faraday's constant, and ΔE^0 ($E^0_{Al/Al(OH)_4^-} - E^0_{Pb/Pb(OH)_4^{2-}}$) is the galvanic cell potential of Equation (9)), is −1047.82 kJ/mol, implying that cementation of dissolved Pb^{2+} from ZPLRs by Al powder is thermodynamically feasible. The distribution of Pb among the cementation product, solution (i.e., dissolved Pb but uncemented), and undissolved Pb from ZPLRs for different leaching times and NaOH concentrations are shown in Figure 4. Figure 4a shows that the amount of dissolved but uncemented Pb decreased with an increase in leaching time of up to 30 min, where almost 100% of the dissolved Pb was cemented by Al metal powder. This entails that dissolved Pb from ZPLRs was cemented as described in the chemical reaction represented by Equation (9). However, a leaching period longer than 30 min led to an increase in the Pb remaining in the solution. This trend could be attributed to the redissolution of cemented Pb after all the Al has been dissolved and consumed by cementation and other side reactions.

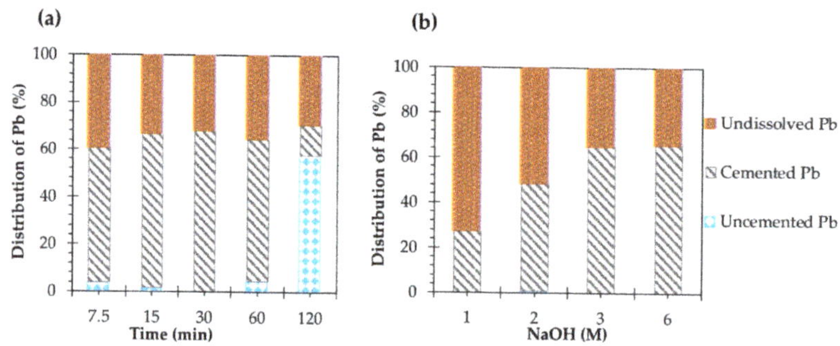

Figure 4. Effects of (**a**) leaching time and (**b**) NaOH concentration on cementation and distribution of Pb in leaching pulp when Al powder was added during ZPLR leaching (for interpretation of the references to color in this figure legend, the reader is referred to the web version of this article).

The effects of the NaOH concentration on the Pb distribution for concurrent cementation of dissolved Pb at 30 min of leaching time is shown in Figure 4b. As expected, the amount of Pb dissolved from ZPLRs and cemented by 0.25 g of Al powder increased with the increase in the NaOH concentration. The cemented Pb increased from 27% for 1 M NaOH to 66.1% for 6 M NaOH, with no dissolved Pb remaining in the solution.

3.2.2. Effects of Time and NaOH Concentration on Cementation and Distribution of Zn in Leaching Pulp

The half-reactions represented by Equations (10) and (11) that add up to the overall chemical reaction depicted in Equation (12) were thermodynamically expected when 0.25 g of Al powder was added during the leaching of ZPLRs in 3 M NaOH solution.

$$Al + 4OH^- \rightarrow Al(OH)_4^- + 3e^- \quad -2.35\ V \quad (10)$$

$$Pb + 4OH^- \rightarrow Pb(OH)_4^{2-} + 2e^- \quad -1.285\ V \quad (11)$$

$$2Al + 3Zn(OH)_4^{2-} \rightarrow 2Al(OH)_4^- + 4OH^- + 3Zn \quad (12)$$

Equation (12) is the cementation reaction of dissolved Zn as $ZnOH)_4{}^{2-}$ from ZPLRs is cemented by Al whose standard Gibbs free energy change, ΔG^0 (i.e., $\Delta G^0 = -nF\Delta E^0$, n is number of electrons transferred, F is Faraday's constant, and ΔE^0 ($E^0_{Al/Al(OH)_4^-} - E^0_{Zn/Zn(OH)_4^{2-}}$) is the galvanic cell potential of Equation (9)), is −616.57 kJ/mol. This means that the cementation of dissolved Zn^{2+} from ZPLRs by Al powder is thermodynamically favorable. However, the results show that little Zn was cemented by Al metal powder from 7.5 up to 120 min using various NaOH concentrations because most of the dissolved Zn remained in solution (Figure 5a,b). This could mean that there were some counter-reactions to the cementation reaction. These reactions could arise from co-dissolved elements in the leachate and/or solid residues.

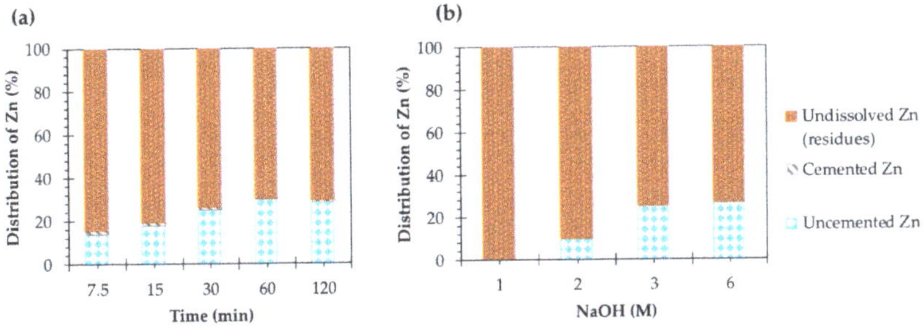

Figure 5. Effects of (**a**) leaching time and (**b**) NaOH concentration on cementation and distribution of Zn in leaching pulp when Al powder was added during ZPLR leaching (for interpretation of the references to color in this figure legend, the reader is referred to the web version of this article).

When the cementation product was analyzed by SEM-EDX, it was shown that both Pb and Zn were cemented but the intensity for Zn was much lower than Pb (Figure 6).

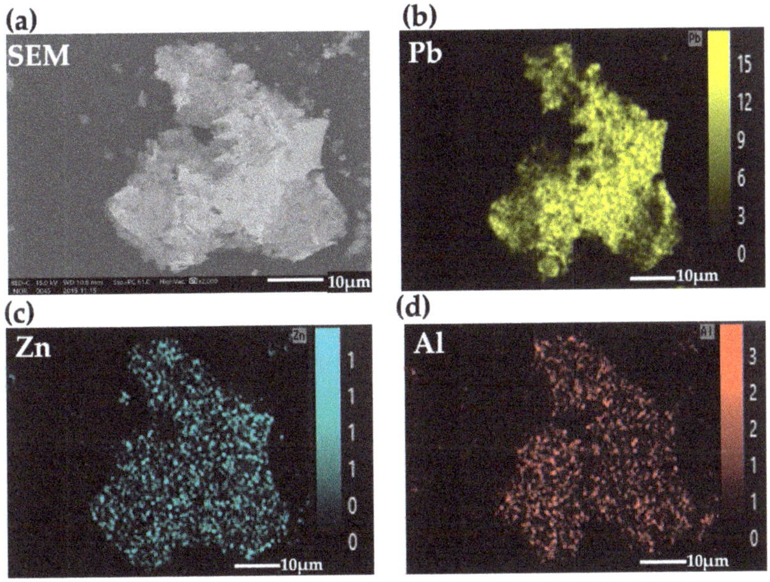

Figure 6. SEM-EDX of cementation product when 2.5 g ZPLRs were leached in 3 M NaOH solution for 30 min with the addition of 0.25 g Al powder: (**a**) SEM microphotography, (**b**) EDX map of Pb, (**c**) EDX map of Zn, and (**d**) EDX map of Al (for interpretation of the references to color in this figure legend, the reader is referred to the web version of this article).

To investigate the effects of co-dissolved elements and solid residues on the cementation of dissolved Zn from leaching solution using Al metal powder, simulated (model) 3 M NaOH solutions containing both 8 mM Pb^{2+} and 10 mM Zn^{2+} and filtrate (to eliminate solid residues interference) after the initial addition of Al powder during ZPLR leaching, respectively, were used. The model solution was prepared by dissolving $ZnCl_2$ and $PbCl_2$ (Wako Pure Chemical Industries, Ltd., Osaka, Japan) in 3 M NaOH. For the model solution,

0.15 g of Al metal powder was added to cement both Pb and Zn. For the filtrate after the concurrent cementation of dissolved Pb and Zn experiments, 0.1 g of Al metal powder was added to the cement residual Zn in the presence of other co-dissolved elements from ZPLRs. Figure 7a shows that 100% of Pb and 100% of Zn in the model solution were cemented out of the solution by Al metal powder. This confirms the thermodynamic feasibility discussed above and that Al metal powder can cement Zn. Likewise, the dissolved Zn in the filtrate after the concurrent cementation experiment was recovered (around 96.9%) by the second portion of Al metal powder added in the filtrate (Figure 7b). This implies that the co-dissolved elements from ZPLRs and from cementation experiments do not affect the cementation of Zn in NaOH solution. The results, however, highlight that solid residue could possibly interfere with and suppress the cementation of Zn during the concurrent cementation of dissolved Pb and Zn experiments.

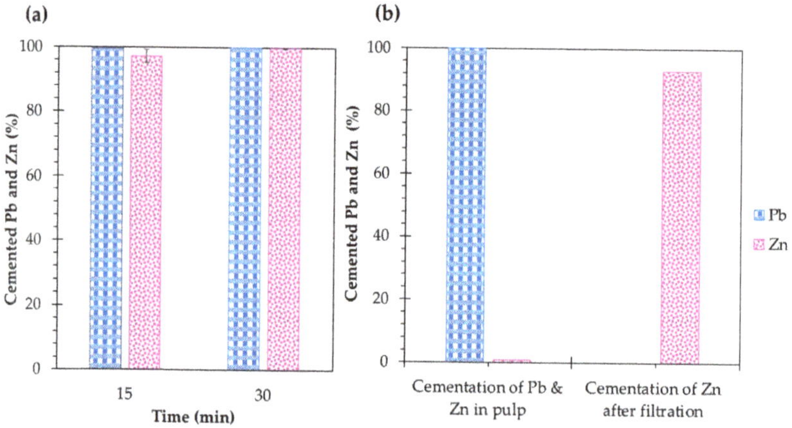

Figure 7. Amounts of Pb and Zn cemented out from (**a**) 3 M NaOH model solution containing Pb and Zn ions and treated for 15 and 30 min, and (**b**) concurrent cementation of dissolved Pb and Zn ions and cementation of Zn in the filtrate solution after concurrent cementation experiment (for interpretation of the references to color in this figure legend, the reader is referred to the web version of this article).

To investigate the effects of minerals in the solid residues from ZPLRs that could affect the cementation of dissolved Zn from leaching pulp using Al metal powder, 2.5 g (to maintain the same S/L ratio) of the three most abundant minerals in the ZPLRs we used. This included SiO_2 (quartz), Fe_2O_3 (hematite), and Fe_3O_4 (magnetite), and each mineral was added in a model solution of 3 M NaOH solutions containing both 8 mM Pb^{2+} and 10 mM Zn^{2+}. To mimic the ZPLR concurrent cementation experiments, 0.25 g of Al metal powder was added to the solution and shaken for 30 min in a temperature-controlled bash shaker. For all of the three solid residues, around 98% of Pb was cemented by Al metal powder (Figure 8). However, Zn cementation was slightly reduced by SiO_2 (93%) and significantly suppressed by Fe_2O_3 (28.5%) and Fe_3O_4 (27.9%). The slight decrease in cementation in the case of Zn in 2.5 g could be ascribed to dissolved silicate anions that exhibit the properties of nanoparticles suspension (colloid) hence affecting viscosity and the transportation of metal ions on the surface of the Al metal and cementation Al ions away from the surface of Al metals [40]. Meanwhile, the significant suppression by Fe_2O_3 and Fe_3O_4 could be attributed to preferential consumption of electrons from Al metal powder to reduce Fe^{3+} to Fe^{2+} (i.e., $Fe(OH)_3$ to $Fe(OH)_2$), whose standard electrode potential in basic solution is around 0.54 V. This deduction is also supported by other works highlighting the participation of Fe_2O_3 and Fe_3O_4 in electrochemical reactions such as pyrite dissolution, arsenite oxidation to arsenate, and the recovery of gold ions from chloride solutions [41,42]

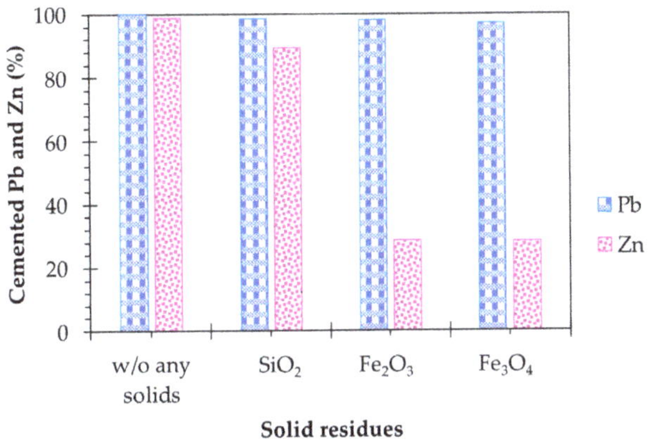

Figure 8. Effects of SiO_2, Fe_2O_3, and Fe_3O_4 on cementation of Pb and Zn when 2.5 of a given mineral was mixed in 3 M NaOH solution containing 8 mM of Pb^{2+} and Zn^{2+} for 30 min (for interpretation of the references to color in this figure legend, the reader is referred to the web version of this article).

The concurrent cementation of both dissolved Pb and Zn can be applied for the remediation of Pb-Zn mine waste materials or Pb-Zn contaminated soil that do not contain a substantial amount of iron oxide. The most toxic metal, Pb, can be cemented in pulp with less suppression by Fe oxides. Meanwhile, the dissolved Zn can be recovered after filtration by either cementation using Al or precipitation as ZnS [21].

4. Conclusions

This study investigated Pb and Zn removal from ZPLRs in alkaline solution by the concurrent cementation of dissolved Pb and Zn in leaching pulp. The findings are summarized as below:

(1) Pb and Zn removal efficiencies were affected by the leaching time, the NaOH concentration, and the S/L ratio. The Pb and Zn removal efficiencies were 62.2% and 27.1%, respectively, when 2.5 g/50 mL (S/L) of ZPLRs were leached in a 3 M NaOH solution for 30 min.

(2) The amounts Pb (62.2%) and Zn (27.1%) that were removed from ZPLRs using NaOH solution correlated and corroborated with the mobile phase fraction (i.e., Pb and Zn bound to water, exchangeable, and carbonates) approximated by sequential extraction.

(3) Around 100% of the dissolved Pb was cemented by Al metal powder for the concurrent cementation of dissolved Pb in the leaching pulp.

(4) The dissolved Zn was not cemented out in leaching pulp by the addition of Al metal powder. However, around 96.9% was cemented by Al after filtration. The suppression of cementation by Al metal was attributed to solid residues, in particular Fe oxides.

(5) The concurrent cementation of both dissolved Pb and Zn in alkaline leaching pulp has the potential to be applied for the remediation of Pb-Zn mine wastes or Pb-Zn contaminated soil, provided they contain minimal amounts of iron oxides, which were found to suppress Zn cementation.

Author Contributions: Conceptualization, M.S.; methodology, M.S., T.F. and R.H.; formal analysis, M.S., M.I. (Mayumi Ito), N.H., S.J., I.P. and C.B.T.; investigation, M.S.; writing—original draft preparation, M.S.; writing—review and editing, M.S., M.I. (Mayumi Ito), N.H., C.B.T., S.J., I.P., I.N., T.S. and T.I.; supervision, M.I. (Mayumi Ito) and N.H.; project administration, M.I. (Mayumi Ishizuka), S.N. and H.N.; funding acquisition, M.I. (Mayumi Ishizuka)., S.N. and H.N. All authors have read and agreed to the published version of the manuscript.

Funding: This work was supported partly by the Japan Society for the Promotion of Science (JSPS) CORE to CORE program (M.I. (Mayumi Ishizuka)), Hokkaido University SOUSEI-TOKUTEI Specific

Research Projects (M.I. (Mayumi Ishizuka)), JSPS Bilateral Open Partnership Joint Research Projects (grant number: JPJSBP120209902) (S.N.), The Japan Prize Foundation (S.N.), and Grants-in-Aid for Scientific Research from the Ministry of Education, and Culture, Sports, Science and Technology of Japan (grant number 20K20633) (S.N.).

Data Availability Statement: Data is available on request because of the restrictions, as the research is ongoing.

Conflicts of Interest: The authors declare no conflict of interest.

References

1. Dahan, A.M.E.; Alorro, R.D.; Pacaña, M.L.C.; Baute, R.M.; Silva, L.C.; Tabelin, C.B.; Resabal, V.J.T. Hydrochloric Acid Leaching of Philippine Coal Fly Ash: Investigation and Optimisation of Leaching Parameters by Response Surface Methodology (RSM). *Sustain. Chem.* **2022**, *3*, 76–90. [CrossRef]
2. Park, I.; Kanazawa, Y.; Sato, N.; Galtchandmani, P.; Jha, M.K.; Tabelin, C.B.; Jeon, S.; Ito, M.; Hiroyoshi, N. Beneficiation of Low-Grade Rare Earth Ore from Khalzan Buregtei Deposit (Mongolia) by Magnetic Separation. *Minerals* **2021**, *11*, 1432. [CrossRef]
3. Tabelin, C.B.; Park, I.; Phengsaart, T.; Jeon, S.; Villacorte-Tabelin, M.; Alonzo, D.; Yoo, K.; Ito, M.; Hiroyoshi, N. Copper and Critical Metals Production from Porphyry Ores and E-Wastes: A Review of Resource Availability, Processing/Recycling Challenges, Socio-Environmental Aspects, and Sustainability Issues. *Resour. Conserv. Recycl.* **2021**, *170*, 105610. [CrossRef]
4. Hund, K.; La Porta, D.; Fabregas, T.P.; Laing, T.; Drexhage, J. *World Bank Minerals for Climate Action—The Mineral Intensity of the Clean Energy Transition 2020*; World Bank Publications: Washington, DC, USA, 2020.
5. Tabelin, C.B.; Dallas, J.; Casanova, S.; Pelech, T.; Bournival, G.; Saydam, S.; Canbulat, I. Towards a Low-Carbon Society: A Review of Lithium Resource Availability, Challenges and Innovations in Mining, Extraction and Recycling, and Future Perspectives. *Miner. Eng.* **2021**, *163*, 106743. [CrossRef]
6. Aikawa, K.; Ito, M.; Segawa, T.; Jeon, S.; Park, I.; Tabelin, C.B.; Hiroyoshi, N. Depression of Lead-Activated Sphalerite by Pyrite via Galvanic Interactions: Implications to the Selective Flotation of Complex Sulfide Ores. *Miner. Eng.* **2020**, *152*, 106367. [CrossRef]
7. Mufalo, W.; Tangviroon, P.; Igarashi, T.; Ito, M.; Sato, T.; Chirwa, M.; Nyambe, I.; Nakata, H.; Nakayama, S.; Ishizuka, M. Solid-Phase Partitioning and Leaching Behavior of Pb and Zn from Playground Soils in Kabwe, Zambia. *Toxics* **2021**, *9*, 248. [CrossRef]
8. Behnajady, B.; Moghaddam, J. Selective Leaching of Zinc from Hazardous As-Bearing Zinc Plant Purification Filter Cake. *Chem. Eng. Res. Des.* **2017**, *117*, 564–574. [CrossRef]
9. Tabelin, C.B.; Silwamba, M.; Paglinawan, F.C.; Mondejar, A.J.S.; Duc, H.G.; Resabal, V.J.; Opiso, E.M.; Igarashi, T.; Tomiyama, S.; Ito, M.; et al. Solid-Phase Partitioning and Release-Retention Mechanisms of Copper, Lead, Zinc and Arsenic in Soils Impacted by Artisanal and Small-Scale Gold Mining (ASGM) Activities. *Chemosphere* **2020**, *260*, 127574. [CrossRef]
10. Abo Atia, T.; Spooren, J. Microwave Assisted Chloride Leaching of Zinc Plant Residues. *J. Hazard. Mater.* **2020**, *398*, 122814. [CrossRef]
11. Ho, G.D.; Tabelin, C.B.; Tangviroon, P.; Tamamura, S.; Igarashi, T. Effects of Cement Addition on Arsenic Leaching from Soils Excavated from Projects Employing Shield-Tunneling Method. *Geoderma* **2021**, *385*, 114896. [CrossRef]
12. Igarashi, T.; Herrera, P.S.; Uchiyama, H.; Miyamae, H.; Iyatomi, N.; Hashimoto, K.; Tabelin, C.B. The Two-Step Neutralization Ferrite-Formation Process for Sustainable Acid Mine Drainage Treatment: Removal of Copper, Zinc and Arsenic, and the Influence of Coexisting Ions on Ferritization. *Sci. Total Environ.* **2020**, *715*, 136877. [CrossRef]
13. Tabelin, C.B.; Igarashi, T.; Villacorte-Tabelin, M.; Park, I.; Opiso, E.M.; Ito, M.; Hiroyoshi, N. Arsenic, Selenium, Boron, Lead, Cadmium, Copper, and Zinc in Naturally Contaminated Rocks: A Review of Their Sources, Modes of Enrichment, Mechanisms of Release, and Mitigation Strategies. *Sci. Total Environ.* **2018**, *645*, 1522–1553. [CrossRef] [PubMed]
14. Rao, S.; Wang, D.; Liu, Z.; Zhang, K.; Cao, H.; Tao, J. Selective Extraction of Zinc, Gallium, and Germanium from Zinc Refinery Residue Using Two Stage Acid and Alkaline Leaching. *Hydrometallurgy* **2019**, *183*, 38–44. [CrossRef]
15. Fattahi, A.; Rashchi, F.; Abkhoshk, E. Reductive Leaching of Zinc, Cobalt and Manganese from Zinc Plant Residue. *Hydrometallurgy* **2016**, *161*, 185–192. [CrossRef]
16. Koleini, S.M.J.; Mehrpouya, H.; Saberyan, K.; Abdolahi, M. Extraction of Indium from Zinc Plant Residues. *Miner. Eng.* **2010**, *23*, 51–53. [CrossRef]
17. Jeon, S.; Tabelin, C.B.; Takahashi, H.; Park, I.; Ito, M.; Hiroyoshi, N. Interference of Coexisting Copper and Aluminum on the Ammonium Thiosulfate Leaching of Gold from Printed Circuit Boards of Waste Mobile Phones. *Waste Manag.* **2018**, *81*, 148–156. [CrossRef]
18. Jeon, S.; Tabelin, C.B.; Park, I.; Nagata, Y.; Ito, M.; Hiroyoshi, N. Ammonium Thiosulfate Extraction of Gold from Printed Circuit Boards (PCBs) of End-of-Life Mobile Phones and Its Recovery from Pregnant Leach Solution by Cementation. *Hydrometallurgy* **2020**, *191*, 105214. [CrossRef]
19. Choi, S.; Yoo, K.; Alorro, R.D.; Tabelin, C.B. Cementation of Co Ion in Leach Solution Using Zn Powder Followed by Magnetic Separation of Cementation-Precipitate for Recovery of Unreacted Zn Powder. *Miner. Eng.* **2020**, *145*, 106061. [CrossRef]

20. Phengsaart, T.; Ito, M.; Hamaya, N.; Tabelin, C.B.; Hiroyoshi, N. Improvement of Jig Efficiency by Shape Separation, and a Novel Method to Estimate the Separation Efficiency of Metal Wires in Crushed Electronic Wastes Using Bending Behavior and "Entanglement Factor". *Miner. Eng.* **2018**, *129*, 54–62. [CrossRef]
21. Liu, Q.; Zhao, Y.; Zhao, G. Production of Zinc and Lead Concentrates from Lean Oxidized Zinc Ores by Alkaline Leaching Followed by Two-Step Precipitation Using Sulfides. *Hydrometallurgy* **2011**, *110*, 79–84. [CrossRef]
22. Şahin, M.; Erdem, M. Cleaning of High Lead-Bearing Zinc Leaching Residue by Recovery of Lead with Alkaline Leaching. *Hydrometallurgy* **2015**, *153*, 170–178. [CrossRef]
23. Huang, Y.; Geng, Y.; Han, G.; Cao, Y.; Peng, W.; Zhu, X.; Zhang, T.; Dou, Z. A Perspective of Stepwise Utilization of Hazardous Zinc Plant Purification Residue Based on Selective Alkaline Leaching of Zinc. *J. Hazard. Mater.* **2020**, *389*, 122090. [CrossRef] [PubMed]
24. Seyed Ghasemi, S.M.; Azizi, A. Alkaline Leaching of Lead and Zinc by Sodium Hydroxide: Kinetics Modeling. *J. Mater. Res. Technol.* **2018**, *7*, 118–125. [CrossRef]
25. Zhang, Y.; Feng, X.; Qian, L.; Luan, J.; Jin, B. Separation of Arsenic and Extraction of Zinc and Copper from High-Arsenic Copper Smelting Dusts by Alkali Leaching Followed by Sulfuric Acid Leaching. *J. Environ. Chem. Eng.* **2021**, *9*, 105997. [CrossRef]
26. Jeon, S.; Bright, S.; Park, I.; Tabelin, C.B.; Ito, M.; Hiroyoshi, N. The Effects of Coexisting Copper, Iron, Cobalt, Nickel, and Zinc Ions on Gold Recovery by Enhanced Cementation via Galvanic Interactions between Zero-Valent Aluminum and Activated Carbon in Ammonium Thiosulfate Systems. *Metals* **2021**, *11*, 1352. [CrossRef]
27. Jeon, S.; Bright, S.; Park, I.; Tabelin, C.B.; Ito, M.; Hiroyoshi, N. A Simple and Efficient Recovery Technique for Gold Ions from Ammonium Thiosulfate Medium by Galvanic Interactions of Zero-Valent Aluminum and Activated Carbon: A Parametric and Mechanistic Study of Cementation. *Hydrometallurgy* **2022**, *208*, 105815. [CrossRef]
28. Silwamba, M.; Ito, M.; Hiroyoshi, N.; Tabelin, C.B.; Hashizume, R.; Fukushima, T.; Park, I.; Jeon, S.; Igarashi, T.; Sato, T.; et al. Recovery of Lead and Zinc from Zinc Plant Leach Residues by Concurrent Dissolution-Cementation Using Zero-Valent Aluminum in Chloride Medium. *Metals* **2020**, *10*, 531. [CrossRef]
29. Silwamba, M.; Ito, M.; Tabelin, C.B.; Park, I.; Jeon, S.; Takada, M.; Kubo, Y.; Hokari, N.; Tsunekawa, M.; Hiroyoshi, N. Simultaneous Extraction and Recovery of Lead Using Citrate and Micro-Scale Zero-Valent Iron for Decontamination of Polluted Shooting Range Soils. *Environ. Adv.* **2021**, *5*, 100115. [CrossRef]
30. Choi, S.; Jeon, S.; Park, I.; Ito, M.; Hiroyoshi, N. Addition of Fe_3O_4 as Electron Mediator for Enhanced Cementation of Cd^{2+} and Zn^{2+} on Aluminum Powder from Sulfate Solutions and Magnetic Separation to Concentrate Cemented Metals from Cementation Products. *J. Environ. Chem. Eng.* **2021**, *9*, 106699. [CrossRef]
31. Silwamba, M.; Ito, M.; Hiroyoshi, N.; Tabelin, C.B.; Fukushima, T.; Park, I.; Jeon, S.; Igarashi, T.; Sato, T.; Nyambe, I. Detoxification of Lead-Bearing Zinc Plant Leach Residues from Kabwe, Zambia by Coupled Extraction-Cementation Method. *J. Environ. Chem. Eng.* **2020**, *8*, 104197. [CrossRef]
32. Orhan, G. Leaching and Cementation of Heavy Metals from Electric Arc Furnace Dust in Alkaline Medium. *Hydrometallurgy* **2005**, *78*, 236–245. [CrossRef]
33. Dutra, A.J.B.; Paiva, P.R.P.; Tavares, L.M. Alkaline Leaching of Zinc from Electric Arc Furnace Steel Dust. *Miner. Eng.* **2006**, *19*, 478–485. [CrossRef]
34. Zhang, Y.; Deng, J.; Chen, J.; Yu, R.; Xing, X. Leaching of Zinc from Calcined Smithsonite Using Sodium Hydroxide. *Hydrometallurgy* **2013**, *131–132*, 89–92. [CrossRef]
35. Tabelin, C.B.; Resabal, V.J.T.; Park, I.; Villanueva, M.G.B.; Choi, S.; Ebio, R.; Cabural, P.J.; Villacorte-Tabelin, M.; Orbecido, A.; Alorro, R.D.; et al. Repurposing of Aluminum Scrap into Magnetic Al0/ZVI Bimetallic Materials: Two-Stage Mechanical-Chemical Synthesis and Characterization of Products. *J. Clean. Prod.* **2021**, *317*, 128285. [CrossRef]
36. Seng, S.; Tabelin, C.B.; Kojima, M.; Hiroyoshi, N.; Ito, M. Galvanic Microencapsulation (GME) Using Zero-Valent Aluminum and Zero-Valent Iron to Suppress Pyrite Oxidation. *Mater. Trans.* **2019**, *60*, 277–286. [CrossRef]
37. Jeon, S.; Tabelin, C.B.; Takahashi, H.; Park, I.; Ito, M.; Hiroyoshi, N. Enhanced Cementation of Gold via Galvanic Interactions Using Activated Carbon and Zero-Valent Aluminum: A Novel Approach to Recover Gold Ions from Ammonium Thiosulfate Medium. *Hydrometallurgy* **2020**, *191*, 105165. [CrossRef]
38. Djokić, S.S. Cementation of Copper on Aluminum in Alkaline Solutions. *J. Electrochem. Soc.* **1996**, *143*, 1300. [CrossRef]
39. Bethke, C.M. *The Geochemist's Workbench—A User's Guide to Rxn, Act2, Tact, React, and Gtplot*; University of Illinois: Urbana, IL, USA, 2002.
40. Yang, X.; Zhu, W.; Yang, Q. The Viscosity Properties of Sodium Silicate Solutions. *J. Solut. Chem.* **2008**, *37*, 73–83. [CrossRef]
41. Calderon, A.R.M.; Alorro, R.D.; Tadesse, B.; Yoo, K.; Tabelin, C.B. Evaluation of Maghemite-Rich Iron Oxide Composite Prepared from Magnetite as Adsorbent for Gold from Chloride Solution. *JOM* **2019**, *71*, 4639–4646. [CrossRef]
42. Tabelin, C.B.; Corpuz, R.D.; Igarashi, T.; Villacorte-Tabelin, M.; Ito, M.; Hiroyoshi, N. Hematite-Catalysed Scorodite Formation as a Novel Arsenic Immobilisation Strategy under Ambient Conditions. *Chemosphere* **2019**, *233*, 946–953. [CrossRef]

MDPI
St. Alban-Anlage 66
4052 Basel
Switzerland
www.mdpi.com

Minerals Editorial Office
E-mail: minerals@mdpi.com
www.mdpi.com/journal/minerals

www.ingramcontent.com/pod-product-compliance
Lightning Source LLC
Chambersburg PA
CBHW061613210926
51850CB00009B/2314
9783036597171